North American
Wildland Plants

North American Wildland Plants

A Field Guide

James Stubbendieck,
Stephan L. Hatch,
L. M. Landholt

Illustrated by Kelly L. Rhodes Hays,
Bellamy Jansen, and Debra Meier
Maps by Kathleen Lonergan-Orr

University of Nebraska Press
Lincoln and London

Library of Congress Cataloging-in-Publication Data
Stubbendieck, James L.
North American wildland plants: a field guide / James Stubbendieck, Stephan L. Hatch, and L. M. Landholt.
p.   cm.
"Sixth edition of North American range plants"—CIP publr. info.
Includes bibliographical refernces and index.
ISBN 0-8032-4306-5 (cloth: alk. paper)—ISBN 0-8032-9306-2 (pbk.: alk. paper)
1. Range plants—North America—Identification.
2. Forage plants—North America—Identification.
3. Range plants—North America.   4. Forage plants—North America.   I. Hatch, Stephan L., 1945–
II. Landholt, L. M., 1975–   III. Stubbendieck, James L. North American range plants.   IV. Title.
SB193.3N67S88   2003
581.7′4′097—dc21
2003042698

Contents

Acknowledgments

The authors wish to acknowledge Kathie J. Diller and Charles H. Butterfield for their contributions to *North American Range Plants*, the predecessor of this book. Kelly L. Rhodes Hays, Debra Meier, and Bellamy Jansen prepared the illustrations.

Juan M. Martìnez-Reyna and Juvenal Gutiérrez-Castillo are recognized for furnishing the Mexican common names. Others contributing to this book and its predecessor include Kelly Allred, Val Anderson, William T. Barker, Alan A. Beetle, Margaret R. Bolick, Kylie Faulk, T. Mark Hart, Marshall Hervey, Barry Irving, Stanley D. Jones, Robert B. Kaul, Gary E. Larson, Elizabeth Manrique, Catherine Mills, Daniel Nosal, Linda L. Rader, Jesús Valdéz-R., Walter H. Schacht, Susan Schuckert, Susan Tunnell, and J. K. Wipff.

Kathleen Lonergan-Orr receives special recognition for her tireless work to revise the distribution maps and expertise in handling the details associated with the process of preparing the manuscript for this book.

North American
Wildland Plants

Introduction

A comprehensive reference containing the important characteristics of the most important wildland plants of North America is critical for ecologists, range managers, land managers, and other natural resource professionals. In addition, university students and range plant identification teams needed a single, primary resource for learning about important wildland plant species. *North American Range Plants* was developed to meet these needs and was first published in 1981. Subsequent editions (1982, 1986, 1992, 1997) included changes in nomenclature, refinement of distributions, additional information on each of the species, and new illustrations. The illustrations were prepared to highlight general and specific characteristics to aid in the identification of the featured range plants. The fifth edition reflected changing attitudes toward riparian areas and wetlands. As a result of increased concern and interest, about 10 percent of the species included in this book occur on these sites. *North American Wildland Plants*, a new title, includes many nomenclatural changes, and the illustrations have been labeled to accentuate specific characters. The title change reflects the importance of plants across ecosystems and the multiple uses of the plant resources within these ecosystems. This field guide will aid both individuals with limited botanical knowledge and natural resources professionals to identify wildland plants.

The two hundred species in this book were selected because of their abundance, desirability, or noxious properties. The list of plant species was developed over the course of nearly fifty years by faculty from the colleges and universities with rangeland management and ecology programs and by coaches of range plant identification teams. The formal list is now the Master Plant List for the International Range Plant Identification Contest sponsored by the Society for Range Management (445 Union Boulevard, Suite 230, Lakewood CO 80228).

Plant species descriptions in this book include characteristics for their identification, a labeled illustration highlighting specific parts of a typical plant, and a refined general distribution map for North America. Each species description includes nomenclature; life span; origin; season of growth; inflorescence, flower or spikelet, or other reproductive parts; vegetative parts; and growth characteristics. Forage value for wildlife and livestock is estimated. Brief notes are included on habitat; livestock losses; and historic, food, and medicinal uses. Information on historic, food, and medicinal uses was gathered from numerous sources and is presented as a point of interest and to broaden readers' appreciation of the plants. It is strongly emphasized that these plant species should not be used for these purposes.

Grasses (POACEAE family) are described first and are aligned by tribe, genus, and specific epithet in alphabetical order by rank. Grass-like plants (CYPERACEAE and JUNCACEAE families) are next. All other families follow in alphabetical order by rank for family, genus, and specific epithet with the exception of the ASTERACEAE family, which follows the same format as the POACEAE family.

The grass (POACEAE) and composite (ASTERACEAE) families are treated by tribe to help the reader relate to smaller groups within these large, complex families. Recognition of species within tribes builds a concept of tribal characteristics. When an unknown species of either family is encountered, knowledge of tribal alignments below family may aid identification, thus reducing the time required for making an identification using a diagnostic key.

The classification system in *Gould's Grasses of Texas* (Department of Rangeland Ecology and Management, 2002) was followed for the grass tribal names. The tribal classification of the composites follows Cronquist's *Vascular Flora of the Southeastern United States* (University of North Carolina Press, 1980).

Numerous authoritative floristic treatments from the wildland areas of North America were consulted for species names and authorities. Selected synonyms, noting other names for the same species, are included on the illustration page for each species to help clarify the species concept used in this text. The synonyms will help in finding additional information from other floristic treatments.

Common and alternative common names are given for the plants, but they may not include the common name used in a particular area. Common names were restricted to two words, sometimes resulting in long and cumbersome words. Spanish common names are listed for the appropriate taxa and may exceed two words.

The origin of each species is given as native or introduced. Origins of introduced taxa are given parenthetically. Many species are known to be introduced, while others are thought to have been. *Poa pratensis* L. is an example of a species listed as introduced but may be native to North America.

Season of growth is listed as cool, warm, or evergreen. Cool-season plants complete most of their growth in the fall, winter, and spring, whereas warm-season plants grow most when temperatures are the highest in the summer. The evergreen plants retain their ability to grow whenever climatic conditions are suitable.

Plant characteristics for each species are separated into categories to help in making comparisons between species. These characteristics are intended to be useful to the amateur botanist. Conservative characteristics, those that are not greatly influenced by the environment, should be the basis for identification. These may include floral, spikelet, leaf, and inflorescence type but may vary with the species. Pubescence, ligule lengths, and awn lengths are highly variable characteristics, and primary importance should not be placed on these when identifying grasses. Presence or absence of rhizomes is another variable characteristic that is somewhat dependant upon moisture and other features of the habitat.

Forage values of the plants discussed in this book are relative values that vary with the type of animal utilizing the particular plant species. Values are determined on the basis of palatability, nutrient content, and the amount of forage produced by the plant species. These values may vary with the climatic conditions, the part of North America in which the plant is growing, when the forage is consumed, associated plant species, and the age class of each animal species utilizing the forage.

Losses due to poisonous plants, one of the major problems facing the livestock industry, are included in these plant descriptions. Annual losses on wildlands amount to millions of dollars, with the effects of poisonous plants varying from slightly reduced rate of weight gain to death of the animal. Losses that are easy to document, such as death, are not as economically significant as the losses wherein growth rate or milk production is reduced. The brief mention of livestock losses in this book gives the animals affected and the type of poison, commonly referred to as the poisonous principle, contained in the plant species.

In addition to the selected synonyms, this book includes a glossary, list of authorities, and list of selected references. This supplementary information will give the student, professional natural resource manager, and anyone else interested in plants a more complete knowledge of plants and a starting place in the literature to seek additional information. The index is comprehensive, including all scientific and common names used in the text.

The information contained in *North American Wildland Plants* is by no means complete. The authors have settled for brevity with the expectation that this book will be a starting point for those interested in wildland plant identification. Plant taxonomists and extension personnel in each locality can provide additional information on plant species of interest.

Wildland Plants

Most wildland plant species are classified as annuals or perennials. Annuals complete their life cycle in one growing season, while perennials generally live three or more years. Herbaceous perennials have aerial stems that die back to the soil level each year while the underground parts remain alive. Perennial grasses and forbs are in this category. Woody perennials have aerial stems that remain alive throughout the year, although they may become dormant for part of the year. Trees and shrubs are in this category. Biennial is a third life-span category. Biennials require two growing seasons to complete their life cycle. Growth during the first year is only vegetative, and seed is produced during the second growing season. Relatively few plants fit into this category.

ORIGIN
Wildland plants that originated in North America are native. Introduced refers to plants that have been brought into North America from another continent and were adapted to conditions here. Several introduced species are valuable forage plants that were intentionally introduced for that purpose. Some introduced species were brought in for various reasons (e.g. landscaping), escaped, and are now troublesome weeds. Some species were accidentally introduced through contaminated crop seed, packing material, or ballast.

CLASSIFICATION
Botanical nomenclature refers to a system of naming plants. Plants are described and grouped according to their structure, particularly the flowering or other reproductive parts. The classification system from general to specific is:

Kingdom (Plant)
  Division (Phylum)
    Class
      Order
        Family
          Tribe
            Genus
              Specific epithet

We will be concerned only with the last four parts of the classification system:
A. Family
  A plant family is the basic division of plant orders. Morphological character-

istics or similarities determine the family to which a plant belongs. Flowering characteristics are extremely important in the classification of families. All grasses have similar flowers and belong to the same family, POACEAE. Numbers of petals, sepals, stamens, pistils, and other flowering parts are the basic divisions. All family names of vascular plants have a standard ending—ACEAE.

B. Tribe

A plant family may be divided into tribes. In this book, the POACEAE and ASTERACEAE are the only families for which tribes are recognized. An example is the ANDROPOGONEAE tribe of the POACEAE family. All tribe names of vascular plants have the standard ending—EAE.

C. Scientific Name

There is only one correct scientific name for each species. The scientific name consists of two main parts. The first part is the genus, and the second is the specific epithet. The authority is added for completeness and accuracy.

1. Genus

Classification of plants into genera (plural of genus) is based on similarities in flowering and/or morphological and non-morphological characteristics, although, with more specific divisions. An example is the genus *Schizachyrium*, which is part of the ANDROPOGONEAE tribe of the POACEAE family. The first letter of the genus is capitalized, and the word is underlined or italicized.

2. Specific Epithet

The second part of the scientific name is the specific epithet. A species is the kind of plant and is named by the genus and specific epithet. This classification is based on differences in flowering and/or morphological and non-morphological characteristics that distinguish the plant from related species. An example is the specific epithet *scoparium* for the species *Schizachyrium scoparium*, which differs from all other species of *Schizachyrium* in morphological characters. The specific epithet is lowercased, and the word is underlined or italicized.

3. Authority

The scientific name, for reasons of completeness and accuracy, is followed by the abbreviation or whole name of the person or persons who first applied that name to the plant. For example, (Michx.) Nash are the authorities for *Schizachyrium scoparium*. The French botanist Andre Michaux (1746–1802) first described and applied the specific epithet to that species of plant, and the American agrostologist George Nash (1864–1921) later transferred the epithet to the genus *Schizachyrium*. A list of authorities follows the glossary.

4. Common name

Common names have been given to many species of plants. Common names are usually simple and often descriptive of the plant, honor some person, or give a geographical location. Little bluestem is a common name of *Schizachyrium scoparium* (Michx.) Nash. Common names are useful only in areas using the same language. For example, the common name used for this species in Mexico is popotillo colorado. Even within the same language, one species may have several common names. Little bluestem is called prairie beardgrass in some parts of North America and popotillo colorado in Mexico. An additional weakness of common names is that one common name may be applied to several species.

Only one correct scientific name exists for each plant. Nevertheless, the name for a given plant will change if that plant is reclassified or if it is discovered that another valid name for it was published earlier. Although date of publication is absolute, the assignment of rank and position in the classification process is a matter of taxonomic opinion, which is often annoying to the layman. An example of a scientific name change is that of little bluestem, which was formerly *Andropogon scoparius* Michx. and is now *Schizachyrium scoparium* (Michx.) Nash. This seems to be a recent change to most of us, but it was first published by Nash in 1903. The relatively recent name change is due to a consensus of taxonomic opinion favoring the Nash classification rather than that of Michaux. Names other than the correct one are synonyms and can be found listed below the accepted scientific and common names on the page of each species illustration.

A summary of the classification system would be:

Family: POACEAE
   Tribe: ANDROPOGONEAE
      Genus: *Schizachyrium*
         Species: *Schizachyrium scoparium* (Michx.) Nash.

PLANT GROUPS

Wildland plants are divided into grasses, grass-like plants, forbs, and woody plants. These can be easily distinguished by certain characteristics. Figure 1 presents a comparison of plant groups.

Grasses have either hollow or solid stems with nodes. Leaves are two-ranked, sheathing, and have parallel veins. Flowers are small, inconspicuous, and occur in spikelets.

Grass-like plants resemble grasses but generally have solid or pithy stems without elongated internodes. Leaf veins are parallel, but the leaves are two-ranked or three-ranked. Stems are often triangular, and the flowers are small and inconspicuous.

Forbs are herbaceous plants other than grasses and grass-like plants. They usually have solid stems and generally have broad leaves with netted venation. Flowers are often large, colored, and showy, although they may be small and inconspicuous.

Woody plants have secondary growth of their aerial stems, which live throughout the year, although they may be dormant part of the time. Leaves are often broad and net veined. Flowers are often showy, but they may be inconspicuous. Both trees and shrubs fit into this category.

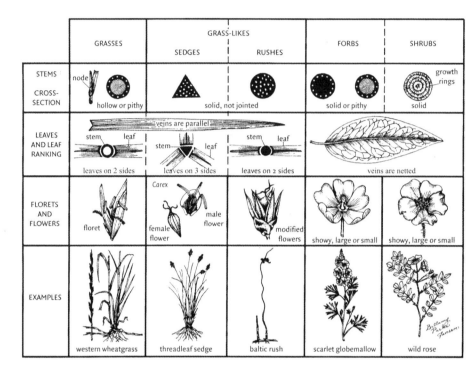

Figure 1. Comparison of plant groups

MORPHOLOGY OF GRASSES
Figures 2 through 5 are a series of drawings illustrating various morphological features of the grass plant. See the glossary for definitions of terms.

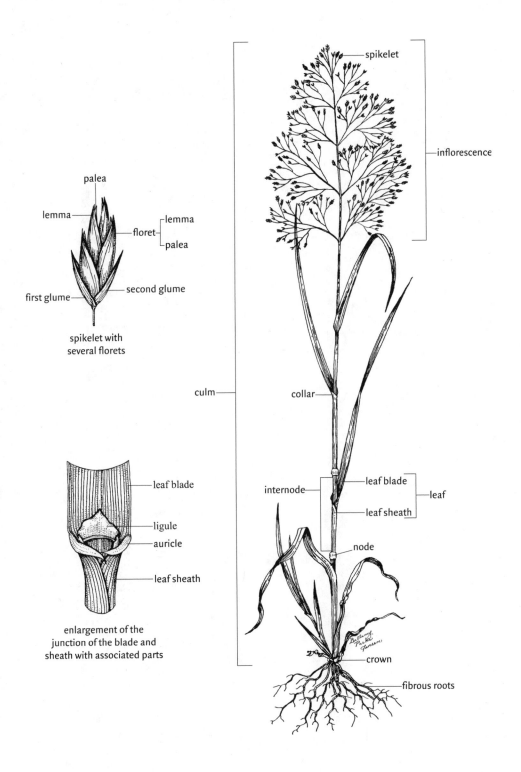

palea

lemma

floret { lemma / palea

first glume — — second glume

spikelet with several florets

leaf blade

ligule

auricle

leaf sheath

enlargement of the junction of the blade and sheath with associated parts

spikelet

inflorescence

culm

collar

internode — — leaf blade

leaf

leaf sheath

node

crown

fibrous roots

Figure 2. Grass plant and spikelet

9

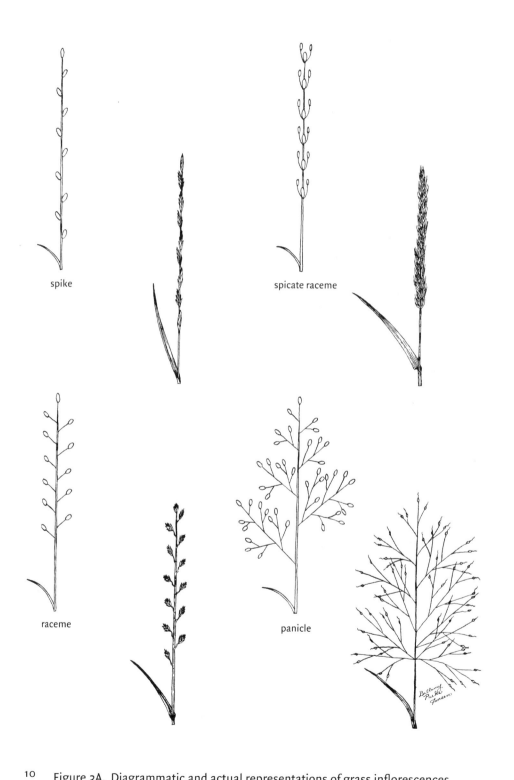

spike

spicate raceme

raceme

panicle

Figure 3A. Diagrammatic and actual representations of grass inflorescences

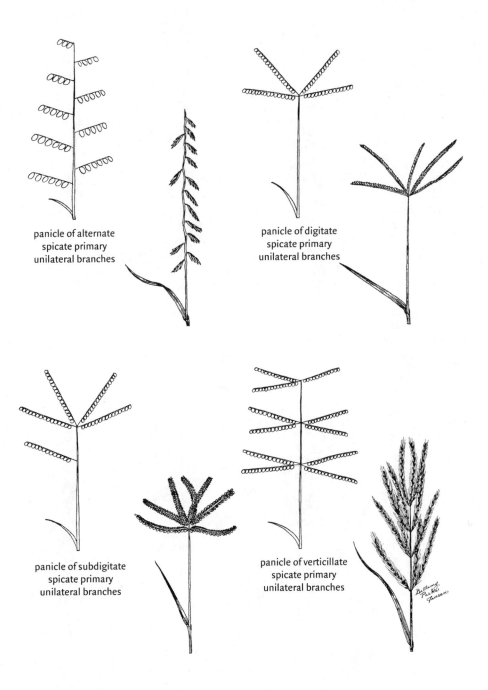

panicle of alternate
spicate primary
unilateral branches

panicle of digitate
spicate primary
unilateral branches

panicle of subdigitate
spicate primary
unilateral branches

panicle of verticillate
spicate primary
unilateral branches

Figure 3B. Diagrammatic and actual representations of grass inflorescences   11

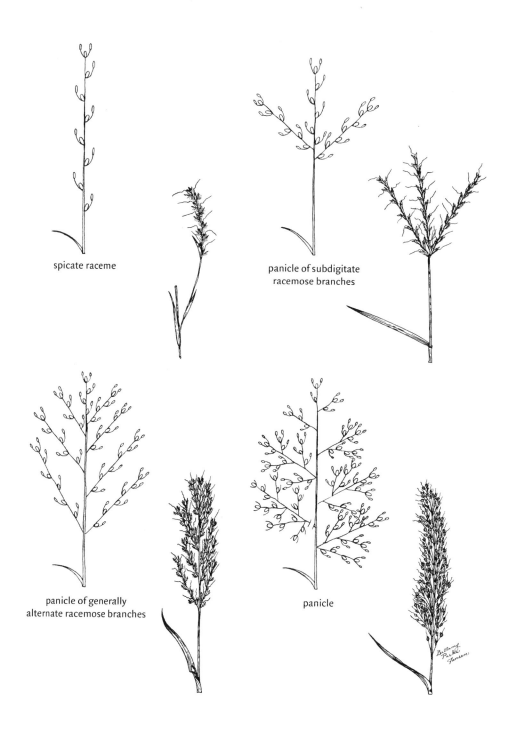

spicate raceme

panicle of subdigitate
racemose branches

panicle of generally
alternate racemose branches

panicle

Figure 3C. Diagrammatic and actual representations of grass inflorescences
of the ANDROPOGONEAE

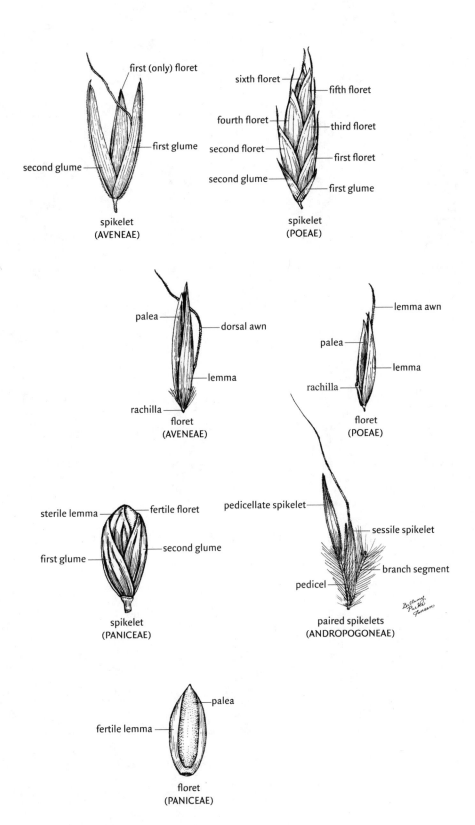

Figure 4. Grass spikelets and representative florets

13

absent

membranous

ciliate

ciliate membrane

LIGULE TYPE

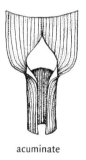
acuminate

acute

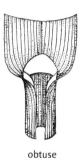
obtuse

truncate

LIGULE APEX SHAPE

entire

notched

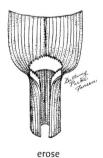
erose

LIGULE MARGIN

Figure 5. Grass ligule types, apex shapes, and margins

MORPHOLOGY OF GRASS-LIKE PLANTS

Figure 6 illustrates the morphological features of grass-like plants. See the glossary for definitions of terms.

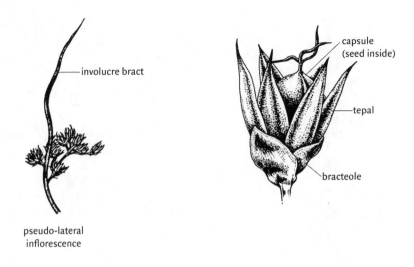

JUNCACEAE

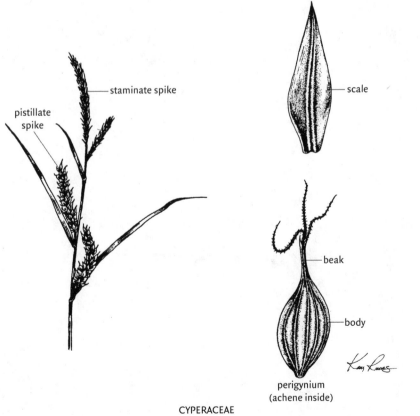

CYPERACEAE

Figure 6. Grass-like inflorescences and flowers

MORPHOLOGY OF FORBS AND WOODY PLANTS

Figures 7 through 19 are a series of drawings illustrating the various morphological features of forbs and woody plants. See the glossary for definitions of terms.

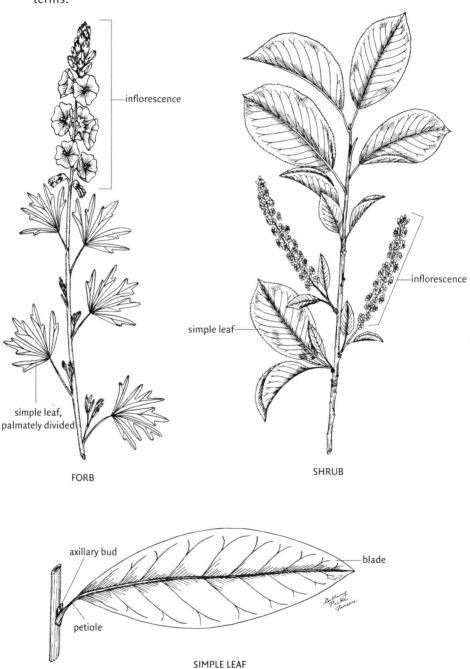

FORB

SHRUB

SIMPLE LEAF

Figure 7. Forb and woody plant leaf parts and a comparison of forbs and shrubs

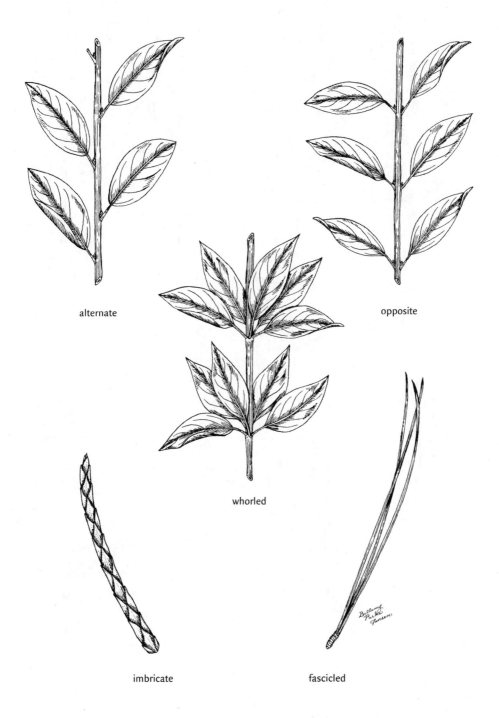

alternate

opposite

whorled

imbricate

fascicled

Figure 8. Forb and woody plant leaf arrangement

17

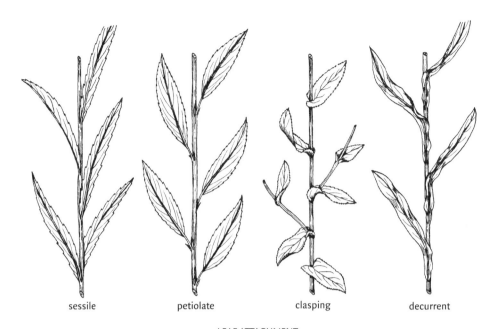

sessile   petiolate   clasping   decurrent

LEAF ATTACHMENT

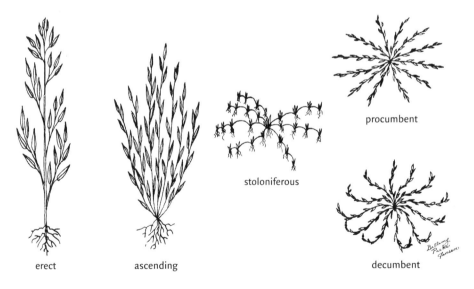

procumbent

stoloniferous

erect   ascending   decumbent

STEM HABITAT TYPES

Figure 9. Forb and woody plant leaf attachments and stem habit types

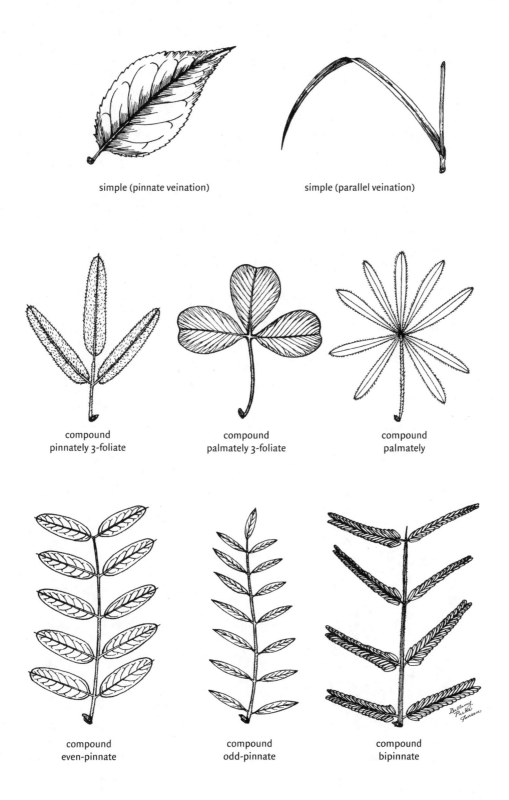

simple (pinnate veination)   simple (parallel veination)

compound
pinnately 3-foliate

compound
palmately 3-foliate

compound
palmately

compound
even-pinnate

compound
odd-pinnate

compound
bipinnate

Figure 10. Forb and woody plant types of simple and compound leaves
(the bud at the base of the petiole which defines the start of the leaf)   19

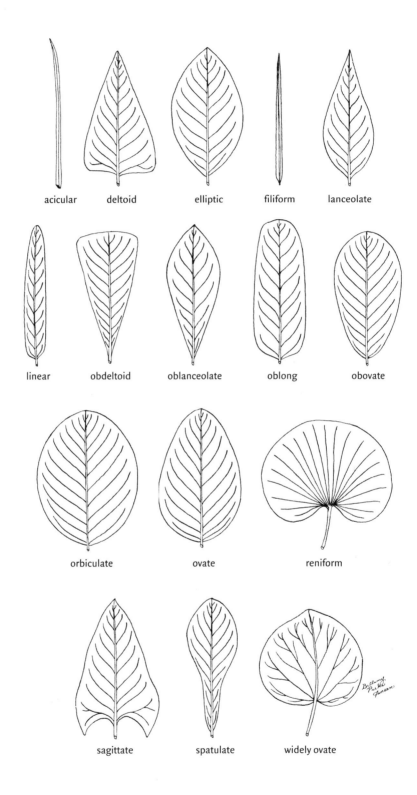

acicular   deltoid   elliptic   filiform   lanceolate

linear   obdeltoid   oblanceolate   oblong   obovate

orbiculate   ovate   reniform

sagittate   spatulate   widely ovate

    Figure 11. Forb and woody plant leaf shapes

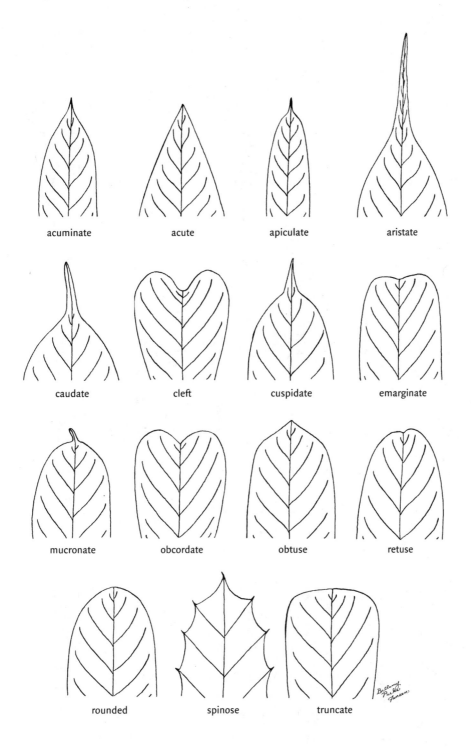

acuminate acute apiculate aristate

caudate cleft cuspidate emarginate

mucronate obcordate obtuse retuse

rounded spinose truncate

Figure 12. Forb and woody plant leaf apices

21

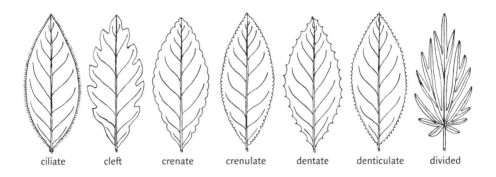

ciliate cleft crenate crenulate dentate denticulate divided

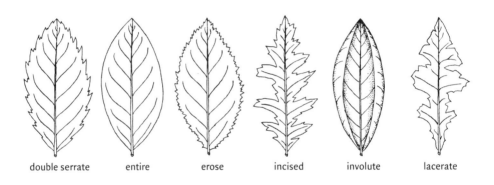

double serrate entire erose incised involute lacerate

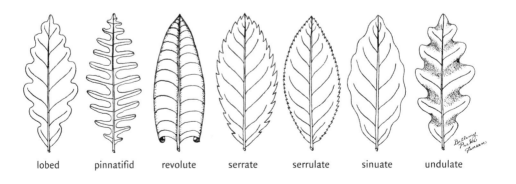

lobed pinnatifid revolute serrate serrulate sinuate undulate

Figure 13. Forb and woody plant leaf margins

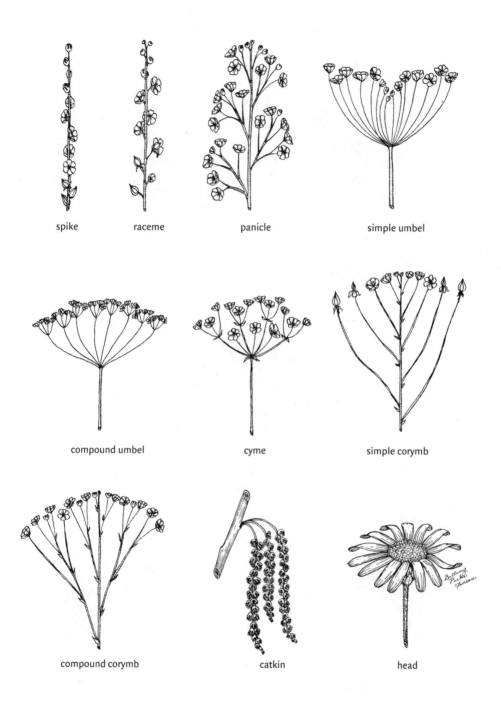

spike   raceme   panicle   simple umbel

compound umbel   cyme   simple corymb

compound corymb   catkin   head

Figure 14. Forb and woody plant inflorescence types

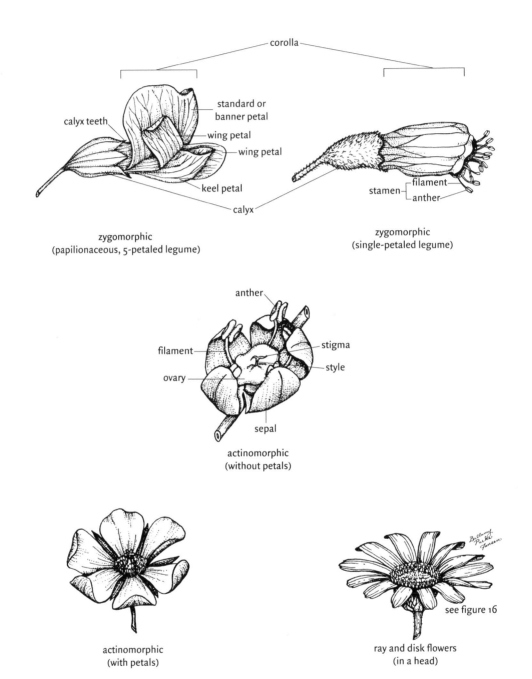

corolla

calyx teeth

standard or
banner petal

wing petal

wing petal

keel petal

calyx

zygomorphic
(papilionaceous, 5-petaled legume)

stamen — filament
— anther

zygomorphic
(single-petaled legume)

anther

filament

ovary

stigma

style

sepal

actinomorphic
(without petals)

actinomorphic
(with petals)

see figure 16

ray and disk flowers
(in a head)

Figure 15. Selected types of forb and woody plant flowers

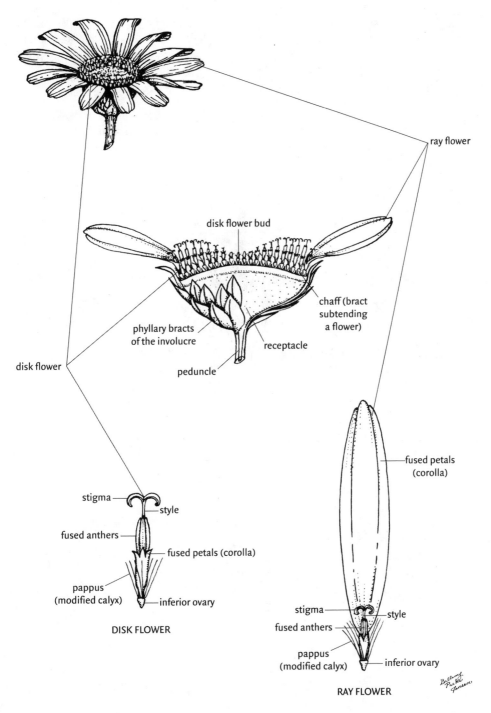

ray flower

disk flower bud

chaff (bract
subtending
a flower)

phyllary bracts
of the involucre

receptacle

disk flower

peduncle

stigma

style

fused anthers

fused petals (corolla)

pappus
(modified calyx)

inferior ovary

DISK FLOWER

fused petals
(corolla)

stigma

style

fused anthers

pappus
(modified calyx)

inferior ovary

RAY FLOWER

Figure 16. A forb and woody plant head inflorescence,
longitudinal-section, with ray and disk flowers

25

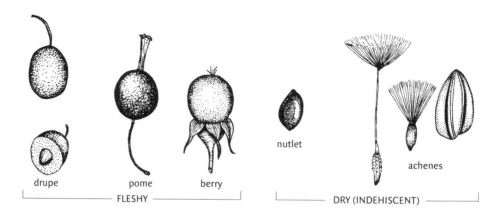

drupe   pome   berry   nutlet   achenes

FLESHY   DRY (INDEHISCENT)

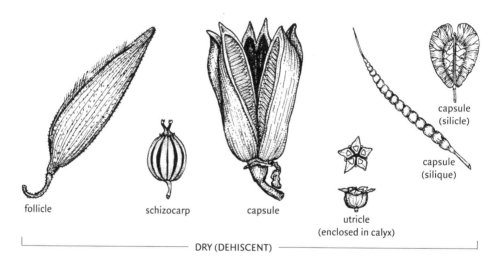

follicle   schizocarp   capsule   utricle (enclosed in calyx)   capsule (silicle)   capsule (silique)

DRY (DEHISCENT)

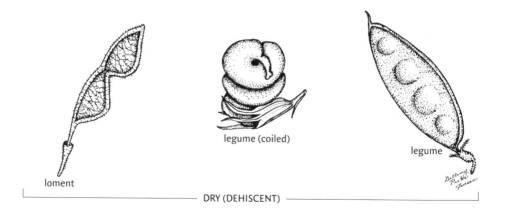

loment   legume (coiled)   legume

DRY (DEHISCENT)

Figure 17. Forb and woody plant fruits

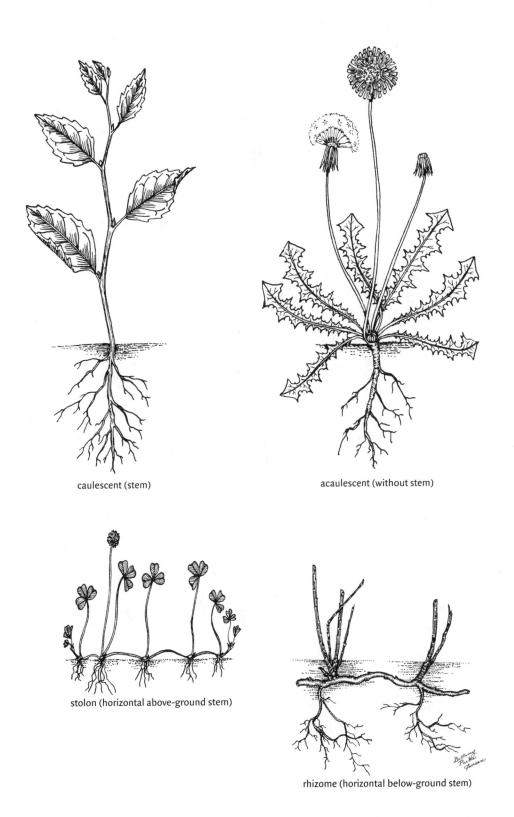

caulescent (stem)

acaulescent (without stem)

stolon (horizontal above-ground stem)

rhizome (horizontal below-ground stem)

Figure 18. Caulescent, acaulescent, and modified stems

27

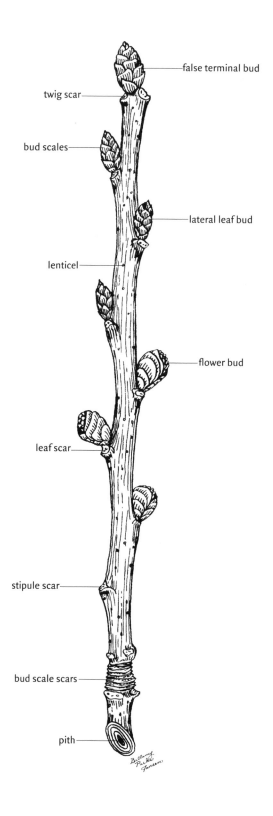

false terminal bud

twig scar

bud scales

lateral leaf bud

lenticel

flower bud

leaf scar

stipule scar

bud scale scars

pith

Figure 19. Twig parts

Grasses

Big bluestem
Andropogon gerardii Vitman

SYN = *A. furcatus* Muhl., *A. hallii* Hack., *A. gerardii* Vitman var. *chrysocomas* (Nash)
Fern., *A. gerardii* Vitman var. *paucipilus* (Nash) Fern.

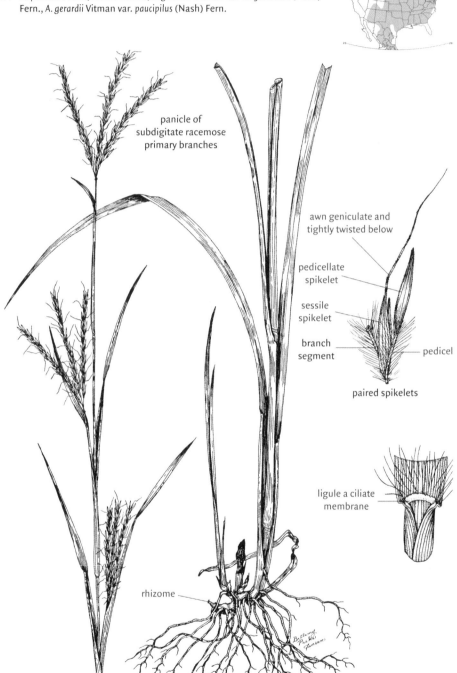

panicle of
subdigitate racemose
primary branches

awn geniculate and
tightly twisted below

pedicellate
spikelet

sessile
spikelet

branch
segment

pedicel

paired spikelets

ligule a ciliate
membrane

rhizome

Tribe:	ANDROPOGONEAE
Species:	*Andropogon gerardii* Vitman
Common Name:	Big bluestem (popotillo gigante, turkeyfoot)
Life Span:	Perennial
Origin:	Native
Season:	Warm

INFLORESCENCE CHARACTERISTICS

type: panicle of 2–6 (commonly 3) digitate or subdigitate racemose primary branches (4–11 cm long), long-exserted, terminal and axillary, fewer than 10 per culm, often purplish, sometimes yellowish

spikelets: paired; pedicellate and sessile spikelets nearly equal in length; lower spikelet sessile and perfect (7–10 mm long); pedicellate spikelet neuter or staminate

glumes: glumes of sessile spikelet subequal (5–11 mm long), first glume slightly grooved or dished; glumes of pedicellate spikelet not grooved (4–10 mm long)

awns: lemma of sessile spikelet awned; awn (1–2 cm long), geniculate and tightly twisted below; pedicellate spikelet awnless

VEGETATIVE CHARACTERISTICS

growth habit: rhizomatous, sometimes appearing cespitose

culms: erect (0.5–2.5 m tall), robust, sparingly branched toward summit, glabrous, glaucous

sheaths: compressed, purplish at base, lower sheaths sometimes villous, margins hyaline

ligules: ciliate membrane (0.4–2.5 mm long)

blades: flat to involute (5–50 cm long, 2–10 mm wide), lower blades often villous; margins scabrous

GROWTH CHARACTERISTICS: grows rapidly from midspring to early fall, many leaves produced in late spring and early summer, growing points stay near ground level until late summer, reproduces primarily from rhizomes

FORAGE VALUE: excellent and highly palatable to all classes of livestock when grazed or consumed in hay, commonly selected by livestock in preference to other grasses on summer range, becomes coarse late in the season

HABITAT: upland and lowland prairies, open woods, and wet overflow sites; adapted to all soil textures; most abundant in lowland prairies; frequently seeded in prairie restorations and for forage production

Broomsedge bluestem
Andropogon virginicus L.

SYN = *A. perangustatus* Nash

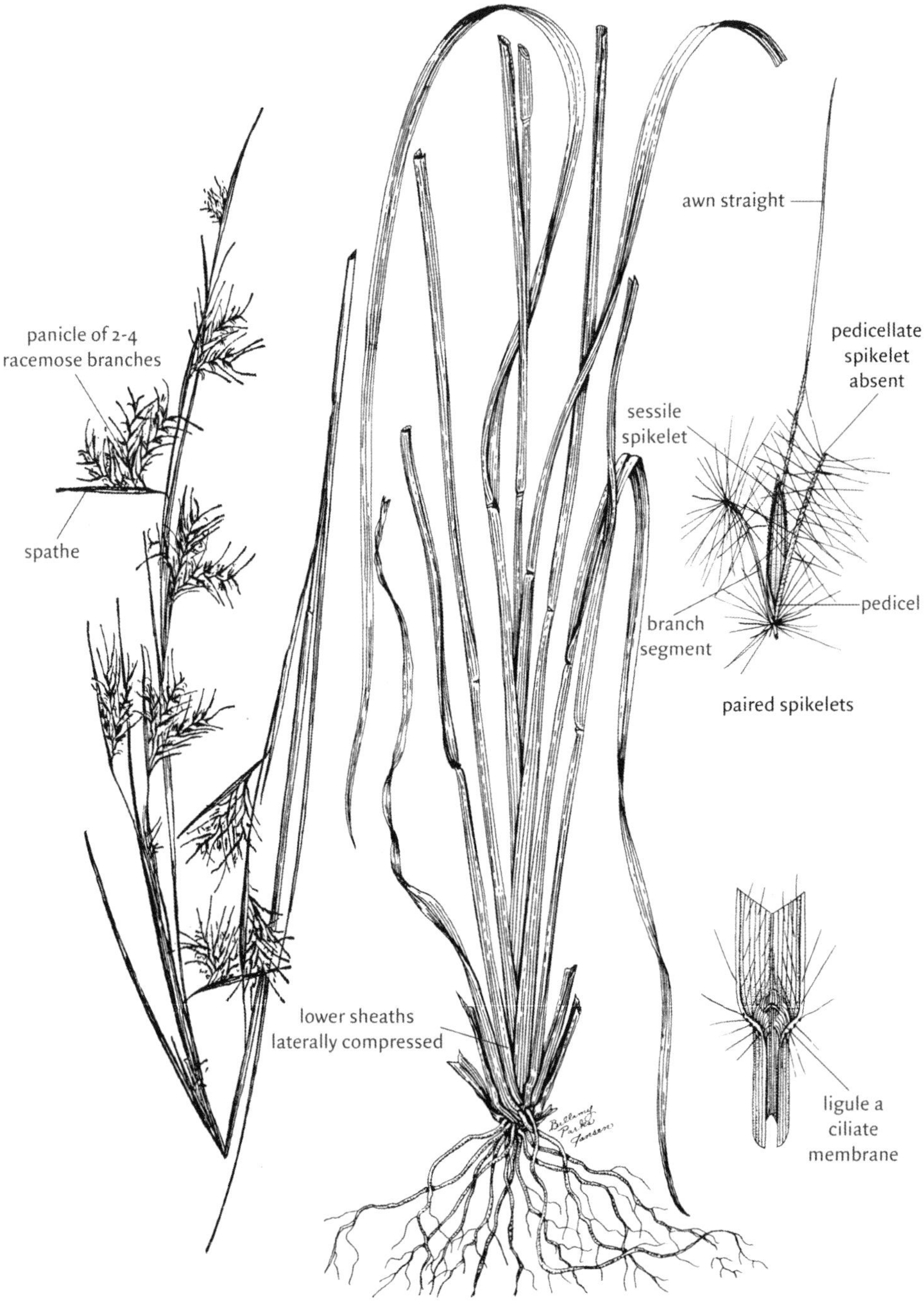

awn straight

panicle of 2-4
racemose branches

pedicellate
spikelet
absent

sessile
spikelet

spathe

pedicel

branch
segment

paired spikelets

lower sheaths
laterally compressed

ligule a
ciliate
membrane

Tribe:	ANDROPOGONEAE
Species:	*Andropogon virginicus* L.
Common Name:	Broomsedge bluestem (popotillo, broomsedge)
Life Span:	Perennial
Origin:	Native
Season:	Warm

INFLORESCENCE CHARACTERISTICS

type: panicle of 2–4 (commonly 2) racemose branches (2–3 cm long), numerous, broom-like appearance; bases of panicle branches enclosed in an inflated, tawny spathe (3–6 cm long, 2–5 mm wide)

spikelets: paired; lower spikelet sessile and perfect (3–4 mm long); pedicellate spikelet absent; pedicels villous

glumes: acuminate (2.5–4 mm long), green to yellowish

awns: upper lemma of sessile spikelet with delicate, straight awn (1–2 cm long)

VEGETATIVE CHARACTERISTICS

growth habit: cespitose

culms: erect (0.5–1.5 m tall), branched above, slender, sulcate on one side, glabrous or with a few short hairs; basal nodes flat

sheaths: imbricate, lower sheaths laterally compressed, strongly keeled; glabrous, scabrous, or pilose; margins hairy; usually wider than the blades

ligules: ciliate membrane (0.3–1 mm long), truncate

blades: flat or folded (8–45 cm long, 2–5 mm wide), tan to straw-colored at maturity, midrib prominent, glabrous to pilose adaxially and near collar

GROWTH CHARACTERISTICS: starts growth when daytime temperatures average 16–17°C, produces seeds mostly from August or September until frost, reproduces from seeds and tillers, grows in infertile soils, not shade tolerant; may rapidly increase with improper grazing; indicator of early stages of plant succession

FORAGE VALUE: poor for livestock and wildlife, except in early growth stages during spring and early summer, nearly unpalatable when mature; may be important for wildlife habitat

HABITAT: open ground, old fields, open woods, lowlands, and sterile hills; sandy to rocky moist soils; most common on improperly grazed rangeland and abandoned fields; will invade improved pastures

Silver bluestem
Bothriochloa laguroides (DC.) Herter

SYN = B. *longipaniculata* (Gould) Allred & Gould, B. *saccharoides* (Sw.) Rydb.,
     Andropogon saccharoides Sw.

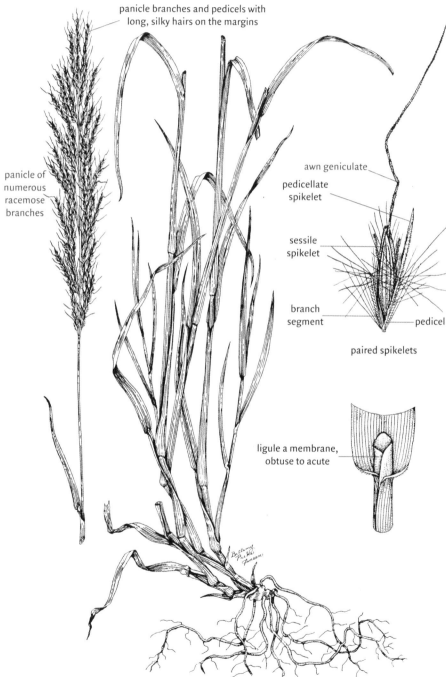

panicle branches and pedicels with
long, silky hairs on the margins

panicle of
numerous
racemose
branches

awn geniculate

pedicellate
spikelet

sessile
spikelet

branch
segment

pedicel

paired spikelets

ligule a membrane,
obtuse to acute

Tribe:	ANDROPOGONEAE
Species:	*Bothriochloa laguroides* (DC.) Herter
Common Name:	Silver bluestem (popotillo plateado, silver beardgrass)
Life Span:	Perennial
Origin:	Native
Season:	Warm

INFLORESCENCE CHARACTERISTICS

type: panicle of 6 to numerous racemose branches, long-exserted, elongate (7–15 cm long), terminal or axillary, usually fewer than 5 per culm, silvery white in color; branches erect to ascending (2–4 cm long), fringed on the margins with long silky hairs; branch joints and pedicels long-villous; pedicels sulcate, dumbbell-shaped in cross-section

spikelets: paired; lower spikelet sessile and perfect (3–4 mm long); pedicellate spikelet (1.5–3 mm long) neuter, narrow

glumes: unequal (2.5–4.5 mm long), firm but papery; first 2-keeled; second 1-keeled, 3-nerved

awns: upper lemma of sessile spikelet with delicate awn (8–25 mm long); awn geniculate

VEGETATIVE CHARACTERISTICS

growth habit: cespitose

culms: erect to decumbent (0.6–1.3 m tall), branched below, sulcate

sheaths: glabrous, keeled near collar; collar usually with a few long hairs on margin; hairs may extend up the leaf margins

ligules: membranous (1–4 mm long), obtuse to acute, erose to entire

blades: flat (2–25 cm long, 3–9 mm wide), linear, acuminate, glaucous; midrib prominent; margins white

GROWTH CHARACTERISTICS: starts growth in spring when daytime temperatures reach 21–24°C, inflorescences emerge 3–4 weeks later; produces abundant seeds, reproduces from seeds and tillers, seedlings must be protected from grazing to enhance establishment

FORAGE VALUE: fair for all classes of livestock and wildlife, only lightly grazed following maturity

HABITAT: prairies, rocky slopes, and roadsides; adapted to a broad range of soil textures, does not grow well on moist sites

Tanglehead
Heteropogon contortus (L.) Beauv. *ex* Roemer & Schultes

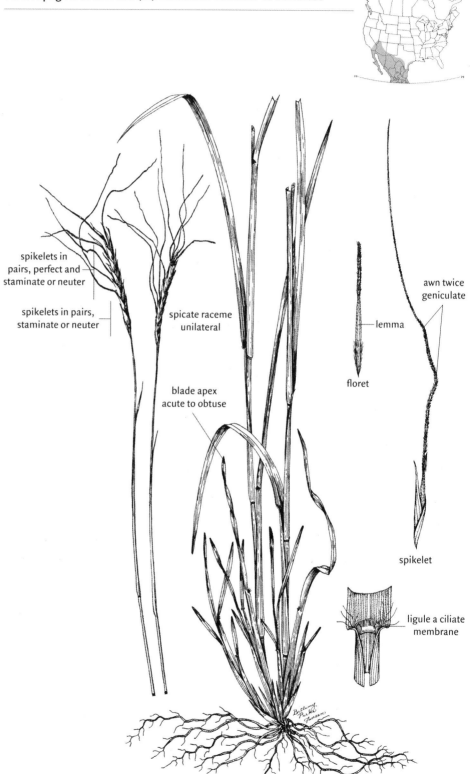

spikelets in pairs, perfect and staminate or neuter

spikelets in pairs, staminate or neuter

spicate raceme unilateral

blade apex acute to obtuse

lemma

floret

awn twice geniculate

spikelet

ligule a ciliate membrane

Tribe:	ANDROPOGONEAE
Species:	*Heteropogon contortus* (L.) Beauv. *ex* Roemer & Schultes
Common Name:	Tanglehead (barba negra, zacate colorado)
Life Span:	Perennial
Origin:	Native
Season:	Warm

INFLORESCENCE CHARACTERISTICS

type: unilateral spicate raceme (3–8 cm long, excluding awns), terminal, few

spikelets: paired, imbricate; sessile spikelets perfect (5–8 mm long); pedicellate spikelets staminate or neuter (7–10 mm long); both spikelets of the few to several pairs at the base of the inflorescence staminate or neuter

glumes: sessile spikelet glumes rounded, brownish, hispid; pedicellate spikelet glumes thin, green, hispid to glabrous

awns: upper lemma of sessile spikelet awned; awn twice-geniculate (5–12 cm long), hispid (hairs 0.5–1 mm long), dark brown to black at maturity, tangled with other awns; lemmas of pedicellate spikelet awnless

VEGETATIVE CHARACTERISTICS

growth habit: cespitose

culms: erect (20–80 cm tall), flat, branched at the base and at upper nodes

sheaths: compressed-keeled; margins glandular; collar with short hairs

ligules: ciliate membrane (1 mm long), acute to truncate

blades: flat (6–20 cm long, 3–7 mm wide); apex acute; midvein prominent adaxially; margins white-glandular, usually ciliate; apex and base red at maturity

GROWTH CHARACTERISTICS: starts growth in early spring, produces inflorescences June through November, low seed production; relatively easy to establish from seeds, reproduces from seeds and tillers

LIVESTOCK LOSSES: awns may be troublesome, especially to sheep

FORAGE VALUE: fair to good for cattle and horses before maturity; little value to sheep due to coarseness

HABITAT: open, dry rocky hills and canyons, usually in sandy soils; most abundant on heavily grazed rangeland

Little bluestem
Schizachyrium scoparium (Michx.) Nash

SYN = *Andropogon divergens* A. S. Hitchc., *A. littoralis* Nash, *A. scoparius* Michx.

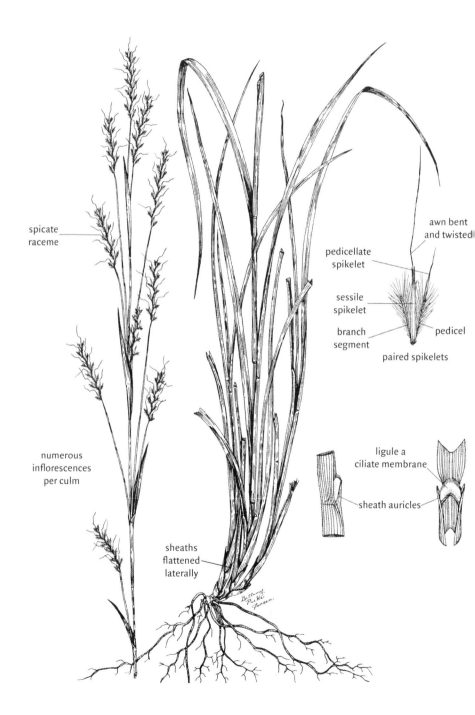

spicate
raceme

awn bent
and twisted

pedicellate
spikelet

sessile
spikelet

branch
segment

pedicel

paired spikelets

numerous
inflorescences
per culm

ligule a
ciliate membrane

sheath auricles

sheaths
flattened
laterally

Tribe:	ANDROPOGONEAE
Species:	*Schizachyrium scoparium* (Michx.) Nash
Common Name:	Little bluestem (popotillo colorado, prairie beardgrass)
Life Span:	Perennial
Origin:	Native
Season:	Warm

INFLORESCENCE CHARACTERISTICS

type: spicate raceme (2.5–5 cm long), several per culm, jointed, breaking apart as spikelet pairs with the rachis; peduncle included in sheath

spikelets: paired; sessile spikelet perfect (6–8 mm long); pedicellate spikelet staminate or neuter; rachis and pedicels pilose

glumes: glumes of sessile spikelet thickened (2.5–4 mm long), glabrous to scabrous, rounded on the back, firm

awns: upper lemma of sessile spikelet awned; awn bent and twisted (9–16 mm long); lemmas of pedicellate spikelet awnless or with a short awn

other: spicate racemes have a zig-zag pattern at maturity

VEGETATIVE CHARACTERISTICS

growth habit: cespitose with or without short rhizomes

culms: erect to decumbent (0.4–1.5 m tall), slender to robust, flat, branching above; base leafy, green to purplish, glaucous

sheaths: keeled, flattened laterally, glabrous to rarely pubescent

ligules: ciliate membrane (1–3 mm long), usually truncate

blades: linear (8–25 cm long, 2–6 mm wide), acute, glabrous to hispid, scabrous adaxially and on margins

GROWTH CHARACTERISTICS: starts growth in late spring, inflorescences appear in midsummer, matures in early fall, seeds mature October to November; reproduces from tillers, rhizomes, and seeds

FORAGE VALUE: good while immature for all classes of livestock; after inflorescences mature, forage is fair for cattle and horses but too coarse for sheep or goats; an important component of upland hay

HABITAT: prairies, open woods, and dry hills in all soil textures; most conspicuous with season-long, moderate grazing which allows spot grazing leaving some plants ungrazed; frequently used in seed mixtures for revegetation

Indiangrass
Sorghastrum nutans (L.) Nash

SYN = *S. avenaceum* (Michx.) Nash

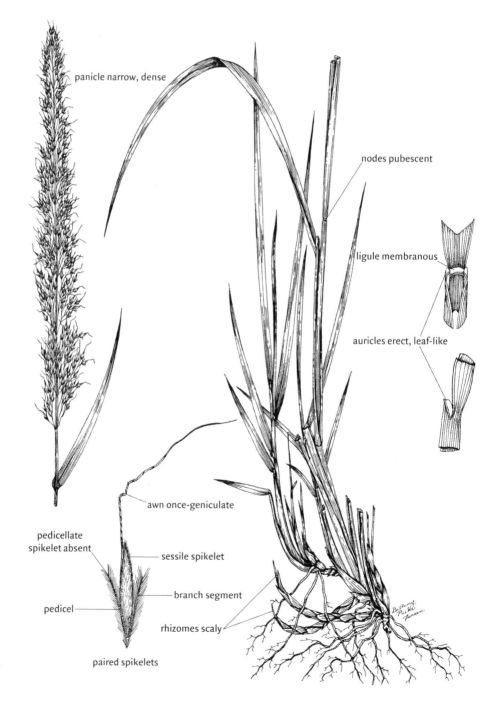

panicle narrow, dense

nodes pubescent

ligule membranous

auricles erect, leaf-like

awn once-geniculate

pedicellate spikelet absent

sessile spikelet

branch segment

pedicel

rhizomes scaly

paired spikelets

Tribe:	ANDROPOGONEAE
Species:	*Sorghastrum nutans* (L.) Nash
Common Name:	Indiangrass (zacate indio, yellow indiangrass)
Life Span:	Perennial
Origin:	Native
Season:	Warm

INFLORESCENCE CHARACTERISTICS

type: panicle (15–30 cm long), dense, yellowish or tawny; apex of branchlets, branch joints, and pedicels grayish-hirsute

spikelets: paired; sessile spikelet perfect (6–8 mm long); pedicellate spikelet absent and represented only by the hairy pedicel

glumes: subequal (5.7–7.5 mm long), leathery, tawny or yellowish, first hirsute with edges inflexed over the second

awns: upper lemma of perfect spikelet awned; once-geniculate (1.2–1.8 cm long), tightly twisted below the bend, loosely twisted above

VEGETATIVE CHARACTERISTICS

growth habit: cespitose with short, scaly rhizomes

culms: erect (1–2 m tall), robust to slender; nodes pubescent

sheaths: round or sometimes flattened, not keeled, glabrous to rarely pilose

ligules: membranous

blades: flat or somewhat keeled (5–40 cm long, 5–10 mm wide), constricted at the base, pointed, midvein conspicuous abaxially

other: sheath extended upward to form erect auricles (2–7 mm long) flanking the ligule

GROWTH CHARACTERISTICS: starts growth in midspring from short rhizomes, matures from September to November, also reproduces from seeds

FORAGE VALUE: excellent, palatable to cattle and horses throughout the summer, but does not cure well and is generally considered only moderately palatable after maturity and fair forage for winter grazing; produces good hay if cut before maturity

HABITAT: prairies, bottomlands, open woods, and meadows (moderately salt tolerant) in all soil textures; withstands occasional flooding; sometimes grown and managed with fertilization and irrigation in pure stands; a common component of seed mixtures for revegetation

Eastern gamagrass
Tripsacum dactyloides (L.) L.

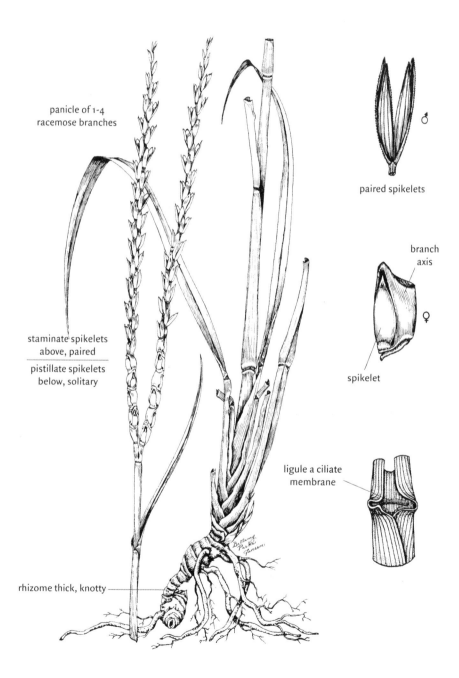

panicle of 1-4
racemose branches

♂

paired spikelets

branch
axis

♀

spikelet

staminate spikelets
above, paired

pistillate spikelets
below, solitary

ligule a ciliate
membrane

rhizome thick, knotty

Tribe:	ANDROPOGONEAE
Species:	*Tripsacum dactyloides* (L.) L.
Common Name:	Eastern gamagrass (zacate maicero, maicillo)
Life Span:	Perennial
Origin:	Native
Season:	Warm

INFLORESCENCE CHARACTERISTICS

type: panicle (12–25 cm long) of 1–4 racemose branches, terminal (occasionally axillary)

spikelets: unisexual; staminate spikelets above, paired, 2-flowered (7–11 mm long), sessile or one slightly pedicellate in 2 rows on the branch, often breaking apart at maturity; pistillate spikelets below (7–10 mm long), solitary, bead-like, embedded in branch, breaking into single spikelet segments at maturity

glumes: staminate spikelet glumes equal (5.5–10 mm long), somewhat pyriform, coriaceous, keeled; pistillate spikelet glumes equal (5–8.5 mm long), indurate, shiny, often embedded in branch

awns: none

other: plants monoecious; pistillate portion of inflorescence one-fourth or less of the entire length

VEGETATIVE CHARACTERISTICS

growth habit: rhizomatous; rhizomes thick, knotty, forming extensive colonies

culms: erect (1.5–3 m tall), decumbent below, stout, solid, slightly flattened, glabrous

sheaths: rounded, smooth, glabrous, usually shorter than the internodes

ligules: ciliate membrane (0.4–2.5 mm long), truncate

blades: flat (30–75 cm long, 1–3 cm wide); midrib prominent; margins scabrous

GROWTH CHARACTERISTICS: most growth is in the spring and early summer, stays green until late fall; produces seeds from July to September, although few seeds are produced; most reproduction is from rhizomes; usually one of the first species to be eliminated by improper grazing

FORAGE VALUE: excellent for all classes of livestock throughout the growing season, foliage breaks down rapidly and it is not dependable for winter grazing

HABITAT: swales, stream banks, and well-drained grasslands; most abundant in fertile soils; does not tolerate standing water for long periods; occasionally seeded in pastures

Prairie threeawn
Aristida oligantha Michx.

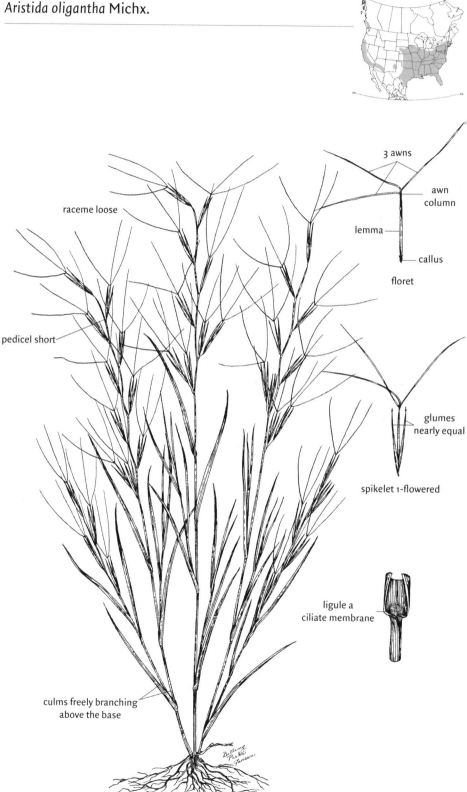

3 awns

awn column

lemma

callus

floret

raceme loose

glumes nearly equal

pedicel short

spikelet 1-flowered

ligule a ciliate membrane

culms freely branching above the base

Tribe:	ARISTIDEAE
Species:	*Aristida oligantha* Michx.
Common Name:	Prairie threeawn (tres barbas anual, oldfield threeawn)
Life Span:	Annual
Origin:	Native
Season:	Warm

INFLORESCENCE CHARACTERISTICS

type: raceme or rarely a panicle (5–20 cm long), loose

spikelets: 1-flowered, widely spaced; lemma firm (6–28 mm long, excluding the awn); callus pilose; pedicel short, scabrous, or pubescent

glumes: nearly equal (2–3 cm long), tapering to awn-like points; first glume 3- to 7-nerved; second glume usually 1-nerved

awns: lemma awn column branches into 3 awns, awns divergent (4–7 cm long), nearly equal or central awn longest; first glume awnless or short-awned; second glume mucronate

VEGETATIVE CHARACTERISTICS

growth habit: cespitose

culms: erect to decumbent (15–80 cm long), much-branched at lower nodes and freely branching above, glabrous to scabrous, base often purplish

sheaths: rounded on back, glabrous or slightly scabrous or pubescent on the sides of the collar

ligules: ciliate membrane (0.1–0.5 mm long)

blades: flat or loosely involute (10–25 cm long, 1–4 mm wide), involute at the tips, reduced upwards, glabrous to scabrous

GROWTH CHARACTERISTICS: seeds germinate in spring, completes life cycle in 2–3 months, seeds may be rapidly spread by wind and animals; an indicator of disturbance or improper grazing

LIVESTOCK LOSSES: long awns are tough, brittle, and can cause injury to eyes, nostrils, and mouths of livestock; awns also may decrease fleece value; losses may occur while animals are grazing or eating contaminated hay

FORAGE VALUE: fair to poor for a brief period in early growth stages, otherwise worthless; its presence in hay greatly reduces the value of the hay

HABITAT: disturbed areas in dry soils, most common on sandy and sandy calcareous soils of deteriorated rangeland, waste ground, and abandoned fields

Purple threeawn
Aristida purpurea Nutt.

SYN = *A. fendleriana* Steud., *A. longiseta* Steud., *A purpurea* Nutt. var. *longiseta* (Steud.)
Vasey, *A. purpurea* Nutt. var. *nealleyi* (Vasey) Allred, *A. purpurea* var. *fendleriana*
(Steud.) Vasey, *A. purpurea* Nutt. var. *wrightii* (Nash) Allred, *A. roemeriana*
Scheele, *A. wrightii* Nash

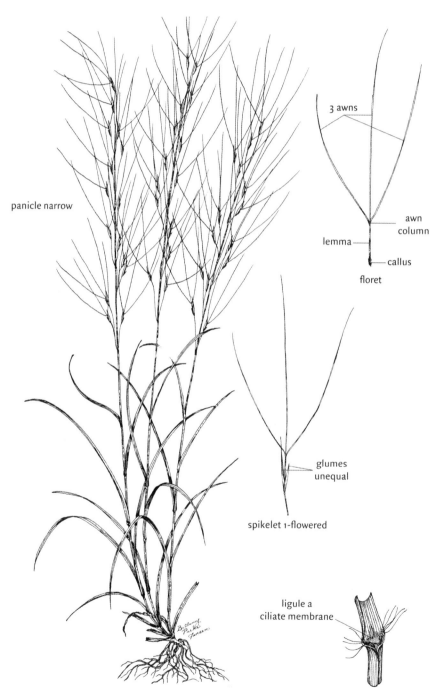

panicle narrow

3 awns

awn
column

lemma

callus

floret

glumes
unequal

spikelet 1-flowered

ligule a
ciliate membrane

Tribe:	ARISTIDEAE
Species:	*Aristida purpurea* Nutt.
Common Name:	Purple threeawn (tres barbas púrpura, red threeawn, dogtowngrass)
Life Span:	Perennial
Origin:	Native
Season:	Warm

INFLORESCENCE CHARACTERISTICS

type: panicle (2–30 cm long), occasionally reduced to a raceme with 3 to 8 spikelets, often purplish to reddish, narrow; branches erect or somewhat flexuous

spikelets: 1-flowered; lemma firm (8–15 mm long, excluding the awn); callus short-bearded

glumes: unequal, broad; first glume about one-half as long (4–15 mm long) as the second (8–26 mm long)

awns: lemma awn column divides into 3 nearly equal awns; awns divergent (4–10 cm long)

VEGETATIVE CHARACTERISTICS

growth habit: cespitose

culms: erect (10–80 cm tall), glabrous

sheaths: glabrous to weakly scabrous; collar often pubescent with a tuft of long, soft hairs on both sides

ligules: ciliate membrane (0.1–0.6 mm long)

blades: involute (2–30 cm long, 1–2 mm wide), curved; apex acute; scabrous adaxially, sometimes hirsute; mostly basal

GROWTH CHARACTERISTICS: starts growth in late spring, strong competitor; produces abundant seeds, reproduces from seeds and tillers; an indicator of range deterioration

LIVESTOCK LOSSES: awns may decrease fleece value; awns may also cause irritation and abscesses in the mouths and nostrils and damage eyes of grazing animals

FORAGE VALUE: poor (rarely fair) for livestock and wildlife, grazed only in early growth stages before awn development, worthless in winter

HABITAT: rangeland, prairies, wastelands, and disturbed sites; in soils of all textures, especially dry sandy soils; most abundant on abused rangeland

Redtop
Agrostis stolonifera L.

SYN = *A. alba* var. *palustris* (Huds.) Pers., *A. alba* var. *stolonifera* (L.) J. G. Smith

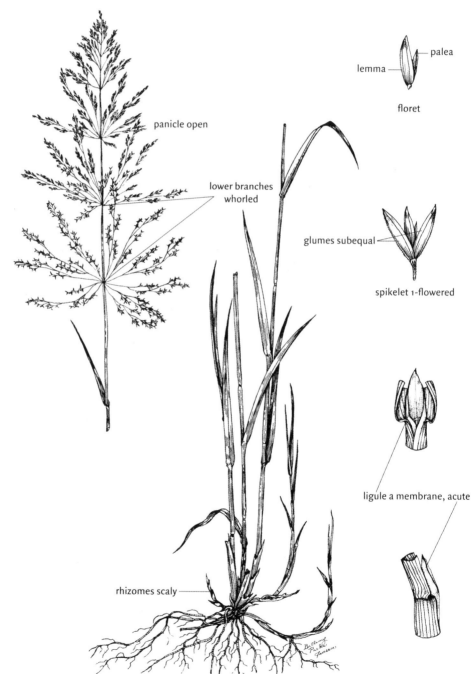

palea

lemma

floret

panicle open

lower branches
whorled

glumes subequal

spikelet 1-flowered

ligule a membrane, acute

rhizomes scaly

48

Tribe:	AVENEAE
Species:	*Agrostis stolonifera* L.
Common Name:	Redtop (zacate punta café, zacate de piedras castillitos, creeping bentgrass, redtop bent)
Life Span:	Perennial
Origin:	Introduced (from Europe)
Season:	Cool

INFLORESCENCE CHARACTERISTICS

type: panicle (5–30 cm long, 5–50 mm wide), open, pyramidal to ovate; branches spreading or ascending to appressed, densely flowered, purplish to reddish; lower branches whorled, not densely flowering at base; flowering nodes reddish

spikelets: 1-flowered; lemma blunt-tipped (1.2–2.5 mm long); palea one-half to two-thirds as long as the lemma

glumes: subequal (1.5–3 mm long), acute, glabrous, keel sometimes scabrous, longer than the lemma

awns: none (lemma rarely awned)

VEGETATIVE CHARACTERISTICS

growth habit: rhizomatous, occasionally with stolons; rhizomes scaly, forming a sod

culms: erect (0.2–1.5 m tall), sometimes decumbent at the base

sheaths: round, glabrous, frequently purplish to reddish

ligules: membranous (1–7 mm long), acute, erose to entire

blades: flat (4–25 cm long, 1–10 mm wide), rolled in the bud, narrow; apex acute; midvein prominent abaxially; margins smooth to scabrous

GROWTH CHARACTERISTICS: starts growth in early spring, flowers in early summer, seeds mature by August, reproduces from rhizomes, stolons, and seeds

FORAGE VALUE: good to very good for cattle and horses, fair to good for sheep, fair for elk; commonly a component of meadow hay, hay quality is acceptable if cut no later than the early flowering stage after which quality rapidly declines

HABITAT: pastures, low ground, and moist meadows; can withstand flooding for extended periods, widely naturalized, will grow on acidic soils, moderately salt tolerant, adapted to a wide range of soil and climatic conditions; commonly seeded in pastures, moist meadows, and irrigated pastures

Slender oat
Avena barbata Pott ex Link

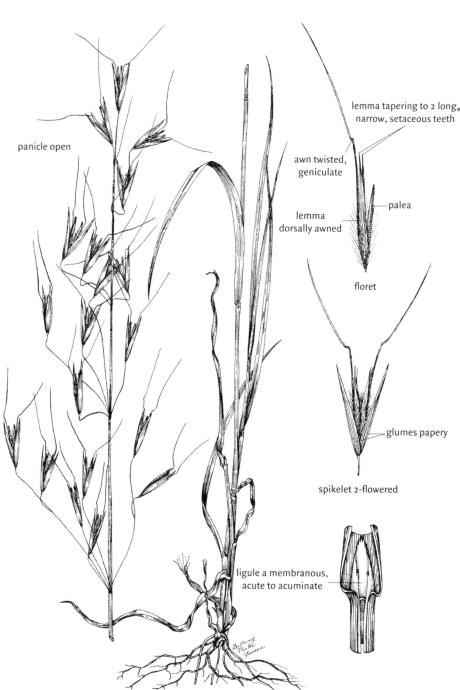

panicle open

lemma tapering to 2 long, narrow, setaceous teeth

awn twisted, geniculate

lemma dorsally awned

palea

floret

glumes papery

spikelet 2-flowered

ligule a membranous, acute to acuminate

Tribe:	AVENEAE
Species:	*Avena barbata* Pott ex Link
Common Name:	Slender oat (avena salvaje, avenilla)
Life Span:	Annual
Origin:	Introduced (from Europe)
Season:	Cool

INFLORESCENCE CHARACTERISTICS

type: panicle (20–40 cm long), open, loose

spikelets: 2-flowered on curved and capillary pedicels, drooping; lemmas with stiff red hairs to the middle, tapering to 2 long and narrow, setaceous teeth (3–4 mm long); pedicels curved, smaller in diameter than those of *Avena fatua*

glumes: subequal (2–2.5 cm long), papery, glabrous, acute, 7-nerved; nerves prominent

awns: lemmas dorsally awned; awn twisted (3–4 cm long), geniculate, stout, reddish-brown to black

VEGETATIVE CHARACTERISTICS

growth habit: cespitose, sometimes solitary culms

culms: erect (0.3–1.2 m tall), slender as compared to *Avena fatua*

sheaths: rounded to somewhat keeled, glabrous or nearly so

ligules: membranous (4 mm long), acute to acuminate, erose

blades: flat to slightly keeled (10–40 cm long, 5–10 mm wide), scabrous on both sides; margins often pilose

GROWTH CHARACTERISTICS: germinates in late fall or early winter, most growth is in early spring during which it grows more rapidly than associated species; flowers March to June; trampling by grazing animals after seeds have matured helps to plant seeds

FORAGE VALUE: good for all classes of livestock during the winter and spring growth period, low quality and palatability after it matures in summer; produces hay with acceptable forage quality if cut before maturity, although cutting at that time will eliminate seed production; considered to be a weed on disturbed sites

HABITAT: foothill rangeland, fields, road rights-of-way, disturbed areas, and waste places; adapted to a wide variety of soils; most abundant in dry, coarse-textured soils

Wild oat
Avena fatua L.

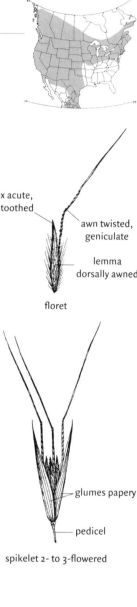

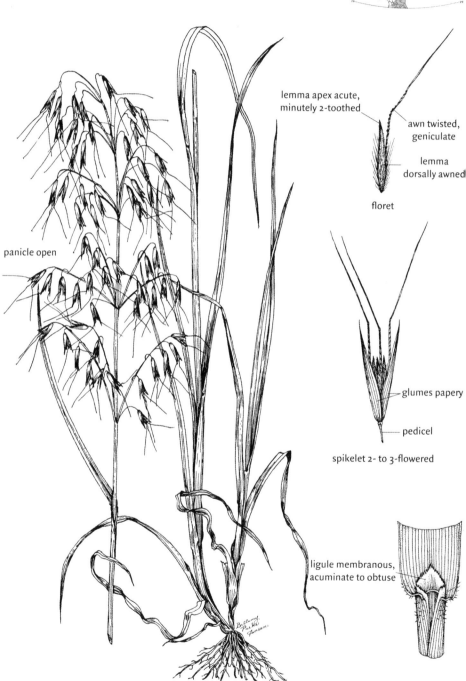

lemma apex acute,
minutely 2-toothed

awn twisted,
geniculate

lemma
dorsally awned

floret

panicle open

glumes papery

pedicel

spikelet 2- to 3-flowered

ligule membranous,
acuminate to obtuse

Tribe:	AVENEAE
Species:	*Avena fatua* L.
Common Name:	Wild oat (avena silvestre, avena guacha, avena loca)
Life Span:	Annual
Origin:	Introduced (from Europe)
Season:	Cool

INFLORESCENCE CHARACTERISTICS

type: panicle (10–30 cm long) or rarely a raceme, loose, open, usually 8–35 spikelets; branches unequal, horizontally spreading to ascending; spikelets pendulous

spikelets: 2- or 3-flowered, rarely 4-flowered; lemmas firm (lowermost lemma 1.5–2 cm long), apex with long hairs and minutely 2-toothed, acute; palea thin, slightly shorter than lemma

glumes: subequal (1.5–2.5 cm long), papery, acute, glabrous; first 7-nerved, second 9-nerved

awns: lemma dorsally awned; awn twisted (3–4 cm long), geniculate, stout, reddish-brown to black

VEGETATIVE CHARACTERISTICS

growth habit: cespitose, sometimes solitary culms

culms: erect (0.3–1.3 m tall), stout, glabrous

sheaths: rounded to somewhat keeled, open; margins broad and thick, glabrous to pubescent; collar pilose on front margins

ligules: membranous (1.5–5 mm long), acuminate to obtuse, erose

blades: flat when mature (10–30 cm long, 5–15 mm wide), scabrous on both sides; margins glabrous to pilose (especially near the base)

GROWTH CHARACTERISTICS: germinates in early winter, most growth in early spring, flowers March to May, seeds are set in June, reproduces from seeds; highest production is on moist and rich soils

FORAGE VALUE: good to excellent for grazing by all classes of livestock until after the florets are shed and the herbage dies, produces good quality hay if cut before maturity; commonly makes up the majority of early forage for grazing on annual rangeland

HABITAT: valleys and open slopes of foothill ranges; cultivated soils and disturbed soils in waste places and road rights-of-way

Bluejoint
Calamagrostis canadensis (Michx.) Beauv.

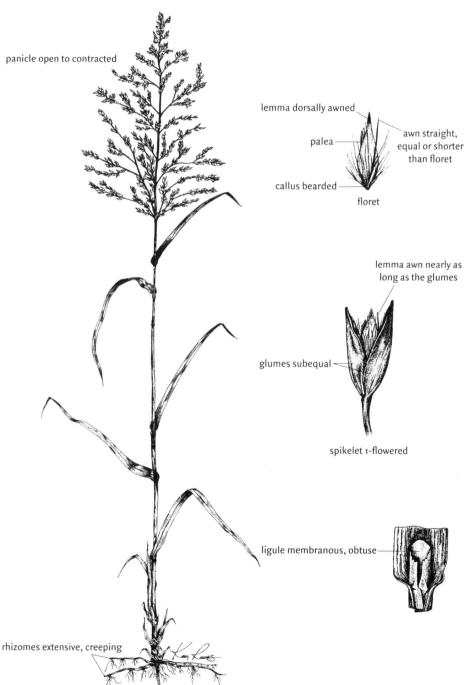

panicle open to contracted

lemma dorsally awned

awn straight, equal or shorter than floret

palea

callus bearded

floret

lemma awn nearly as long as the glumes

glumes subequal

spikelet 1-flowered

ligule membranous, obtuse

rhizomes extensive, creeping

Tribe:	AVENEAE
Species:	*Calamagrostis canadensis* (Michx.) Beauv.
Common Name:	Bluejoint (meadow pinegrass, Canadian reedgrass, marsh pinegrass)
Life Span:	Perennial
Origin:	Native
Season:	Cool

INFLORESCENCE CHARACTERISTICS

type: panicle (7–20 cm long, 1–15 cm wide), open to somewhat contracted, often nodding, pale or purple in color; branches mostly visible

spikelets: 1-flowered (2–4.5 mm long); lemma (1.5–4 mm long) nearly as long as the glumes, apex 2- to 4-toothed; callus bearded, hairs (1–2.5 mm long) often as long as the lemma

glumes: subequal (1.5–4.5 mm long), second slightly shorter than the first; acute or acuminate, keel scabrous

awns: lemma dorsally awned from the middle or below; awn slender, usually straight (0.2–2.2 mm long), equal to or shorter than the floret, nearly as long as the glumes

VEGETATIVE CHARACTERISTICS

growth habit: weakly cespitose to solitary, with extensively creeping rhizomes

culms: erect (0.7–1.3 m tall), stout, glabrous

sheaths: round, smooth, open, glabrous to scabrous; collar yellowish, constricted, glabrous to scabrous

ligules: membranous (3–7 mm long), obtuse, entire to erose-ciliate or lacerate

blades: flat to slightly involute (8–40 cm long, 2–7 mm wide), often drooping, glabrous to scabrous, widely spaced ridges and furrows

other: a highly variable species with several varieties

GROWTH CHARACTERISTICS: flowers June to August; reproduces by seeds, rhizomes, and tillers

FORAGE VALUE: fair to good for cattle and horses in the spring, fair for wildlife; palatability decreases in the summer; makes good quality hay if cut early

HABITAT: marshes, sloughs, meadows, and other wet places

Pine reedgrass
Calamagrostis rubescens Buckl.

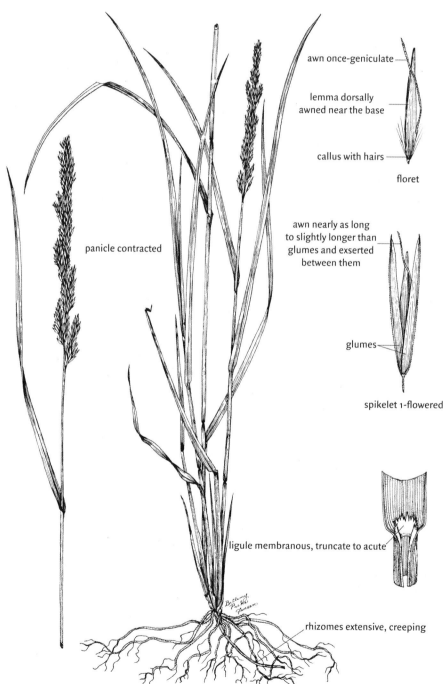

awn once-geniculate

lemma dorsally awned near the base

callus with hairs

floret

panicle contracted

awn nearly as long to slightly longer than glumes and exserted between them

glumes

spikelet 1-flowered

ligule membranous, truncate to acute

rhizomes extensive, creeping

Tribe:	AVENEAE
Species:	*Calamagrostis rubescens* Buckl.
Common Name:	Pine reedgrass (pinegrass)
Life Span:	Perennial
Origin:	Native
Season:	Cool

INFLORESCENCE CHARACTERISTICS

type: panicle (7–18 cm long), contracted, dense to occasionally loose, interrupted, pale green to purplish

spikelets: 1-flowered; lemma pale (2.5–5 mm long), nearly as long as the glumes, smooth to scaberulous; callus with tuft of hairs (about 1 mm long); palea shorter than the lemma

glumes: equal to subequal (3–6 mm long), narrow, acuminate, glabrous to scabridulous; first glume 1-nerved; second 3-nerved, lateral nerves obscure

awns: lemma dorsally awned from near base; awn delicate (2.5–4.5 mm long), exserted at side between the glumes, nearly as long to slightly longer than glumes, once-geniculate

VEGETATIVE CHARACTERISTICS

growth habit: cespitose or solitary, with extensively creeping rhizomes; leaves mostly basal

culms: erect (0.4–1 m tall), slender, smooth; nodes may be dark; basal internodes may be reddish

sheaths: smooth, distinctly veined, often purplish at base, generally glabrous, collar pubescent although sometimes obscurely so

ligules: membranous (1–4 mm long), truncate to acute, erose

blades: flat (8–35 cm long, 2–6 mm wide) or involute at the apex, ascending with curved or drooping apexes; scabrous to pilose

GROWTH CHARACTERISTICS: starts growth in early spring, flowers in July or August, reproduces from seeds and rhizomes; remains green late into the fall; quickly declines with excessive use

FORAGE VALUE: poor to fair for sheep, fair for cattle and horses, less desirable for big game although elk and deer graze it in the spring when it is immature and most palatable; herbage becomes unpalatable with maturity

HABITAT: open to dense pine woods, prairies, meadows, and stream banks in dry to moderately moist soils; will not persist in open sunlight

Tufted hairgrass
Deschampsia cespitosa (L.) Beauv.

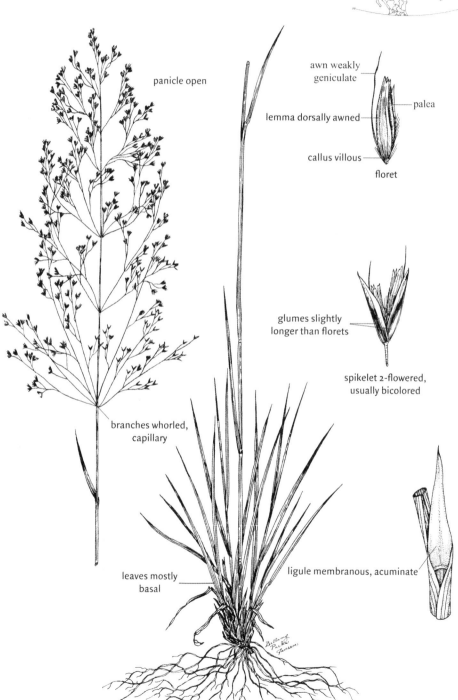

panicle open

awn weakly geniculate

palea

lemma dorsally awned

callus villous

floret

glumes slightly longer than florets

spikelet 2-flowered, usually bicolored

branches whorled, capillary

leaves mostly basal

ligule membranous, acuminate

Tribe:	AVENEAE
Species:	*Deschampsia cespitosa* (L.) Beauv.
Common Name:	Tufted hairgrass (zacate de montaña, saltandpeppergrass)
Life Span:	Perennial
Origin:	Native
Season:	Cool

INFLORESCENCE CHARACTERISTICS

type: panicle (10–25 cm long), open to narrow, loose and often nodding, shiny black and light brown when mature; branches whorled, capillary

spikelets: 2-flowered (rarely 3-flowered), pale or dark purplish-black, usually bicolored (hence the name saltandpeppergrass); lemmas purplish-black at base (2.5–4.5 mm long), membranaceous with erose or lacerate apex, apex usually light brown; callus villous; paleas about equaling the lemmas

glumes: subequal, slightly longer than florets, lanceolate, acute, glabrous or scaberulous; first (2–5 mm long) 1-nerved; second (2.5–5.2 mm long) 1- to 3-nerved

awns: lemma dorsally awned from near the base; awn weakly geniculate (2–6 mm long), length varies from short to twice as long as spikelet

VEGETATIVE CHARACTERISTICS

growth habit: strongly cespitose; leaves mostly basal

culms: erect (0.2–1.2 m tall), smooth

sheaths: keeled, open, glabrous to scabrous, veins prominent

ligules: membranous (3–10 mm long), acuminate, entire to lacerate

blades: flat or folded (5–30 cm long, 1–4 mm wide), firm, constricted at collar; margins scabrous; ridged and scabrous adaxially, glabrous abaxially

GROWTH CHARACTERISTICS: starts growth early in the spring, flowers from July to September, seeds mature August to September, reproduces from seeds and tillers

FORAGE VALUE: good to excellent for all classes of livestock and fair to good for wildlife, produces good quality hay

HABITAT: wet meadows, prairies, stream banks, ditches, and open bogs; as well as in the spruce–fir zone; adapted to a wide range of soil textures; occasionally occurs in moderately saline and alkaline soils

Spike oat
Helictotrichon hookeri (Scribn.) Henr.

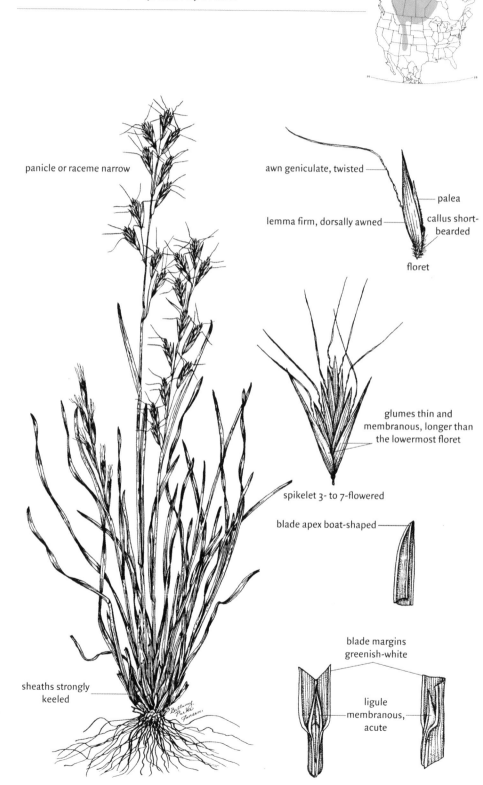

panicle or raceme narrow

awn geniculate, twisted

palea

lemma firm, dorsally awned

callus short-bearded

floret

glumes thin and membranous, longer than the lowermost floret

spikelet 3- to 7-flowered

blade apex boat-shaped

blade margins greenish-white

ligule membranous, acute

sheaths strongly keeled

Tribe:	AVENEAE
Species:	*Helictotrichon hookeri* (Scribn.) Henr.
Common Name:	Spike oat
Life Span:	Perennial
Origin:	Native
Season:	Cool

INFLORESCENCE CHARACTERISTICS

type: panicle (5–12 cm long) or raceme, narrow, long-exserted at maturity; branches erect or ascending, upper branches with a single spikelet, lower branches with 2 spikelets

spikelets: 3- to 7-flowered; lemmas firm (1–1.2 cm long), 5-nerved, brown, toothed at the apex, rounded on the back; callus short-bearded; paleas well developed, shorter than the lemmas; rachilla villous

glumes: subequal (8–15 mm long), apex acute, thin and membranous; first 3-nerved; second 3- to 5-nerved

awns: lemma dorsally awned; awn (1–1.8 cm long) from slightly above the middle of the abaxial side of the lemma, geniculate (often twice), twisted

VEGETATIVE CHARACTERISTICS

growth habit: cespitose; leaves mostly basal

culms: erect (10–75 cm tall), glabrous, smooth or scaberulous

sheaths: strongly keeled, open to the base, glabrous

ligules: membranous (2–7 mm long), acute, erose to lacerate, sometimes entire, whitish

blades: flat or involute (4–20 cm long, 1–4 mm wide), folded in the bud, twisted with age; apex boat-shaped; midrib prominent and whitish abaxially; scabrous abaxially; slightly pubescent adaxially; margins somewhat thickened, greenish-white

GROWTH CHARACTERISTICS: starts growth in early spring, flowers June to July, seeds mature July to August, reproduces from seeds and tillers, may regrow in fall if adequate soil moisture is available

FORAGE VALUE: good for cattle and horses, fair for sheep and big game; quality rapidly declines with maturity

HABITAT: hillsides, prairies, foothills, and mountain tops at or near the tree line, in all soil textures; most abundant in moist to moderately dry soils

Junegrass
Koeleria macrantha (Ledeb.) J. A. Schultes

SYN = K. *cristata* Pers., K. *nitida* Nutt., K. *pyramidata* (Lam.) Beauv.

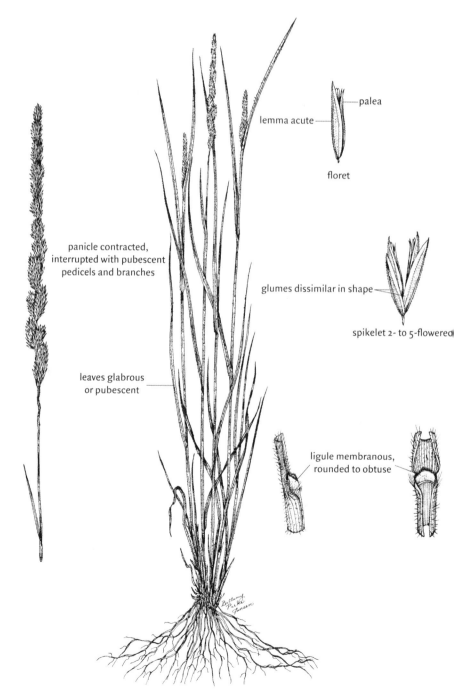

palea

lemma acute

floret

panicle contracted, interrupted with pubescent pedicels and branches

glumes dissimilar in shape

spikelet 2- to 5-flowered

leaves glabrous or pubescent

ligule membranous, rounded to obtuse

Tribe:	AVENEAE
Species:	*Koeleria macrantha* (Ledeb.) J. A. Schultes
Common Name:	Junegrass (zacate de Junio, zacate de cresta, prairie junegrass)
Life Span:	Perennial
Origin:	Native
Season:	Cool

INFLORESCENCE CHARACTERISTICS

type: panicle (3–18 cm long, 1–3 cm wide) dense, contracted, interrupted, tapered toward the apex; branches erect to spreading at anthesis, pubescent; pedicels pubescent; one of the most variable grasses

spikelets: 2- to 5-flowered (4–6 mm long), flattened laterally; lemmas narrow (3–6 mm long), lanceolate, acute (often appearing awn-tipped), mid-nerve scabrous; rachilla pubescent

glumes: subequal (3–6 mm long), dissimilar in shape; first 1-nerved, narrow; second 3-nerved and broadened about the middle, shiny and translucent, shorter than first floret

awns: lemmas very rarely awned

VEGETATIVE CHARACTERISTICS

growth habit: cespitose; leaves mostly basal

culms: erect (20–60 cm tall), may be pubescent below the panicle

sheaths: rounded, distinctly veined, retrorsely pubescent, upper sheaths may be glabrous; collar pilose on margins

ligules: membranous (0.5–1.5 mm long), may be minutely ciliate, rounded to obtuse, erose to entire

blades: flat or involute (3–25 cm long, 1–3 mm wide), folded in bud; apex blunt or prow-shaped; coarsely veined adaxially, glabrous or pubescent abaxially; may have long hairs near collar

GROWTH CHARACTERISTICS: starts growth in early spring, flowers in June and July, produces seed through September, may regrow in the fall if soil moisture is adequate; reproduces from seeds and tillers

FORAGE VALUE: excellent for all classes of livestock, although its forage production is low; good for wildlife in spring and in the fall after curing, less palatable during seed production; hay quality is fair to good

HABITAT: prairies, open woods, foothills, and subalpine areas in all soil textures; normally not found in wetlands

Reed canarygrass
Phalaris arundinacea L.

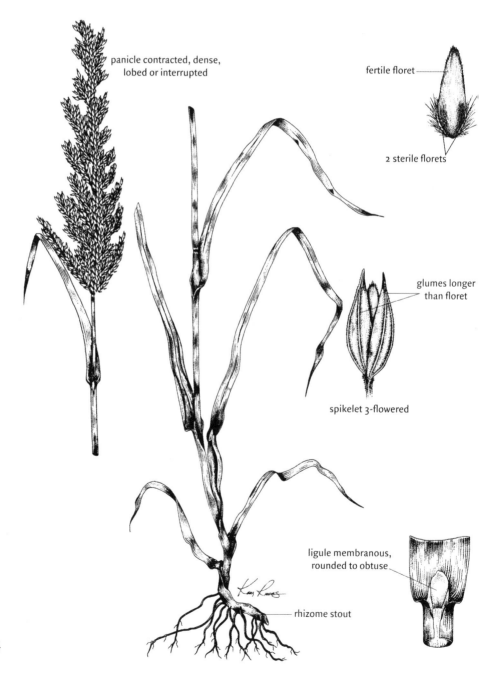

panicle contracted, dense,
lobed or interrupted

fertile floret

2 sterile florets

glumes longer
than floret

spikelet 3-flowered

ligule membranous,
rounded to obtuse

rhizome stout

Tribe:	AVENEAE
Species:	*Phalaris arundinacea* L.
Common Name:	Reed canarygrass
Life Span:	Perennial
Origin:	Native
Season:	Cool

INFLORESCENCE CHARACTERISTICS

type: panicle (6–18 cm long, 1–3 cm wide), contracted, dense, lobed or interrupted; spikelets occurring in clusters on short ascending scabrous branches, reddish or purplish in anthesis, becoming straw-colored

spikelets: 3-flowered; 1 fertile floret (2.7–4.2 mm long) with 2 sterile florets below and unlike, glabrous to appressed-pubescent; fertile lemma ovate (2.5–4.1 mm long), firm and shiny, villous on margins; sterile lemmas scale-like (1–2 mm long), villous, appressed to fertile lemma base

glumes: subequal (3.4–7 mm long), 3-nerved, ovate to lanceolate, greenish-white to purplish becoming straw colored, midnerve scabrous, not winged

awns: none

VEGETATIVE CHARACTERISTICS

growth habit: rhizomatous; rhizomes stout, forming large colonies

culms: erect (0.6–2.1 m tall) to occasionally geniculate below, glabrous

sheaths: round, lower sheaths longer than the internodes, with obvious air chambers, cross-septate, glabrous, sometimes purplish

ligules: membranous (2–8 mm long), rounded to obtuse, entire to lacerate

blades: flat (7–40 cm long, 4–20 mm wide), glabrous to scabrous, midrib prominent abaxially

GROWTH CHARACTERISTICS: reproduces by rhizomes, tillers, and seeds; forms dense colonies, aggressive and may dominate wet areas, often seeded or transplanted for wetland recovery; a form of this species with white striped-leaves is used as an ornamental

FORAGE VALUE: good to fair for livestock, provides good hay if cut prior to maturity, fair for wildlife, provides important nesting cover for waterfowl

HABITAT: wet meadows, marshes, ditches, edges of ponds and streams, and roadside wetlands

Alpine timothy
Phleum alpinum L.

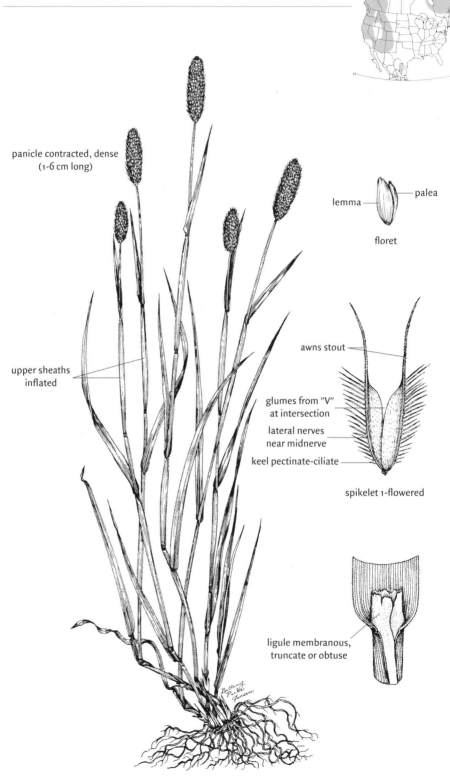

panicle contracted, dense
(1-6 cm long)

upper sheaths
inflated

lemma — palea

floret

awns stout

glumes from "V"
at intersection

lateral nerves
near midnerve

keel pectinate-ciliate

spikelet 1-flowered

ligule membranous,
truncate or obtuse

Tribe:	AVENEAE
Species:	*Phleum alpinum* L.
Common Name:	Alpine timothy (timothy alpino, mountain timothy, wild timothy)
Life Span:	Perennial
Origin:	Native
Season:	Cool

INFLORESCENCE CHARACTERISTICS

type: panicle (1–6 cm long, 8–12 mm wide), contracted, dense, ovoid to cylindric, usually from 1.5 to 3 times as long as wide, often purplish in color, bristly

spikelets: 1-flowered, elliptic, small, somewhat flattened; lemma (1.7–2.5 mm long) shorter than glumes, lance-ovate, glabrous to puberulent, apex minutely erose; palea slightly shorter than the lemma

glumes: subequal (3–5 mm long), keeled, pectinate-ciliate on keel, hyaline, gradually tapering to form a "V" where the two glumes intersect; lateral nerves near the midnerve; first glume sometimes with ciliate margins

awns: glumes awned from apex; awn stout (1.5–3 mm long)

VEGETATIVE CHARACTERISTICS

growth habit: usually cespitose; base of culm may be decumbent, sometimes creeping, and forming a sod

culms: erect to occasionally decumbent at base (15–60 cm tall), glabrous, internodes visible

sheaths: upper sheaths inflated, glabrous, prominently veined, open; collar yellowish; auricles rare, small, blunt to rounded

ligules: membranous (0.5–4 mm long), truncate or obtuse, erose

blades: flat (2–15 cm long, 3–8 mm wide), rolled in bud, tapering; scabrous abaxially and on margins

GROWTH CHARACTERISTICS: starts growth in early spring, flowers June to August, remains green throughout the summer, reproduces from seeds and tillers; requires occasional rest from heavy grazing to sustain productivity

FORAGE VALUE: good to excellent for all classes of livestock, as well as for elk and deer; most valuable in summer

HABITAT: mountain meadows, bogs, and moist woods in deep, poorly drained soils to deep stony loam soils; generally above 1,250 m elevation

Timothy
Phleum pratense L.

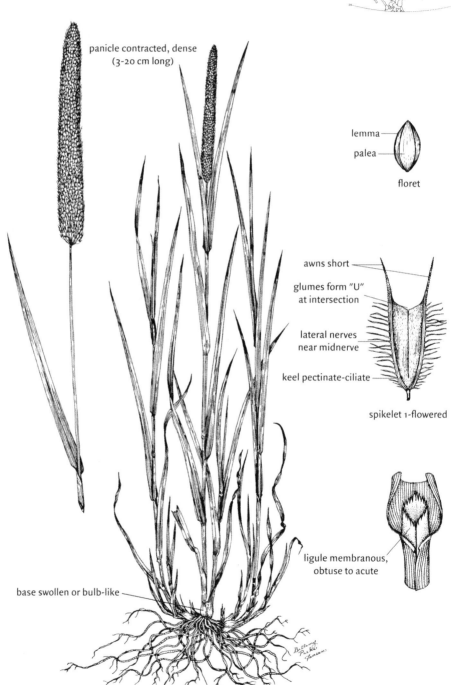

panicle contracted, dense
(3-20 cm long)

lemma

palea

floret

awns short

glumes form "U"
at intersection

lateral nerves
near midnerve

keel pectinate-ciliate

spikelet 1-flowered

ligule membranous,
obtuse to acute

base swollen or bulb-like

Tribe:	AVENEAE
Species:	*Phleum pratense* L.
Common Name:	Timothy (zacate cola de gato, common timothy)
Life Span:	Perennial
Origin:	Introduced (from Europe)
Season:	Cool

INFLORESCENCE CHARACTERISTICS

type: panicle (3–20 cm long, 5–9 mm wide), contracted, dense, cylindrical, several times longer than wide, bristly

spikelets: 1-flowered, elliptic, flattened, small; lemma shorter (1.3–2.5 mm long) than glumes, delicate; usually disarticulating below the glumes

glumes: subequal to equal (2–3.5 mm long, excluding the awns), keeled, pectinate-ciliate on keel, 3-nerved (lateral nerves close to the keel), abrupt taper forming "U" where the two glumes intersect; lateral nerves near the midnerve

awns: nerves of glumes extended into awns; awns short (0.5–1.5 mm long); central nerve of lemma sometimes extended into an awn point

VEGETATIVE CHARACTERISTICS

growth habit: cespitose, rarely solitary

culms: erect (0.5–1.2 m tall), sometimes geniculate below, glabrous; base swollen or bulb-like

sheaths: round, open, glabrous, distinctly veined, often purplish at the base; collar indistinct; auricles rare, small

ligules: membranous (2–5 mm long), obtuse to acute, entire to erose

blades: flat or loosely involute (5–30 cm long, 3–8 mm wide); apex acute; glabrous or scabrous; margins retrorsely scabrous

GROWTH CHARACTERISTICS: starts growth in early spring, flowers May to August, reproduces from seeds and tillers; cold tolerant, poor drought tolerance, responds to nitrogen fertilizer

FORAGE VALUE: good to excellent for all classes of livestock, as well as for deer and elk; produces leafy, palatable hay for cattle and horses and is seeded as a pasture grass; not tolerant of heavy grazing

HABITAT: seeded in pastures and meadows; commonly escaped from cultivation into roadsides, fields, forests, and waste places from the plains to subalpine elevations; most abundant on moist, fertile sites with fine-textured, nonsaline soils

Spike trisetum
Trisetum spicatum (L.) Richter

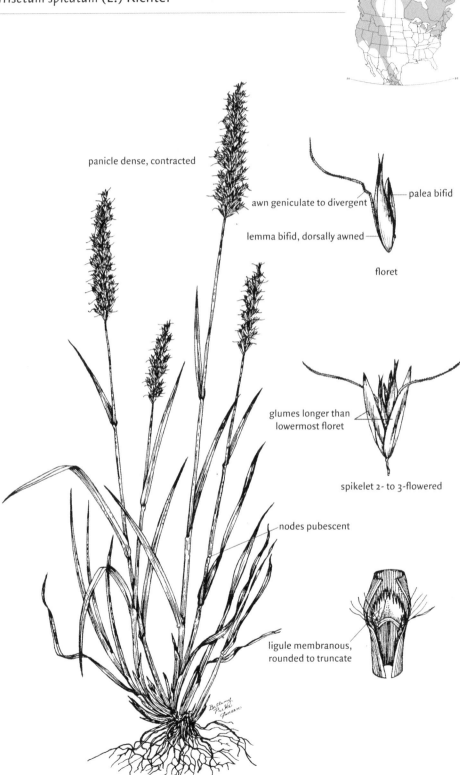

panicle dense, contracted

awn geniculate to divergent

palea bifid

lemma bifid, dorsally awned

floret

glumes longer than lowermost floret

spikelet 2- to 3-flowered

nodes pubescent

ligule membranous, rounded to truncate

Tribe:	AVENEAE
Species:	*Trisetum spicatum* (L.) Richter
Common Name:	Spike trisetum (zacate tres cerdas)
Life Span:	Perennial
Origin:	Native
Season:	Cool

INFLORESCENCE CHARACTERISTICS

type: panicle (2–15 cm long), narrow, dense, contracted, interrupted, purplish-green, bristly

spikelets: 2- to 3-flowered (4–6 mm long); lemmas keeled (3.5–5.5 mm long), 5-nerved, bifid, scaberulous, callus hairs short; paleas bifid; rachilla with short hairs

glumes: unequal, lanceolate, acute; first 1-nerved (3–5 mm long); second 3-nerved (3.5–6 mm long), broader than the first

awns: lemma awned from one-third below the apex; awn conspicuous (4–6 mm long), geniculate to divergent

VEGETATIVE CHARACTERISTICS

growth habit: cespitose; leaves mostly basal

culms: erect (10–70 cm tall); internodes glabrous to downy-pubescent; nodes pubescent

sheaths: keeled at upper end, glabrous to pubescent (except on margins), with scattered long hairs near collar

ligules: membranous (0.5–3 mm long), rounded to truncate, erose or sometimes finely ciliate

blades: flat to folded (3–12 cm long, 2–5 mm wide), glabrous to pubescent, tapering to a blunt apex, distinctly veined

GROWTH CHARACTERISTICS: starts growth in early spring, remains green until August; reproduces from seeds and tillers; low production of viable seeds; seldom occurs in dense stands; does not withstand heavy use

FORAGE VALUE: good for all classes of livestock and wildlife throughout the growing season, cures well, furnishes forage late into the fall if not covered by snow; one of the most important grasses in the mountains

HABITAT: moist to moderately moist alpine and subalpine meadows, bottomlands, and gentle slopes; abundant on well-drained, medium-textured soils

Mountain brome
Bromus carinatus Hook. & Arn.

SYN = *B. marginatus* Nees

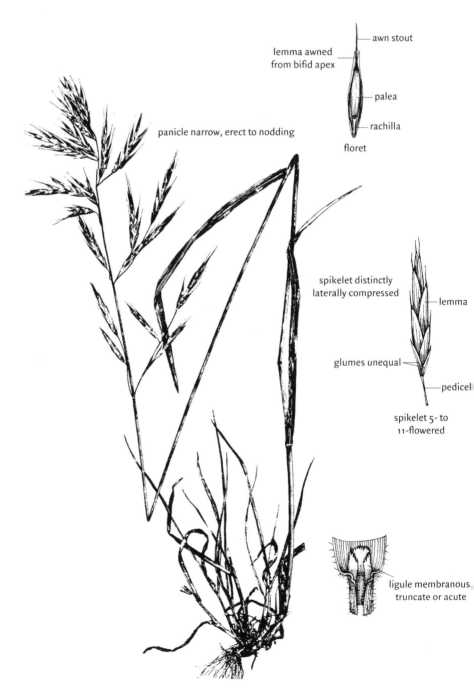

awn stout

lemma awned
from bifid apex

palea

rachilla

floret

panicle narrow, erect to nodding

spikelet distinctly
laterally compressed

lemma

glumes unequal

pedicel

spikelet 5- to
11-flowered

ligule membranous
truncate or acute

Tribe:	BROMEAE
Species:	*Bromus carinatus* Hook. & Arn.
Common Name:	Mountain brome (bromo triste, bromo, California brome)
Life Span:	Perennial
Origin:	Native
Season:	Cool

INFLORESCENCE CHARACTERISTICS

type: panicle (10–30 cm long), narrow, erect to nodding; branches appressed to spreading

spikelets: 5- to 11-flowered (2.5–4 cm long), distinctly laterally compressed; lemmas keeled (11–17 mm long), glabrous to slightly hirsute; apex bifid at maturity

glumes: unequal, strongly keeled; first 3- to 5-nerved (7–11 mm long); second 5- to 7-nerved (9–13 mm long), shorter than the lowermost lemma

awns: lemmas awned; awn stout (4–8 mm long), straight

VEGETATIVE CHARACTERISTICS

growth habit: cespitose

culms: erect (0.3–1.2 m tall), coarse, glabrous to pubescent

sheaths: rounded, retrorsely pubescent to pilose or nearly glabrous, pilose near the throat; margins connate to near the apex

ligules: membranous (1.5–4 mm long), truncate or acute, erose

blades: flat (15–40 cm long, 6–10 mm wide), scabrous to pilose or glabrous

GROWTH CHARACTERISTICS: starts growth in early spring, seeds mature by August, reproduces from seeds and tillers; seeded for rangeland revegetation and erosion control; does not withstand continuous heavy grazing, reducing grazing pressure during the reproductive stages will help to maintain this plant; plants have a relatively short life span

FORAGE VALUE: excellent for cattle, horses, and elk; good for sheep and deer; becoming harsh and fibrous at maturity; seeds furnish food for many kinds of birds and small mammals

HABITAT: mountain slopes, ridge tops, valleys, meadows, and waste places; grows in a broad range of soils; most abundant in moderately moist, well-developed, deep, medium-textured soils and in partial shade; may be found in poorly drained soils

Ripgut brome
Bromus diandrus Roth

SYN = B. rigidus auct. non Roth

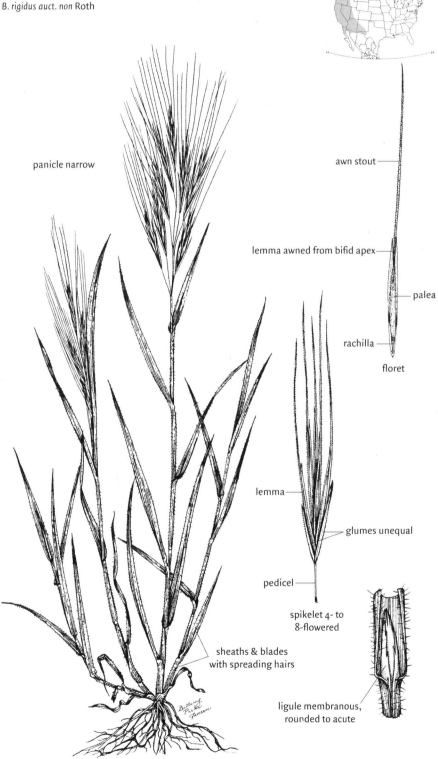

panicle narrow

awn stout

lemma awned from bifid apex

palea

rachilla

floret

lemma

glumes unequal

pedicel

spikelet 4- to 8-flowered

sheaths & blades with spreading hairs

ligule membranous, rounded to acute

Tribe:	BROMEAE
Species:	*Bromus diandrus* Roth
Common Name:	Ripgut brome (bromo frágil)
Life Span:	Annual
Origin:	Introduced (from Europe)
Season:	Cool

INFLORESCENCE CHARACTERISTICS

type: panicle (7–20 cm long), narrow, stout; branches elongate, erect above, spreading below

spikelets: 4- to 8-flowered (3–4 cm long, excluding awns); lemmas (2–3 cm long), glabrous or scabrous, margins broad and hyaline; apex bifid; apical teeth (3–6 mm long)

glumes: unequal; first 1- to 3-nerved (1.3–2.2 cm long); second 3- to 5-nerved (2–2.8 cm long), smooth, margins hyaline, scabrous

awns: lemmas awned; awn stout (3–6 cm long)

VEGETATIVE CHARACTERISTICS

growth habit: cespitose or solitary

culms: ascending to decumbent at base (20–90 cm tall), thick but weak, easily broken after maturity

sheaths: round, pubescent with spreading hairs; margins connate

ligules: membranous (3–5 mm long), rounded to acute, erose

blades: flat (10–20 cm long, 3–8 mm wide), sometimes involute on drying, soft, erect to spreading, pubescent with spreading hairs on both surfaces

GROWTH CHARACTERISTICS: seeds germinate in late fall, grows rapidly in the spring, matures 2–3 months later, reproduces from seeds

LIVESTOCK LOSSES: awns may be injurious to livestock, frequently working into the nostrils and eyes of grazing animals

FORAGE VALUE: excellent in seedling stage and during vigorous vegetative growth but becomes poor for sheep and wildlife and fair for cattle at flowering, worthless at maturity

HABITAT: open ground, waste places, road rights-of-way, field borders, and disturbed sites; adapted to nearly all soil types, most abundant on dry sites

Soft brome
Bromus hordeaceus L.

SYN = *B. mollis* L.

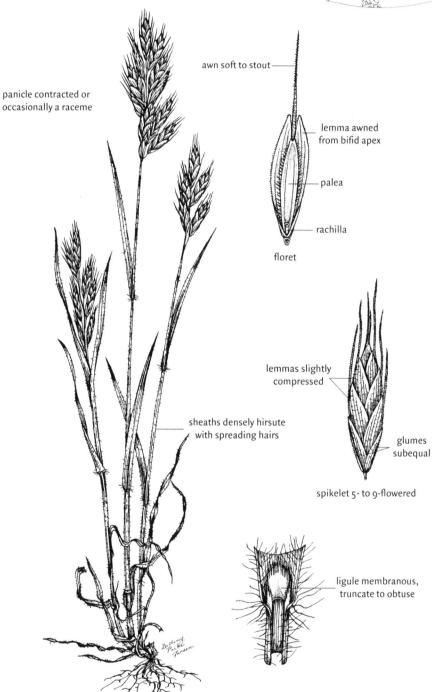

panicle contracted or
occasionally a raceme

awn soft to stout

lemma awned
from bifid apex

palea

rachilla

floret

sheaths densely hirsute
with spreading hairs

lemmas slightly
compressed

glumes
subequal

spikelet 5- to 9-flowered

ligule membranous,
truncate to obtuse

Tribe:	BROMEAE
Species:	*Bromus hordeaceus* L.
Common Name:	Soft brome (bromo, soft chess, bald brome)
Life Span:	Annual
Origin:	Introduced (from Europe)
Season:	Cool

INFLORESCENCE CHARACTERISTICS

type: panicle or occasionally a raceme (3–10 cm long), erect, contracted; pedicels shorter than spikelets, ascending

spikelets: 5- to 9-flowered (1.3–2 cm long, 4–7 mm wide), pubescent to glabrous; lemmas slightly compressed (7–10 mm long, 1.5–2.5 mm wide), obtuse, pilose or scabrous; 7-nerved, prominent; margins hyaline; apex bifid; teeth short (1 mm); palea subequal to the lemma

glumes: subequal, broad, obtuse, coarsely pilose to scabrous, rarely glabrous; first 3- to 5-nerved (4–7.5 mm long); second 5- to 9-nerved (5–8.5 mm long)

awns: lemma awned; awn straight (5–9 mm long), soft to stout

VEGETATIVE CHARACTERISTICS

growth habit: cespitose or solitary

culms: base geniculate (20–80 cm tall), weak; glabrous to retrorsely pubescent, especially at the nodes

sheaths: rounded, densely hirsute with spreading hairs; margins connate

ligules: membranous (0.5–2 mm long), truncate to obtuse, erose

blades: flat (3–15 cm long, 2–5 mm wide), glabrous or sparsely pilose above

GROWTH CHARACTERISTICS: germinates in late fall when moisture is adequate, flowers in early spring, seeds mature in May and June; growth period of about 12 weeks, reproduces from seeds

FORAGE VALUE: excellent for livestock and good for wildlife while immature, good to fair when mature, soft awns allow it to be grazed without injury even after seed maturity, seeds remain on the plants providing good winter grazing

HABITAT: open ground, fields, waste places, and disturbed sites; most abundant on clay loam and sandy soils but grows well on well-drained soils

Smooth brome
Bromus inermis Leyss.

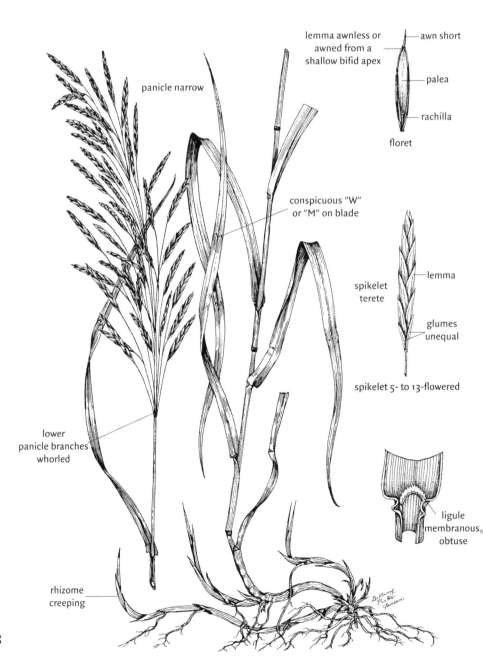

panicle narrow

lemma awnless or
awned from a
shallow bifid apex

awn short

palea

rachilla

floret

conspicuous "W"
or "M" on blade

spikelet
terete

lemma

glumes
unequal

spikelet 5- to 13-flowered

lower
panicle branches
whorled

ligule
membranous
obtuse

rhizome
creeping

Tribe:	BROMEAE
Species:	*Bromus inermis* Leyss.
Common Name:	Smooth brome (bromo suave)
Life Span:	Perennial
Origin:	Introduced (from Europe)
Season:	Cool

INFLORESCENCE CHARACTERISTICS

type: panicle (7–24 cm long), narrow to somewhat open, usually contracted at maturity; branches ascending; lower branches whorled

spikelets: 5- to 13-flowered (1.5–4 cm long, 2.5–7 mm wide), terete, pointed; lemmas rounded on back (9–14 mm long), glabrous to puberulent, greenish to purplish; apex shallowly bifid

glumes: unequal, papery, lanceolate, glabrous to puberulent; first usually 1-nerved (4–8 mm long); second usually 3-nerved (6–10 mm long)

awns: lemmas awnless or with a short awn (1–2 mm)

VEGETATIVE CHARACTERISTICS

growth habit: rhizomatous; rhizomes creeping, forming a sod

culms: erect to rarely decumbent at base (0.4–1.2 m tall), glabrous

sheaths: round, glabrous to rarely pilose, prominently veined; margins connate

ligules: membranous (0.5–2.5 mm long), obtuse, minutely erose-ciliate

blades: flat (15–40 cm long, 4–15 mm wide), glabrous to rarely pilose, margins scabrous; conspicuous "W" or "M" on blade; auricles may be present (to 0.5 mm long)

GROWTH CHARACTERISTICS: starts growth in early spring; flowers May to July; reproduces from seeds, tillers, and rhizomes; responds to fertilization with nitrogen; may regrow and reflower in the fall if moisture is sufficient

FORAGE VALUE: excellent for livestock and wildlife, quality and palatability rapidly decline after inflorescence development, produces excellent hay, regrowth may furnish valuable fall grazing

HABITAT: cultivated as a dryland or irrigated hay and pasture grass; roadsides and waste places on all soil types; often invades native prairies

Downy brome
Bromus tectorum L.

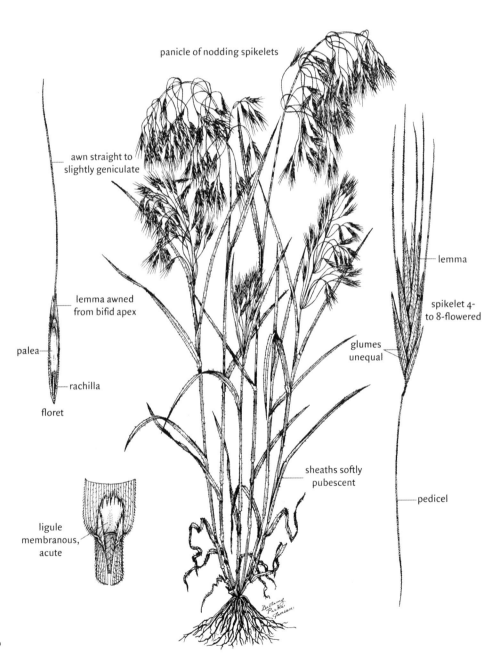

panicle of nodding spikelets

awn straight to
slightly geniculate

lemma awned
from bifid apex

palea

rachilla

floret

ligule
membranous,
acute

sheaths softly
pubescent

lemma

spikelet 4-
to 8-flowered

glumes
unequal

pedicel

Tribe:	BROMEAE
Species:	*Bromus tectorum* L.
Common Name:	Downy brome (bromo velloso, cheatgrass, broncograss, junegrass)
Life Span:	Annual
Origin:	Introduced (from Europe)
Season:	Cool

INFLORESCENCE CHARACTERISTICS

type: panicle (5–20 cm long), dense, much branched, nodding, often purplish; branches and pedicels slender, flexuous

spikelets: 4- to 8-flowered (1.2–2 cm long, 3–6 mm wide excluding awns); lemmas lanceolate (9–15 mm long), rounded on back, glabrous to hirsute; margins thin, membranous; apex bifid; teeth slender (1–3 mm long)

glumes: unequal, first (4–7 mm long), 1-nerved; second (8–11 mm long), 3-nerved, glabrous to hirsute; margins broad, hyaline

awns: lemma awned; awn straight to slightly geniculate (1–1.8 cm long)

VEGETATIVE CHARACTERISTICS

growth habit: cespitose or solitary

culms: erect or decumbent at base (25–60 cm tall), weak

sheaths: round, keeled toward collar, softly pubescent; margins connate

ligules: membranous (2–3.5 mm long), acute, erose to lacerate

blades: flat (5–12 cm long, 2–7 mm wide), glabrous to hispid

GROWTH CHARACTERISTICS: seeds germinate in the late fall or early spring, rapid spring growth, seeds mature about 2 months later, reproduces from seeds; an aggressive weed

LIVESTOCK LOSSES: awns may injure eyes and mouths of grazing animals and contaminate fleece

FORAGE VALUE: fair to good for livestock before the inflorescence emerges, then is practically worthless; deer and pronghorn graze it in the spring while it is actively growing; furnishes food for some upland birds and small mammals

HABITAT: heavily grazed rangeland, roadsides, waste places, and disturbed sites; adapted to a broad range of soil textures, most abundant on dry sites

Sideoats grama
Bouteloua curtipendula (Michx.) Torr.

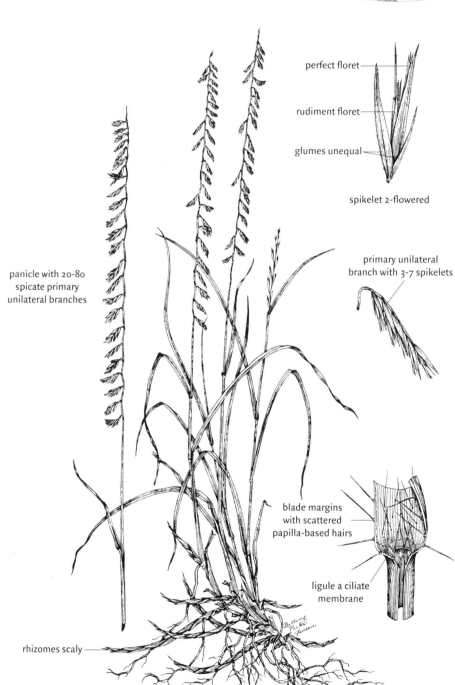

perfect floret

rudiment floret

glumes unequal

spikelet 2-flowered

primary unilateral
branch with 3-7 spikelets

panicle with 20-80
spicate primary
unilateral branches

blade margins
with scattered
papilla-based hairs

ligule a ciliate
membrane

rhizomes scaly

Tribe:	CYNODONTEAE
Species:	*Bouteloua curtipendula* (Michx.) Torr.
Common Name:	Sideoats grama (banderilla, banderita)
Life Span:	Perennial
Origin:	Native
Season:	Warm

INFLORESCENCE CHARACTERISTICS

type: panicle (10–30 cm long) of 20–80 spicate primary unilateral branches; branches (1–3 cm long) distant, pendant; individual branches turned to one side of inflorescence, spicate branches fall as a unit, branch base remaining on culm after disarticulation; spikelets 3–7 per branch, crowded

spikelets: 2-flowered; 1 perfect floret with sterile floret above; fertile lemma (3–6 mm long) 3-nerved; rudiment variable, usually a short lemma with 3 awns

glumes: unequal; first short (2.5–6 mm long), thin; second longer (4–8 mm long), thick, tapering, glabrous or scabrous, purplish

awns: fertile lemma 3-toothed, with 3 awn tips (1–2 mm long); rudimentary lemma with 3 awns (3–6 mm long), unequal

VEGETATIVE CHARACTERISTICS

growth habit: cespitose or rhizomatous; rhizomes scaly; leaves mostly basal

culms: erect (0.2–1 m tall), smooth; nodes purplish

sheaths: round, glabrous below to somewhat pilose above, prominently veined; collar pilose on margin

ligules: ciliate membrane (0.3–0.7 mm long), truncate

blades: flat to subinvolute (2–30 cm long, 2–6 mm wide), linear; usually scabrous adaxially, smooth abaxially; margins with scattered papilla-based hairs

GROWTH CHARACTERISTICS: starts growth in early spring and flowers July to September; reproduces from seeds, tillers, and rhizomes; commonly included in seeding mixtures for revegetation

FORAGE VALUE: good for all classes of livestock and wildlife throughout summer and fall, remains moderately palatable into winter; makes good hay

HABITAT: dry plains, prairies, and rocky hills; most abundant in fine-textured soils, seldom grows in coarse-textured soils, better adapted to calcareous and moderately alkaline soils than to neutral or acid soils

Black grama
Bouteloua eriopoda (Torr.) Torr.

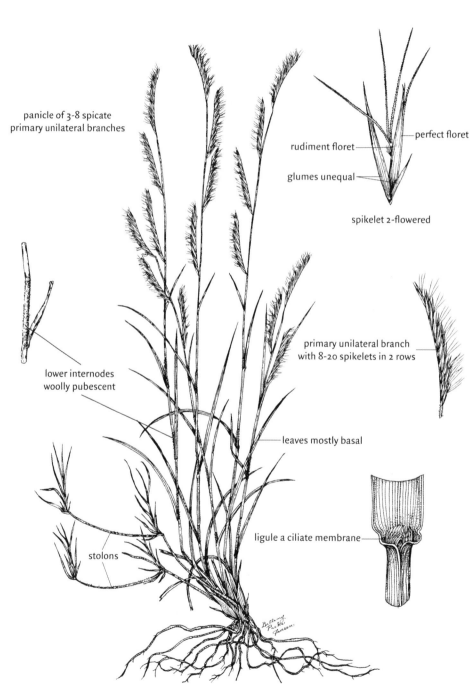

panicle of 3-8 spicate
primary unilateral branches

rudiment floret

perfect floret

glumes unequal

spikelet 2-flowered

primary unilateral branch
with 8-20 spikelets in 2 rows

lower internodes
woolly pubescent

leaves mostly basal

stolons

ligule a ciliate membrane

Tribe:	CYNODONTEAE
Species:	*Bouteloua eriopoda* (Torr.) Torr.
Common Name:	Black grama (navajita negra)
Life Span:	Perennial
Origin:	Native
Season:	Warm

INFLORESCENCE CHARACTERISTICS

type: panicle of 3–8 spicate primary unilateral branches; branches ascending (2–5 cm long), slender, delicate, white-lanate, branch not projecting beyond the spikelet-bearing portion; spikelets 8–20 in 2 rows on each branch, not crowded, somewhat pectinate

spikelets: 2-flowered; 1 perfect floret with 1 rudiment above; fertile lemma (4.5–6.5 mm long) 3-nerved, usually bearded at the base; rudiment may be reduced to 3 awns

glumes: unequal; first shorter (3–6 mm long) than the second (6–9 mm long), acute to acuminate, glabrous to scabrous, purplish

awns: fertile lemma with 3 awns (1.5–3 mm long), central awn longest; rudiment with 3 awns (3.5–8 mm long)

VEGETATIVE CHARACTERISTICS

growth habit: stoloniferous; leaves mostly basal

culms: ascending to decumbent (20–60 cm long), spreading, wiry, slender; internodes arched, lower internodes woolly pubescent; base swollen, knotty, woolly

sheaths: round, shorter than the internodes, glabrous

ligules: ciliate membrane (0.2–0.7 mm long), truncate

blades: flat below, twisted and involute above (2–7 cm long, 0.5–2 mm wide), flexuous; apex acute

GROWTH CHARACTERISTICS: starts growth when sufficient moisture is available in late spring, reproduces mainly from stolons and tillers, low seed viability; frequently grows in nearly pure stands; not tolerant of heavy grazing

FORAGE VALUE: excellent for all classes of livestock and wildlife throughout the year, produces excellent hay during years with adequate moisture

HABITAT: rocky or sandy mesas, dry hills, and dry open prairie from 1,000 to 2,000 m altitude; seldom grows in fine-textured soils

Blue grama
Bouteloua gracilis (Willd. *ex* Kunth) Lag. *ex* Griffiths

SYN = *B. oligostachya* (Nutt.) Torr. *ex* A. Gray

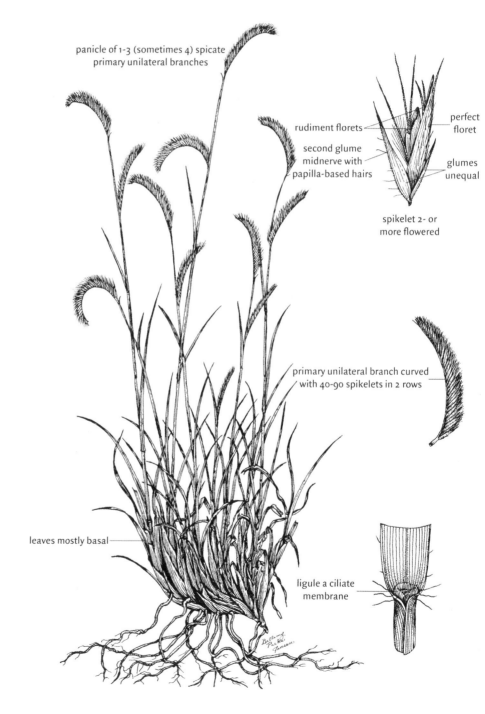

panicle of 1-3 (sometimes 4) spicate primary unilateral branches

rudiment florets

second glume midnerve with papilla-based hairs

perfect floret

glumes unequal

spikelet 2- or more flowered

primary unilateral branch curved with 40-90 spikelets in 2 rows

leaves mostly basal

ligule a ciliate membrane

86

Tribe:	CYNODONTEAE
Species:	*Bouteloua gracilis* (Willd. *ex* Kunth) Lag. *ex* Griffiths
Common Name:	Blue grama (navajita azúl, navajita común)
Life Span:	Perennial
Origin:	Native
Season:	Warm

INFLORESCENCE CHARACTERISTICS

type: panicle of 1–3 (sometimes 4) spicate primary unilateral branches; branches (1.5–5 cm long) curved and spreading at maturity, not extending beyond the spikelet-bearing portion; spikelets 40–90 in 2 rows on each branch, crowded, pectinate

spikelets: 2- or more flowered; 1 perfect floret per spikelet with 1 or more rudiments above; fertile lemma (4.2–5.7 mm long) pubescent on back; rudiments (0.8–3 mm long) highly variable, may be reduced to 1–3 awns

glumes: unequal, 1-nerved; first short (1.5–3.5 mm long), glabrous or with hairs on the nerve; second longer (3.5–6 mm long), papilla-based hairs on the midnerve, may be purplish

awns: fertile lemma with 3 awns (0.5–1.5 mm long), central awn longest; rudimentary lemmas with 1–3 awns (2.5–5 mm long) or may be awnless

VEGETATIVE CHARACTERISTICS

growth habit: cespitose, occasionally with short rhizomes; forming mats; leaves mostly basal

culms: erect (20–60 cm tall), slender, often geniculate below, glabrous

sheaths: round, glabrous to pilose, long pilose at collar

ligules: ciliate membrane (0.1–0.5 mm long), truncate

blades: flat or loosely involute (2–25 cm long, 1–4 mm wide), tapering, puberulent to scabrous above, smooth to slightly scabrous below, sometimes sparingly pilose on both surfaces

GROWTH CHARACTERISTICS: starts growth in May or June, flowers June to October, reproduces primarily from tillers; cannot tolerate shading by taller plants, withstands relatively heavy grazing

FORAGE VALUE: good for all classes of livestock and wildlife, quality is highest when it is green, but it retains much of its value when dry and furnishes fall and winter grazing

HABITAT: open plains, mesas, foothills, and woodlands; in all soil textures, but most abundant in sandy or gravelly soils; not found in wet, poorly drained soils

Hairy grama
Bouteloua hirsuta Lag.

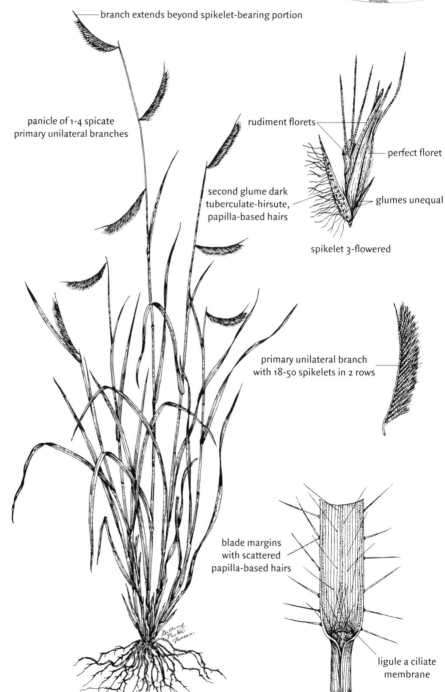

branch extends beyond spikelet-bearing portion

panicle of 1-4 spicate
primary unilateral branches

rudiment florets

perfect floret

second glume dark
tuberculate-hirsute,
papilla-based hairs

glumes unequal

spikelet 3-flowered

primary unilateral branch
with 18-50 spikelets in 2 rows

blade margins
with scattered
papilla-based hairs

ligule a ciliate
membrane

Tribe:	CYNODONTEAE
Species:	*Bouteloua hirsuta* Lag.
Common Name:	Hairy grama (navajita vellucla)
Life Span:	Perennial
Origin:	Native
Season:	Warm

INFLORESCENCE CHARACTERISTICS

type: panicle of 1–4 spicate primary unilateral branches; branches (1–3.5 cm long) ascending to spreading, occasionally curved, extending beyond spikelet-bearing portion in an obvious point (5–6 mm long); spikelets 18–50 in 2 rows on each branch, crowded, pectinate

spikelets: 3-flowered; 1 perfect floret with 2 rudiments above; fertile lemma (2–4.5 mm long), deeply 3-cleft, hairy; lemmas of rudiments highly reduced often to only 1 to 3 awns

glumes: unequal; first short (1.5–3.5 mm long); second longer (3–6 mm long), dark tuberculate-hirsute, papilla-based hairs on the midnerve

awns: second glume awn-tipped; fertile lemma with 3 awns (0.2–2.5 mm long), central awn longest; rudiment lemmas 1- to 3-awned (2–4.5 mm long), may be awnless, dark

VEGETATIVE CHARACTERISTICS

growth habit: cespitose; leaves basal and cauline

culms: erect to spreading (15–70 cm tall), somewhat geniculate below; nodes 4–8

sheaths: round, glabrous to sparsely pilose; margins with papilla-based hairs; collar hairy

ligules: ciliate membrane (0.2–0.4 mm long), truncate

blades: flat or involute (2–18 cm long, 1–3 mm wide), narrow; apex acute; surfaces smooth to scabrous; margins with scattered papilla-based hairs (often on both surfaces as well), thickened

GROWTH CHARACTERISTICS: starts growth in June or as soon as moisture is available, flowers in July and August, reproduces primarily from tillers and seeds

FORAGE VALUE: fair for both livestock and wildlife, palatability is highest late in the growing season, cures well and furnishes winter grazing when not covered by snow

HABITAT: rangeland, prairies, shallow uplands, and rocky ridges; most abundant on dry, loose sands and neutral to slightly calcareous soils

Slender grama
Bouteloua repens (Kunth) Scribn. & Merr.

SYN = *B. filiformis* (Fourn.) Griffiths

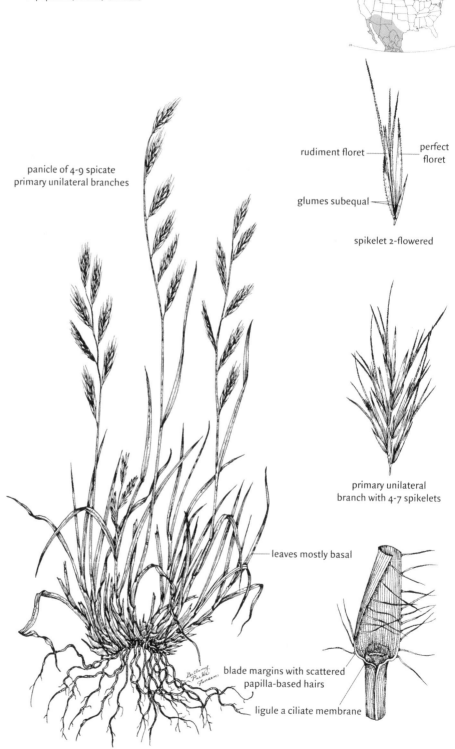

panicle of 4-9 spicate
primary unilateral branches

rudiment floret

perfect floret

glumes subequal

spikelet 2-flowered

primary unilateral
branch with 4-7 spikelets

leaves mostly basal

blade margins with scattered
papilla-based hairs

ligule a ciliate membrane

Tribe:	CYNODONTEAE
Species:	*Bouteloua repens* (Kunth) Scribn. & Merr.
Common Name:	Slender grama (navajita esbelta, navajita rastrera)
Life Span:	Perennial
Origin:	Native
Season:	Warm

INFLORESCENCE CHARACTERISTICS

type: panicle of 4–9 spicate primary unilateral branches; branches (1–2 cm long) flattened, ascending to spreading, extending beyond (2–6 mm) the uppermost spikelets; branches fall as a unit; spikelets typically 4–7 per branch, not crowded, not pectinate

spikelets: 2-flowered; 1 perfect floret below with 1 rudiment above; fertile lemma (3.5–6 mm long) 3-nerved, 3-cleft, glabrous; rudiment lemma well-developed, 3-nerved

glumes: subequal (3–6 mm long), broad, scabrous on the midnerve, otherwise glabrous

awns: fertile lemma short-awned; rudiment lemma with 3 stout awns (3–7 mm long), central awn the longest

VEGETATIVE CHARACTERISTICS

growth habit: cespitose; leaves mostly basal

culms: erect (20–45 cm tall) to ascending, slender, weak

sheaths: round, strongly veined, glabrous; collar sparsely pilose

ligules: ciliate membrane (usually less than 0.3 mm long), truncate

blades: flat (5–16 cm long, 1–3 mm wide), thin; margins with scattered papilla-based hairs below the middle

GROWTH CHARACTERISTICS: starts growth in April when adequate moisture is available, flowers May through November, reproduces from seeds and tillers; withstands extended moderate grazing; may persist where other grasses have been eliminated; relatively short-lived; not highly drought tolerant

FORAGE VALUE: good for livestock and wildlife, cures well and is moderately palatable even when dry; especially valuable in fall and winter

HABITAT: open or brushy pastures, dry slopes, along stream banks, and road rights-of-way; adapted to a broad range of soil textures, most abundant on sandy or rocky soils

Buffalograss
Buchloe dactyloides (Nutt.) Engelm.

spikelet 1-flowered

cluster bur-like,
3-5 spikelets

blades with
papilla-based hairs

ligule a ciliate
membrane

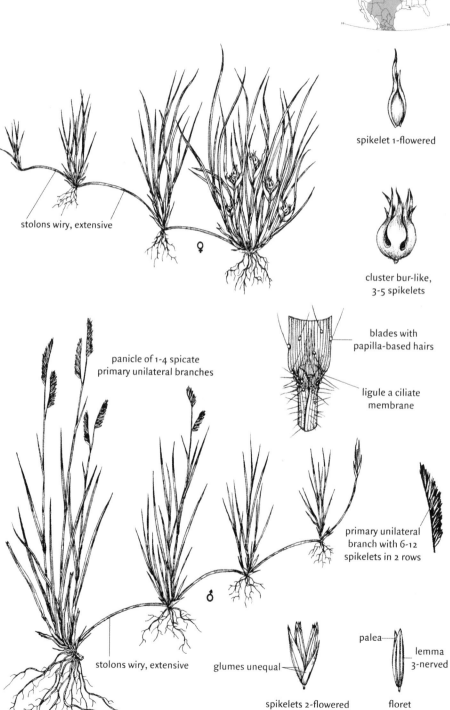

stolons wiry, extensive

♀

panicle of 1-4 spicate
primary unilateral branches

primary unilateral
branch with 6-12
spikelets in 2 rows

♂

stolons wiry, extensive

glumes unequal

spikelets 2-flowered

palea

lemma
3-nerved

floret

Tribe:	CYNODONTEAE
Species:	*Buchloe dactyloides* (Nutt.) Engelm.
Common Name:	Buffalograss (pasto chino, búfalo)
Life Span:	Perennial
Origin:	Native
Season:	Warm

INFLORESCENCE CHARACTERISTICS

type: staminate plants with an elevated panicle of 1–4 spicate primary unilateral branches, branches (6–14 mm long) erect to spreading, spikelets 6–12 in 2 rows on each branch, pectinate; pistillate plants have bur-like clusters (up to 1 cm long) near the mid-culm or within the leaves, spikelets 3–5 per cluster; burs fall as a unit

spikelets: staminate spikelets (4–5.5 mm long) with 2 florets, lemmas longer than the glumes and glabrous; pistillate spikelets usually 1-flowered, yellowish, lemmas 3-nerved and enclosed in glumes

glumes: staminate spikelet glumes unequal (first 1–3 mm long, second 2–4.5 mm long), acute; pistillate glumes unequal, first glume highly reduced, second indurate, enclosing lemma, bearing 3 awn-tipped lobes

awns: second glume of pistillate spikelets awn-tipped

other: plants dioecious or occasionally monoecious

VEGETATIVE CHARACTERISTICS

growth habit: stoloniferous; stolons wiry, extensive, forming mats

culms: culm of staminate plants erect (5–25 cm tall), nodes glabrous; culms of pistillate plants much shorter

sheaths: rounded on the back, open, glabrous except for a few marginal hairs near the collar

ligules: ciliate membrane (0.5–1 mm long), truncate to obtuse, often flanked with longer hairs

blades: flat (1–15 cm long, 1–2.5 mm wide), curly, with papilla-based hairs on both surfaces

GROWTH CHARACTERISTICS: starts growth in midspring when adequate moisture is available, flowers in summer, reproduces from stolons and seeds; can withstand heavy grazing and dry conditions, cannot tolerate shading

FORAGE VALUE: good for all classes of livestock and fair for wildlife, cures well and provides winter grazing when not covered by snow

HABITAT: dry plains on medium- to fine-textured soils, rare on sandy soils, most abundant on heavily used rangeland

Hooded windmillgrass
Chloris cucullata Bisch.

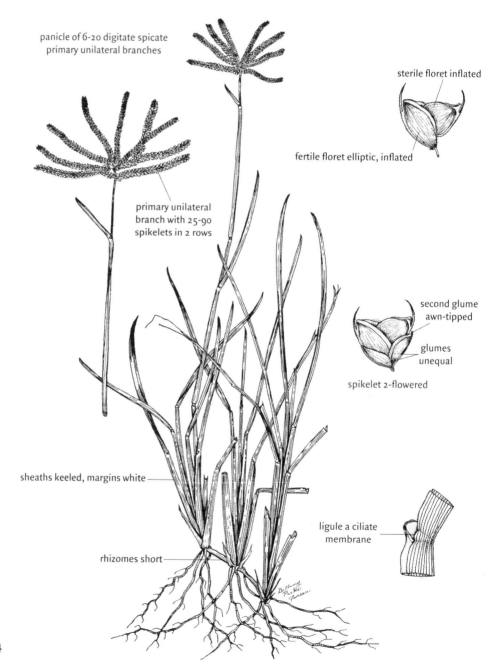

panicle of 6-20 digitate spicate
primary unilateral branches

sterile floret inflated

fertile floret elliptic, inflated

primary unilateral
branch with 25-90
spikelets in 2 rows

second glume
awn-tipped

glumes
unequal

spikelet 2-flowered

sheaths keeled, margins white

ligule a ciliate
membrane

rhizomes short

94

Tribe:	CYNODONTEAE
Species:	*Chloris cucullata* Bisch.
Common Name:	Hooded windmillgrass (pata de gallo arenosa, verdillo papalote)
Life Span:	Perennial
Origin:	Native
Season:	Warm

INFLORESCENCE CHARACTERISTICS

type: panicle of 6–20 digitate or subdigitate spicate primary unilateral branches, brown to tawny at maturity; branches (2–5 cm long) curled; spikelets 25–90 in 2 rows on each branch, crowded

spikelets: 2-flowered; 1 perfect floret below a sterile floret; fertile lemma elliptic (1.5–2.1 mm long); sterile lemma inflated (1–1.5 mm long), cup-shaped, 3-nerved, upper margins inrolled

glumes: unequal, first shorter (0.5–1.5 mm long) than the second (1.5–2 mm long), lanceolate to obovate, membranous, glabrous

awns: second glume awn-tipped, fertile lemma awned (0.3–1.8 mm long) from below the apex, sterile lemma may be awnless or with an awn (0.3–1.5 mm long)

VEGETATIVE CHARACTERISTICS

growth habit: cespitose, rarely with short rhizomes

culms: erect (15–70 cm tall) to ascending, compressed, glabrous

sheaths: keeled, glabrous; margins white

ligules: ciliate membrane (1–2 mm long), truncate

blades: folded (2–25 cm long, 2–4 mm wide), keeled near the base, glabrous to scabrous, midrib white

GROWTH CHARACTERISTICS: starts growth in early spring, stays green until fall, may produce inflorescences several times each year; reproduces primarily from seeds

FORAGE VALUE: fair to good for livestock and wildlife, cures well; provides fair forage in winter but should be supplemented with protein concentrate

HABITAT: pastures, plains, road rights-of-way, lawns, and disturbed areas; most abundant on acid to neutral soils with medium to coarse texture; not adapted to calcareous or clay soils

Bermudagrass
Cynodon dactylon (L.) Pers.

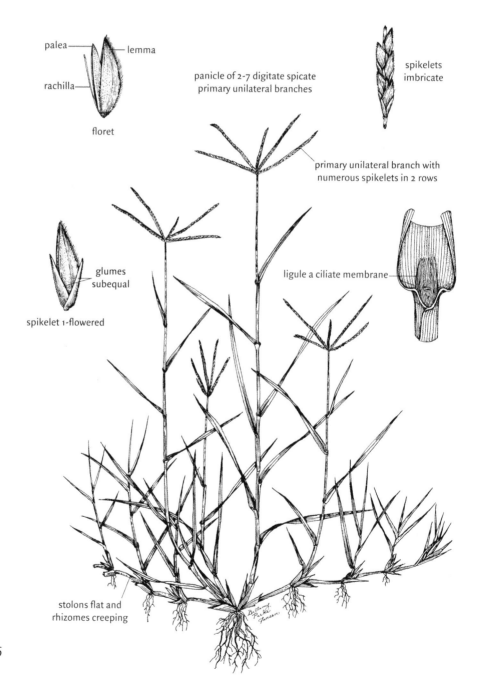

palea — lemma

rachilla

floret

panicle of 2-7 digitate spicate
primary unilateral branches

spikelets
imbricate

primary unilateral branch with
numerous spikelets in 2 rows

glumes
subequal

spikelet 1-flowered

ligule a ciliate membrane

stolons flat and
rhizomes creeping

96

Tribe:	CYNODONTEAE
Species:	*Cynodon dactylon* (L.) Pers.
Common Name:	Bermudagrass (zacate bermuda común, zacate pata de gallo, bermuda, agrarista)
Life Span:	Perennial
Origin:	Introduced (from Africa)
Season:	Warm

INFLORESCENCE CHARACTERISTICS

type: panicle of 2–7 digitate spicate primary unilateral branches; branches (2–6 cm long) ascending to spreading, floriferous to the base; spikelets numerous in 2 rows on each branch, imbricate, glabrous or scabrous

spikelets: 1-flowered (2–2.5 mm long); lemma laterally compressed; rachilla extended behind the palea, occasionally bearing a rudiment

glumes: subequal (1.3–1.8 mm long), lanceolate, 1-nerved

awns: none

VEGETATIVE CHARACTERISTICS

growth habit: rhizomatous and stoloniferous; rhizomes creeping, extensive, forming mats; stolons flat

culms: creeping, weak; only the flowering culms erect (10–50 cm tall), flattened

sheaths: rounded, glabrous except for tufts of hair on either side of the collar and on either side of the ligule

ligules: ciliate membrane (0.2–0.5 mm long), truncate

blades: flat or folded (3–12 cm long, 1–4 mm wide), linear, glabrous or occasionally pilose adaxially

GROWTH CHARACTERISTICS: may grow and flower throughout the year if temperatures and moisture permit; reproduces from seeds, rhizomes, and stolons; pastures are commonly started with vegetative sprigs rather than from seeds

FORAGE VALUE: good for cattle, poor for wildlife; used for pasture in spring and early summer, quality declines in summer; infrequently cut for hay, hay quality is fair

HABITAT: open ground, fields, ditches, waste places, stream banks, along lakes, marshy swales, common lawn grass, and planted in pastures; common in moist saline soils, grows in all soil textures; well adapted to clayey bottom lands that are occasionally flooded

Saltgrass
Distichlis spicata (L.) Greene

SYN = *D. stricta* (Torr.) Rydb.

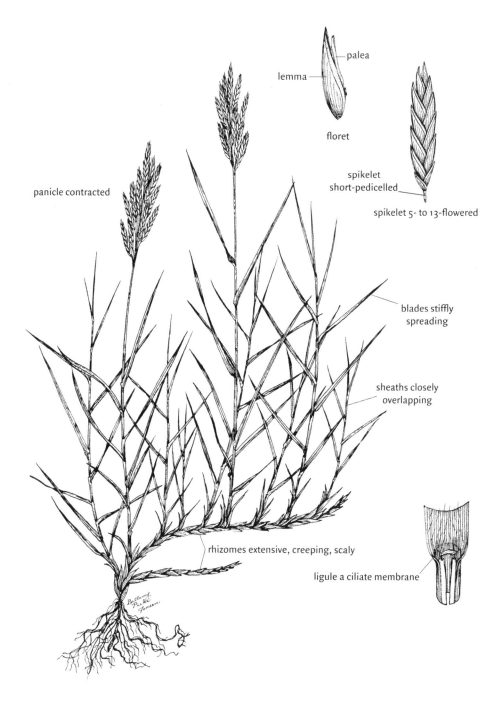

palea

lemma

floret

spikelet
short-pedicelled

spikelet 5- to 13-flowered

panicle contracted

blades stiffly
spreading

sheaths closely
overlapping

rhizomes extensive, creeping, scaly

ligule a ciliate membrane

Tribe:	CYNODONTEAE
Species:	*Distichlis spicata* (L.) Greene
Common Name:	Saltgrass (zacate salado, inland saltgrass)
Life Span:	Perennial
Origin:	Native
Season:	Warm

INFLORESCENCE CHARACTERISTICS

type: panicle (1–7 cm long), contracted; pedicels short

spikelets: unisexual; 5- to 13-flowered (6–17 mm long); staminate spikelets straw-colored, pistillate spikelets green; lemma acute (3–6 mm long), margins yellow and coarse; palea soft, narrowly winged

glumes: unequal, acute, glabrous, first 3- to 9-nerved; second 5- to 11-nerved

awns: none

other: plants dioecious

VEGETATIVE CHARACTERISTICS

growth habit: rhizomatous; rhizomes extensive, creeping, scaly

culms: decumbent to erect (10–40 cm tall), internodes short and numerous, glabrous

sheaths: closely overlapping

ligules: ciliate membrane (0.1–0.5 mm long), truncate, often flanked with long hairs

blades: flat to involute (2–12 cm long, 1–3.5 mm wide), conspicuously ranked (distichous), sharp-pointed, tightly involute at the tip, glabrous, stiffly spreading

other: pistillate plants usually not as tall as staminate plants

GROWTH CHARACTERISTICS: starts growth in early summer, slow rate of growth, remains green until fall, few seeds produced, reproduction mostly from rhizomes, highly resistant to trampling, may aggressively increase when competition from other plants is reduced

LIVESTOCK LOSSES: rumen compaction may develop if cattle are allowed to only graze dried saltgrass in fall or winter; sharp-pointed blades may discourage grazing

FORAGE VALUE: poor for livestock and wildlife, seldom grazed if other grasses are available, heavy grazing pressure will force use

HABITAT: moist alkaline or saline soils on inlands and seashores; frequently the dominant species

Curly mesquite
Hilaria belangeri (Steud.) Nash

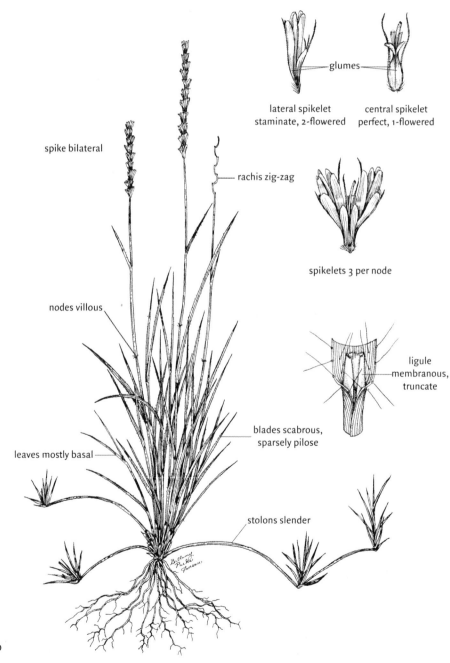

glumes

lateral spikelet
staminate, 2-flowered

central spikelet
perfect, 1-flowered

spike bilateral

rachis zig-zag

spikelets 3 per node

nodes villous

ligule
membranous,
truncate

blades scabrous,
sparsely pilose

leaves mostly basal

stolons slender

Tribe:	CYNODONTEAE
Species:	*Hilaria belangeri* (Steud.) Nash
Common Name:	Curly mesquite (toboso menudo, espiga negra)
Life Span:	Perennial
Origin:	Native
Season:	Warm

INFLORESCENCE CHARACTERISTICS

type: spike (2–3.5 cm long), bilateral, dense, exserted on narrow peduncles; rachis axis flat and strongly angled at each node (zig-zag); inflorescence nodes 4–8

spikelets: spikelets (4.5–6 mm long) 3 per node, central spikelet shortest with 1 perfect floret; 2 lateral spikelets with 2 staminate florets each; lemmas of all florets similar, lanceolate

glumes: lateral spikelet glumes unequal, scabrous, united below, usually shorter than lemmas; outer glume broadened above, notched; central spikelet glumes subequal, glabrous or scabrous, lobed

awns: midrib of glumes extended into short awn; lateral glume awns shorter (less than 1 mm long) than central glume awns (2.5–5 mm long)

VEGETATIVE CHARACTERISTICS

growth habit: cespitose; slender stolons (5–20 cm long) usually present; leaves mostly basal

culms: erect (10–30 cm tall), internodes wiry; nodes villous, especially lower ones

sheaths: round, glabrous

ligules: membranous (0.5–1.5 mm long), truncate, erose to occasionally lacerate

blades: flat or less commonly involute (5–20 cm long, 1–3 mm wide), ascending, scabrous, sparsely pilose

GROWTH CHARACTERISTICS: starts growth in late spring, inflorescenses emerge 1 month or more later, highly drought tolerant; produces a relatively large amount of forage for the plant size; reproduces primarily from stolons, usually occurs in nearly pure stands, does not tolerate shade

FORAGE VALUE: fair for cattle, sheep, goats, deer, and pronghorn; cures well and furnishes forage in the fall and winter

HABITAT: slopes, dry hillsides, and grassy or brushy plains; grows in a wide range of soil textures; most abundant on medium- to fine-textured soils with a pH of 6.8–7.4

Galleta
Hilaria jamesii (Torr.) Benth.

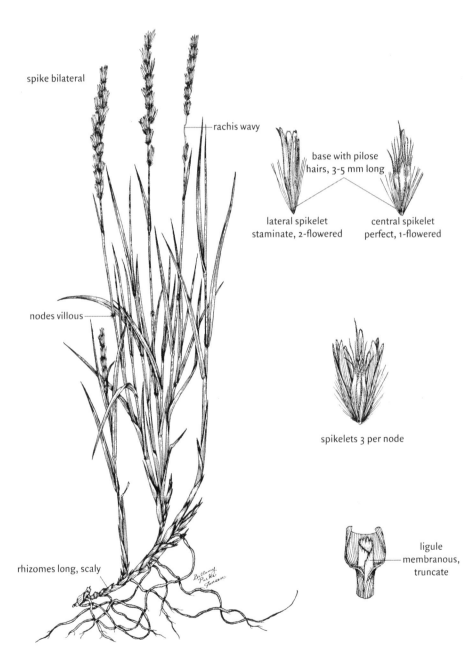

spike bilateral

rachis wavy

base with pilose
hairs, 3-5 mm long

lateral spikelet
staminate, 2-flowered

central spikelet
perfect, 1-flowered

nodes villous

spikelets 3 per node

rhizomes long, scaly

ligule
membranous,
truncate

Tribe:	CYNODONTEAE
Species:	*Hilaria jamesii* (Torr.) Benth.
Common Name:	Galleta
Life Span:	Perennial
Origin:	Native
Season:	Warm

INFLORESCENCE CHARACTERISTICS

type: spike (3–8 cm long), bilateral, dense; rachis wavy; inflorescence nodes 10–25

spikelets: spikelets (6–9 mm long) 3 per node, each node pilose at the base (hairs 3–5 mm long); central spikelet with 1 perfect floret set behind 2 lateral spikelets with 2 staminate florets each; lemma of perfect spikelet (5.8–7.6 mm long) bifid; lemma of lateral spikelets (4.4–7 mm long) ciliate at apex, inrolled

glumes: lateral spikelet glumes subequal, narrowing from the middle upward, not fan-shaped (compare with *Hilaria mutica*); central spikelet glumes subequal, short, cleft

awns: first glume of lateral spikelets awned (3–6 mm long) from back; glumes and lemmas of central spikelet awn-tipped

VEGETATIVE CHARACTERISTICS

growth habit: rhizomatous; rhizomes long, scaly; leaves mostly basal

culms: erect (30–75 cm tall); wiry, glabrous, nodes villous; base usually decumbent

sheaths: round, prominently veined; collar glabrous to villous

ligules: membranous (1–3 mm long), truncate, erose

blades: flat at base (3–15 cm long, 2–5 mm wide), upper two-thirds often involute, scabrous

GROWTH CHARACTERISTICS: grows mainly in summer after sufficient rain, reproduces from rhizomes and seeds, may occur in nearly pure or scattered stands, withstands heavy grazing

FORAGE VALUE: good for cattle, horses, and wildlife and fair for sheep while it is green; worthless to fair for all classes of livestock during the dormant periods; not grazed in fall or winter unless other forage is unavailable

HABITAT: deserts, canyons, open valleys, and dry plains; most abundant on fine-textured soils but it will occur on coarse-textured soils

Tobosa
Hilaria mutica (Buckl.) Benth.

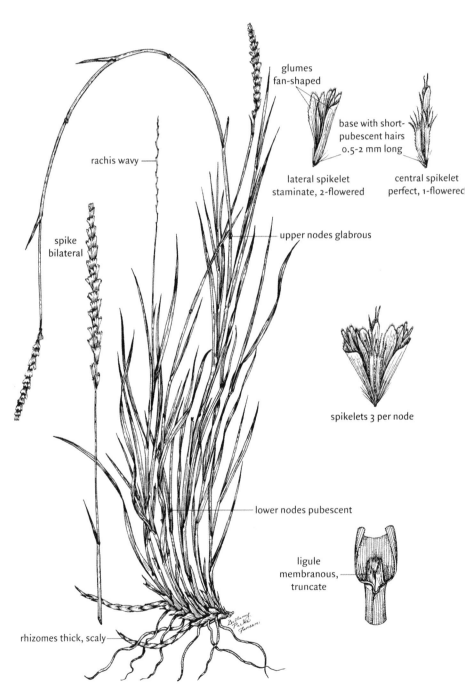

glumes
fan-shaped

base with short-
pubescent hairs
0.5-2 mm long

lateral spikelet
staminate, 2-flowered

central spikelet
perfect, 1-flowered

rachis wavy

upper nodes glabrous

spike
bilateral

spikelets 3 per node

lower nodes pubescent

ligule
membranous,
truncate

rhizomes thick, scaly

Tribe:	CYNODONTEAE
Species:	*Hilaria mutica* (Buckl.) Benth.
Common Name:	Tobosa (toboso)
Life Span:	Perennial
Origin:	Native
Season:	Warm

INFLORESCENCE CHARACTERISTICS

type: spike (4–6 cm long), bilateral, dense; rachis wavy; inflorescence nodes 7–20

spikelets: spikelets (6–9 mm long) 3 per node, short-pubescent to hirsute (hairs 0.5–2 mm long) at the base; central spikelet with 1 perfect floret set behind 2 lateral spikelets with 2 staminate florets each; lemmas entire or irregularly erose and ciliate at apex

glumes: lateral spikelet glumes subequal, fan-shaped, ciliate; central spikelet glumes subequal, narrow, short, cleft

awns: inner glumes of lateral spikelets with a rough or hairy awn (0.5–3 mm long); glumes of central spikelet awn-tipped

VEGETATIVE CHARACTERISTICS

growth habit: rhizomatous; rhizomes thick, scaly

culms: erect (30–75 cm tall), slender; bases decumbent, wiry; lower nodes pubescent, upper nodes glabrous

sheaths: round, glabrous, prominently veined; collar pubescent on margins

ligules: membranous (1–2 mm long), truncate, erose

blades: flat to involute (5–10 cm long, 2–4 mm wide), glabrous (occasionally pubescent adaxially)

GROWTH CHARACTERISTICS: growth starts in late spring or summer after sufficient rain, grows vigorously from thick rhizomes, forms dense and nearly pure to scattered stands, responds quickly to extra moisture during the growing season, will increase with heavy grazing; low seed production

FORAGE VALUE: good to fair for cattle and horses, fair for sheep, poor for wildlife; becomes relatively unpalatable when mature; makes good hay if cut at about the time that the inflorescences appear

HABITAT: dry and rocky slopes, dry upland plains and plateaus, alluvial flats, and swales; most abundant on heavy clay soils, occasionally on sandy to gravelly soils

Tumblegrass
Schedonnardus paniculatus (Nutt.) Trel.

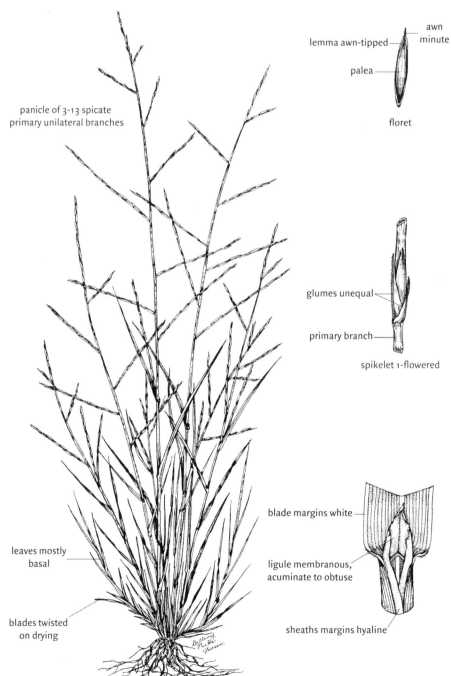

awn
minute

lemma awn-tipped

palea

floret

panicle of 3-13 spicate
primary unilateral branches

glumes unequal

primary branch

spikelet 1-flowered

blade margins white

leaves mostly
basal

ligule membranous,
acuminate to obtuse

blades twisted
on drying

sheaths margins hyaline

Tribe:	CYNODONTEAE
Species:	*Schedonnardus paniculatus* (Nutt.) Trel.
Common Name:	Tumblegrass (Texas crabgrass, wiregrass)
Life Span:	Perennial
Origin:	Native
Season:	Warm

INFLORESCENCE CHARACTERISTICS

type: panicle (30–60 cm long) of 3–13 spicate primary unilateral branches; branches (2–20 cm long) spreading, remote, main axis and branches curving at maturity; entire inflorescence breaks off at the base and is tumbled by the wind

spikelets: 1-flowered, widely spaced to only slightly imbricate, imbedded in and appressed to branches, slender, sessile; lemma 3-nerved (3–5 mm long), glabrous to scabrous

glumes: unequal; second glume as long as lemma, 1-nerved, lanceolate, acuminate; first glume shorter

awns: lemma and second glume awn-tipped, minute

VEGETATIVE CHARACTERISTICS

growth habit: cespitose; leaves mostly basal

culms: erect to ascending (8–70 cm tall), often decumbent at the base, stiffly curving

sheaths: laterally compressed, upper portion keeled, glabrous; margins hyaline

ligules: membranous (1–3 mm long), acuminate to obtuse, erose

blades: flat or folded (2–12 cm long, 1–3 mm wide), twisted on drying, folded in the bud, glabrous; midrib prominent abaxially; margins scabrous, white

GROWTH CHARACTERISTICS: grows from early spring to late fall when moisture is available; flowers throughout the growing period under favorable conditions, reproduces from seeds and tillers; regarded as an indicator of abusive grazing and disturbance

FORAGE VALUE: poor to worthless for livestock and wildlife, seldom grazed, develops little herbage

HABITAT: open prairies, plains, roadsides, and waste places; adapted to a broad range of soil types; most frequent on dry clay or clay loam soils

Alkali cordgrass
Spartina gracilis Trin.

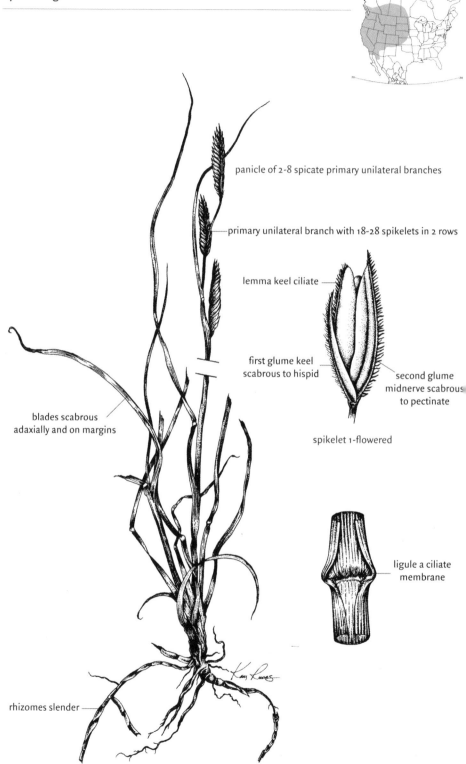

panicle of 2-8 spicate primary unilateral branches

primary unilateral branch with 18-28 spikelets in 2 rows

lemma keel ciliate

first glume keel scabrous to hispid

second glume midnerve scabrous to pectinate

spikelet 1-flowered

blades scabrous adaxially and on margins

ligule a ciliate membrane

rhizomes slender

Tribe:	CYNODONTEAE
Species:	*Spartina gracilis* Trin.
Common Name:	Alkali cordgrass (zacate espinilla)
Life Span:	Perennial
Origin:	Native
Season:	Warm

INFLORESCENCE CHARACTERISTICS

type: panicle of 2–8 spicate primary unilateral branches; branches (2–7 cm long) appressed to ascending, lower branches not floriferous to the base; spikelets 18–28 per branch, imbricate in 2 rows on abaxial side of the branch

spikelets: 1-flowered, sessile, strongly compressed; lemma lanceolate (5.5–8.5 mm long), 3-nerved, blunt, keel ciliate toward apex; palea subequal to lemma, papery

glumes: unequal; first (3.5–6.5 mm long) linear, shorter than lemma, 1-nerved, scabrous to hispid on keel; glume (6–10 mm long) narrowly lanceolate, 3- to 5-nerved, lateral and upper midnerve scabrous to pectinate

awns: glumes awnless to mucronate (to 0.5 mm long)

VEGETATIVE CHARACTERISTICS

growth habit: rhizomatous; rhizomes slender (3–5 mm in diameter), spreading, scaly

culms: erect (30–90 cm tall), solitary, glabrous, terete to slightly flattened

sheaths: rounded to slightly flattened, smooth to striate, glabrous

ligules: ciliate membrane (0.5–1.8 mm long), obtuse to truncate

blades: flat (6–40 cm long, 2.5–6 mm wide), becoming involute when dried, strongly ridged, scabrous adaxially and on margins, glabrous abaxially, attenuate

GROWTH CHARACTERISTICS: flowers June to September; reproduces from seeds, rhizomes, and tillers

FORAGE VALUE: fair to good for cattle in spring and summer, poor for wildlife; palatability rapidly declines as the plants mature and only the leaf tips are eaten, produces fair hay if cut while immature

HABITAT: moist alkaline or saline sites, shores of lakes or ponds, wet meadows, seepage areas, flats, marshes, to dry sandy soils

Prairie cordgrass
Spartina pectinata Link

SYN = *S. michauxiana* A. S. Hitchc.

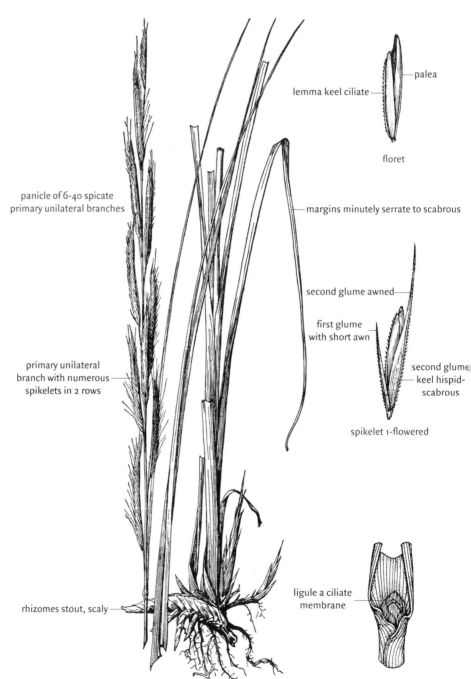

palea

lemma keel ciliate

floret

panicle of 6-40 spicate
primary unilateral branches

margins minutely serrate to scabrous

second glume awned

first glume
with short awn

second glume
keel hispid-
scabrous

primary unilateral
branch with numerous
spikelets in 2 rows

spikelet 1-flowered

ligule a ciliate
membrane

rhizomes stout, scaly

Tribe:	CYNODONTEAE
Species:	*Spartina pectinata* Link
Common Name:	Prairie cordgrass (espartillo, tall marshgrass, sloughgrass)
Life Span:	Perennial
Origin:	Native
Season:	Warm

INFLORESCENCE CHARACTERISTICS

type: panicle of 6–40 spicate primary unilateral branches; branches (4–15 cm long) ascending to occasionally spreading, lower branches not floriferous to the base; spikelets numerous per branch, imbricate in 2 rows on abaxial side of the branch

spikelets: 1-flowered, sessile; lemma (6–9 mm long) laterally compressed, 3-nerved, scabrous on keel, cleft at tip, shorter than palea

glumes: unequal; first (4–7 mm long) 1-nerved, usually as long as floret; second (8–12 mm long) 3-nerved usually exceeding the floret, hispid-scabrous on keel

awns: first glume with short awn (0.9–4 mm long); second glume awned (4–10 mm long), stout

VEGETATIVE CHARACTERISTICS

growth habit: rhizomatous; rhizomes stout (4–10 mm in diameter), widely spreading, scaly, sharp-pointed

culms: erect (1–2.5 m tall), solitary or in small clusters, robust, terete

sheaths: round, may be keeled above, distinctly veined, smooth to slightly striate, usually pubescent only in the throat

ligules: ciliate membrane (2–4 mm long), truncate

blades: flat when green (20–120 cm long, 6–15 mm wide), involute when dry, tapering to a point, glabrous; margins minutely serrate to scabrous

GROWTH CHARACTERISTICS: growth starts in early spring, produces flowers in late summer, reproduces from seeds and rhizomes, may grow in nearly pure stands

FORAGE VALUE: herbage is coarse and furnishes poor to fair forage for cattle and poor forage for wildlife; becomes unpalatable with maturity, produces fair hay if cut while immature

HABITAT: marshy meadows, along swales, ditches, and moist areas; in both fresh and salt water

California oatgrass
Danthonia californica Boland.

SYN = *D. americana* Scribn.

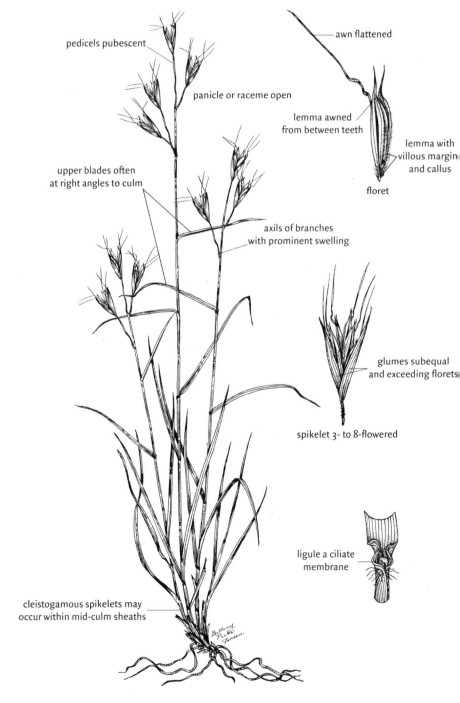

pedicels pubescent

panicle or raceme open

awn flattened

lemma awned
from between teeth

lemma with
villous margin
and callus

floret

upper blades often
at right angles to culm

axils of branches
with prominent swelling

glumes subequal
and exceeding florets

spikelet 3- to 8-flowered

ligule a ciliate
membrane

cleistogamous spikelets may
occur within mid-culm sheaths

Tribe:	DANTHONIEAE
Species:	*Danthonia californica* Boland.
Common Name:	California oatgrass
Life Span:	Perennial
Origin:	Native
Season:	Cool

INFLORESCENCE CHARACTERISTICS

type: panicle or raceme (2–7 cm long), open with 3–5 spreading to reflexed pedicels; pedicels flexuous (1–2 cm long), pubescent; axils of pedicels with prominent swelling

spikelets: 3- to 8-flowered, large (2 cm long); lemmas bifid (9–13 mm long), frequently purplish, mostly glabrous except for villous margins and callus; lemma teeth narrow (2–5 mm long), becoming aristate; cleistogamous spikelets may occur within mid-culm sheaths

glumes: subequal (1.5–2 cm long), usually exceeding all florets, keeled, glabrous or scaberulous

awns: lemma awned from between teeth, lower portion flat, geniculate; terminal segment 5–10 mm long

VEGETATIVE CHARACTERISTICS

growth habit: cespitose; bases retain old, brown sheaths

culms: erect (0.3–1 m tall), robust, glabrous; nodes abruptly constricted

sheaths: round; margins hyaline; throat glabrous to pilose

ligules: ciliate membrane (0.3–1 mm long)

blades: flat to involute (5–25 cm long, 2–5 mm wide), scaberulous, margins glabrous to short-pilose; upper blades often at right-angles to culm

GROWTH CHARACTERISTICS: starts growth in the early spring, flowers May to July, seed matures July to August; reproduces from seeds, tillers, and cleistogamous spikelets

FORAGE VALUE: good to excellent for cattle, horses, and wildlife; somewhat less palatable to sheep and goats; good quality hay if cut before reaching maturity

HABITAT: mountain meadows, open woods, sagebrush hills, coastal prairies; occurs in both open and partially shaded areas; most abundant on dry sites

Timber oatgrass
Danthonia intermedia Vasey

SYN = D. *canadensis* Baum & Findlay

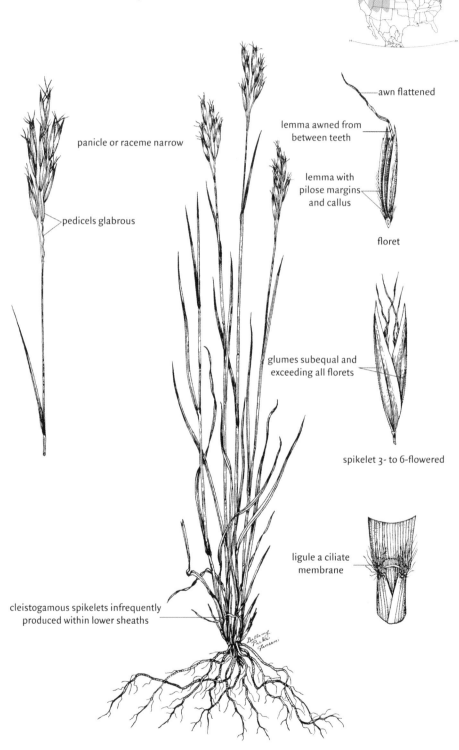

panicle or raceme narrow

pedicels glabrous

awn flattened

lemma awned from between teeth

lemma with pilose margins and callus

floret

glumes subequal and exceeding all florets

spikelet 3- to 6-flowered

ligule a ciliate membrane

cleistogamous spikelets infrequently produced within lower sheaths

Tribe:	DANTHONIEAE
Species:	*Danthonia intermedia* Vasey
Common Name:	Timber oatgrass (timber danthonia)
Life Span:	Perennial
Origin:	Native
Season:	Cool

INFLORESCENCE CHARACTERISTICS

type: panicle or raceme (2–8 cm long), narrow, with 4–12 ascending branches; each branch bearing 1–2 spikelets; pedicels glabrous

spikelets: 3- to 6-flowered, often purplish; lemmas (7–10 mm long) primarily glabrous, although pilose along margins and callus, bifid, teeth acuminate (1.5–2.5 mm long); palea narrowed above, notched at apex; cleistogamous spikelets infrequently produced within lower leaf sheaths

glumes: subequal (1.3–1.8 cm long), usually exceeding all florets, irregularly 3- to 5-nerved, lateral nerves obscure, glabrous

awns: lemma awned from between teeth (awn 7–10 mm long), twisted and somewhat geniculate, flattened

VEGETATIVE CHARACTERISTICS

growth habit: cespitose; leaves mainly basal, with a few persisting brown sheaths at the crown

culms: erect (10–50 cm tall), glabrous

sheaths: round, glabrous (rarely pilose); throat pilose (hairs up to 2 mm long); collar usually pilose

ligules: ciliate membrane (0.3–1 mm long)

blades: flat or involute (5–15 cm long, 2–4 mm wide), ascending if on upper part of culm, generally glabrous, occasionally pubescent abaxially

GROWTH CHARACTERISTICS: starts growth in early spring; seeds mature by September; reproduces from seeds, tillers, and cleistogamous spikelets

FORAGE VALUE: generally good to excellent for livestock, deer, and elk; utilized in the spring; in some cases, has a low relative palatability and is not grazed or only lightly grazed

HABITAT: mountain meadows and bogs (above 1,800 m), as understory in forests at lower elevations; most abundant in medium- to fine-textured soils

Parry oatgrass
Danthonia parryi Scribn.

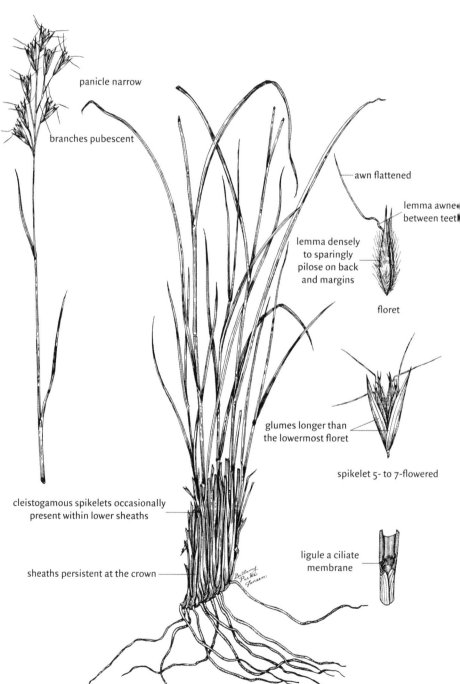

panicle narrow

branches pubescent

awn flattened

lemma awned between teeth

lemma densely to sparingly pilose on back and margins

floret

glumes longer than the lowermost floret

spikelet 5- to 7-flowered

cleistogamous spikelets occasionally present within lower sheaths

ligule a ciliate membrane

sheaths persistent at the crown

Tribe:	DANTHONIEAE
Species:	*Danthonia parryi* Scribn.
Common Name:	Parry oatgrass (Parry danthonia)
Life Span:	Perennial
Origin:	Native
Season:	Cool

INFLORESCENCE CHARACTERISTICS

type: panicle (3–7 cm long), narrow, with 3–8 spikelets on ascending or appressed pedicels; branches usually pubescent

spikelets: 5- to 7-flowered; lemma 11-nerved (1–1.5 cm long), densely to sparingly pilose (hair to 1 cm long) on back and margins, bifid, teeth acuminate (4–5 mm long); palea narrowed above, nearly as long as the lemma; cleistogamous spikelets occasionally present within lower sheaths

glumes: equal to subequal (1.6–2.3 cm long), longer than the lowermost floret, 5-nerved but only the midvein distinct

awns: lemma awned from between teeth; awn twisted (8–15 mm long), geniculate, flattened

VEGETATIVE CHARACTERISTICS

growth habit: cespitose; sheaths persistent and coarse at the crown

culms: erect (20–60 cm tall), stout, glabrous, somewhat enlarged at the base because of numerous overlapping sheaths

sheaths: round, glabrous; margins glabrous or pubescent; throat pilose; collar with a line or ridge

ligules: ciliate membrane (0.3–0.7 mm long)

blades: flat or involute (10–25 cm long, 1–4 mm wide), erect to flexuous, not curled, glabrous; margins scaberulous to scabrous

GROWTH CHARACTERISTICS: starts growth in spring, seed matures in July and August; reproduces from seeds, tillers, and occasional cleistogamous spikelets

FORAGE VALUE: fair for cattle, horses, and wildlife; fair to poor for sheep; quality rapidly declines with maturity; seldom abundant enough to furnish large quantities of forage

HABITAT: open grasslands, open woods, rocky hillsides, and valleys; most abundant at relatively high altitudes (2,000–3,000 m) in coarse-textured soils

Pine dropseed
Blepharoneuron tricholepis (Torr.) Nash

SYN = *Vilfa tricholepis* Torr.

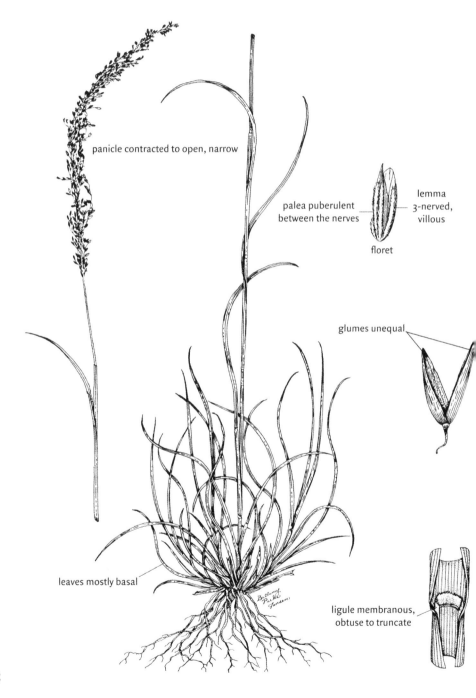

panicle contracted to open, narrow

palea puberulent between the nerves

lemma 3-nerved, villous

floret

glumes unequal

leaves mostly basal

ligule membranous, obtuse to truncate

Tribe:	ERAGROSTEAE
Species:	*Blepharoneuron tricholepis* (Torr.) Nash
Common Name:	Pine dropseed (popotillo del pinar, hairy dropseed)
Life Span:	Perennial
Origin:	Native
Season:	Warm

INFLORESCENCE CHARACTERISTICS

type: panicle (5–22 cm long, 2–5 cm wide), oblong to elliptical, contracted to open, narrow, long-exserted or sometimes enclosed in the uppermost sheath; branches capillary, glabrous, branch bases free of spikelets

spikelets: 1-flowered, with a distinctive greenish-gray or bluish-gray color; lemma obtuse (2–4 mm long), 3-nerved, nerves villous; palea equaling or slightly exceeding the lemma, acute, 2-nerved, puberulent between the nerves

glumes: unequal (1.5–3.2 mm long), broad, rounded on the back, often minutely pointed at the apex, glabrous; first faintly 5-nerved, usually slightly shorter than the second

awns: none

VEGETATIVE CHARACTERISTICS

growth habit: cespitose; leaves mostly basal

culms: erect (20–70 cm tall), slender, glabrous, sometimes purplish

sheaths: rounded on the back, glabrous; margins ciliate

ligules: membranous (0.3–1.9 mm long), obtuse to truncate, erose-dentate to ciliolate

blades: filiform (5–20 cm long, 1–2 mm wide), involute, often strongly flexuous, glabrous or scabrous, margins scabrous

GROWTH CHARACTERISTICS: starts growth in late June or early July, completes growth in September; reproduces from seeds and tillers; generally comprises only a small portion of the vegetation

FORAGE VALUE: palatability and quality of young plants is very good for all classes of livestock, quality rapidly declines with maturity, stems are neglected or only slightly grazed after maturity; generally one of the grasses with the highest forage quality in timbered areas

HABITAT: glades, open slopes, and in dry woodlands at medium to high elevations; adapted to a broad range of soils; most abundant in rocky, moderately dry soils

Prairie sandreed
Calamovilfa longifolia (Hook.) Scribn.

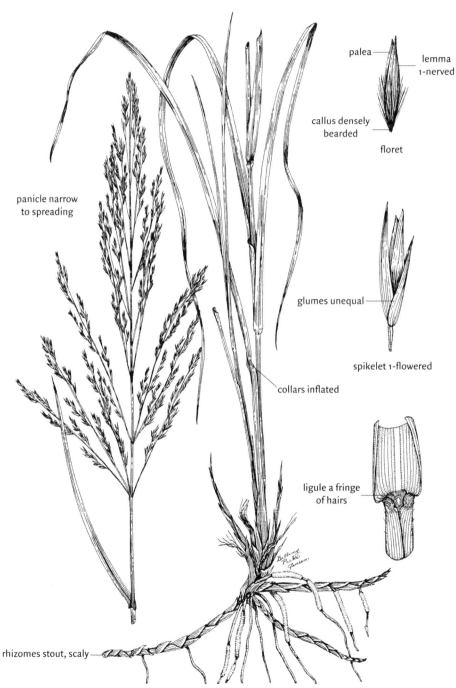

palea

lemma 1-nerved

callus densely bearded

floret

panicle narrow to spreading

glumes unequal

spikelet 1-flowered

collars inflated

ligule a fringe of hairs

rhizomes stout, scaly

Tribe:	ERAGROSTEAE
Species:	*Calamovilfa longifolia* (Hook.) Scribn.
Common Name:	Prairie sandreed (sandgrass)
Life Span:	Perennial
Origin:	Native
Season:	Warm

INFLORESCENCE CHARACTERISTICS

type: panicle (15–40 cm long), narrow to spreading, shiny; branches ascending

spikelets: 1-flowered; lemma acute (4–7 mm long), 1-nerved, usually shorter than the second glume, glabrous, callus densely bearded with hairs usually one-half the length of the lemma; palea firm, slightly shorter than the lemma

glumes: unequal, first (4–7 mm long) shorter than the second (5–8 mm long), acute to acuminate, rigid, 1-nerved

awns: none

VEGETATIVE CHARACTERISTICS

growth habit: rhizomatous; rhizomes spreading, stout, scaly

culms: erect (0.5–1.8 m tall), solitary, robust, glabrous

sheaths: round, glabrous to pubescent; throat and collar pilose (hairs 2–3 mm long), collar inflated

ligules: fringe of hairs (0.5–3 mm long)

blades: flat below (10–60 cm long, 3–12 mm wide), involute above, basal portion keeled, glabrous, margins scabrous

GROWTH CHARACTERISTICS: grows rapidly in late spring and throughout the summer, remains green until frost, reproduces from seeds and rhizomes; drought tolerant and may increase during dry years when associated grasses decline; tolerates relatively heavy grazing; may form large, dense colonies, especially when grazed in winter

FORAGE VALUE: fair for cattle, horses, and wildlife throughout the summer; cures well and provides good standing winter feed, produces hay with acceptable quality if it is not cut too late

HABITAT: prairies, plains, and open woods; most abundant in sand and sandy soils, although it may be locally abundant in deep, medium-textured soils; rarely found on subirrigated or other moist sites

Weeping lovegrass
Eragrostis curvula (Schrad.) Nees

SYN = *E. chloromelas* Steud.

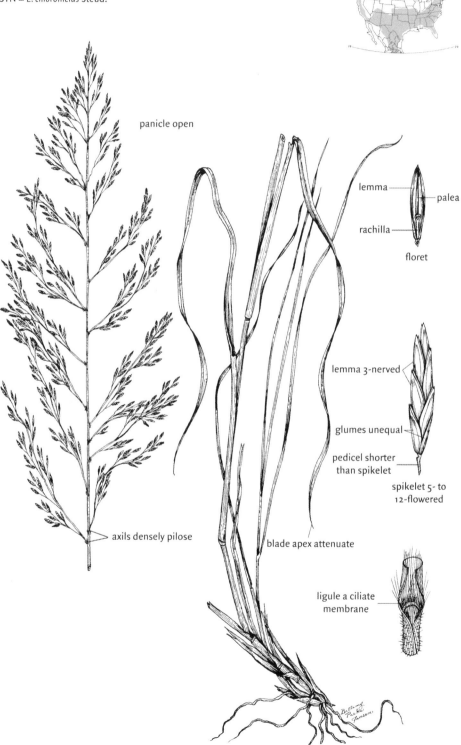

panicle open

lemma

palea

rachilla

floret

lemma 3-nerved

glumes unequal

pedicel shorter
than spikelet

spikelet 5- to
12-flowered

axils densely pilose

blade apex attenuate

ligule a ciliate
membrane

Tribe:	ERAGROSTEAE
Species:	*Eragrostis curvula* (Schrad.) Nees
Common Name:	Weeping lovegrass (zacate llorón, zacate del amor)
Life Span:	Perennial
Origin:	Introduced (from South Africa)
Season:	Warm

INFLORESCENCE CHARACTERISTICS

type: panicle (15–40 cm long, 8–12 cm wide), open, oblong to ovate, eventually drooping or nodding, often lead-gray in color; branches solitary or in pairs, naked at the base, densely pilose in the axils

spikelets: 5- to 12-flowered (6–10 mm long, 1.4–2 mm wide), compressed, short-pediceled; lemmas obtuse (2.2–2.7 mm long), membranous, 3-nerved, lateral nerves conspicuous; paleas scabridulous

glumes: unequal; first acute to acuminate (1.5–2 mm long), 1-nerved; second longer (2–3 mm long), membranous

awns: none

VEGETATIVE CHARACTERISTICS

growth habit: cespitose; leaves mainly basal

culms: erect (0.6–1.5 m tall), glabrous; base often geniculate

sheaths: keeled, shorter than the internodes, basal sheaths hairy abaxially, upper sheaths pilose only at the throat

ligules: ciliate membrane (0.5–1 mm long), white, backed by longer hairs

blades: involute (culm leaves 20–30 cm long, 1–3 mm wide, basal leaves much longer), apex attenuate, flexuous and arching toward the ground, glabrous to scabrous

GROWTH CHARACTERISTICS: starts growth in late spring, produces seeds in spring and fall, reproduces from seeds and tillers, withstands dry conditions, does not tolerate extended periods of temperature below -10°C, tolerates moderate to heavy grazing, planted as a forage grass

FORAGE VALUE: fair for livestock and relatively poor for wildlife; most palatable in spring, may be grazed very little from flowering through dormancy, less palatable than many other seeded species

HABITAT: pastures, sandy fields, waste areas, and road rights-of-ways; sometimes seeded on burned rangeland in the southwest

Sand lovegrass
Eragrostis trichodes (Nutt.) Wood

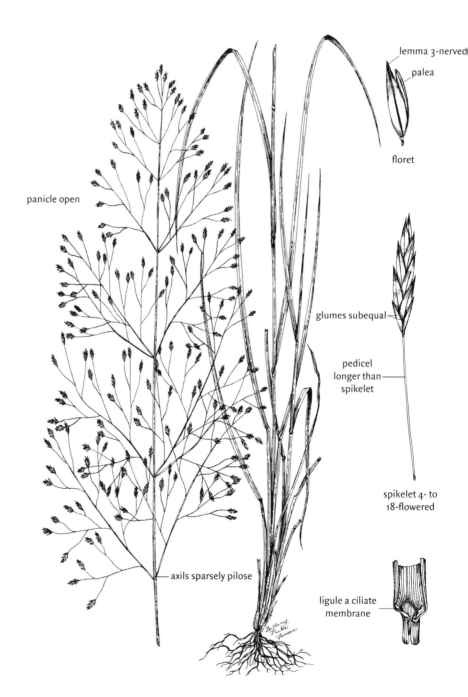

lemma 3-nerved

palea

floret

panicle open

glumes subequal

pedicel longer than spikelet

spikelet 4- to 18-flowered

axils sparsely pilose

ligule a ciliate membrane

Tribe:	ERAGROSTEAE
Species:	*Eragrostis trichodes* (Nutt.) Wood
Common Name:	Sand lovegrass
Life Span:	Perennial
Origin:	Native
Season:	Warm

INFLORESCENCE CHARACTERISTICS

type: panicle (35–55 cm long, 7–30 cm wide), open, diffuse, oblong to ovate, often purple or red, may be one-half the length of the plant; branches in groups of 3 or 4, capillary, sometimes flexuous, sparsely pilose in the axils

spikelets: 4- to 18-flowered (4–10 mm long, 1.5–3.5 mm wide), compressed, purplish, long pediceled; lemmas acute (2.2–3.5 mm long), 3-nerved, nerves conspicuous to obscure; paleas shorter than the lemmas, scabrous on the nerves

glumes: subequal (1.8–4 mm long), acuminate, 1-nerved

awns: none

VEGETATIVE CHARACTERISTICS

growth habit: cespitose

culms: erect (0.3–1.6 m tall), glabrous

sheaths: round, imbricate, occasionally hairy on the back or margins; throat pilose

ligules: ciliate membrane (0.2–0.5 mm long)

blades: flat or somewhat involute (15–50 cm long, 2–8 mm wide), elongate, midrib prominent, usually scabrous adaxially but may be pilose near the ligule

GROWTH CHARACTERISTICS: starts growth as much as 2 weeks earlier than other warm-season grasses, remains green into the fall if moisture is available; reproduces from seeds and tillers; sometimes included in seeding mixtures to furnish quick cover and forage production

FORAGE VALUE: excellent for all classes of livestock and wildlife during summer, fair to good after maturity, cures well, furnishes good grazing in fall and winter, sometimes cut for hay

HABITAT: prairies, open woods, and disturbed sites; most abundant on deep sands and sandy loam soils, although it is sometimes found on fine-textured soils; often included in seeding mixtures

Green sprangletop
Leptochloa dubia (Kunth) Nees

SYN = *Diplachne dubia* (Kunth) Scribn.

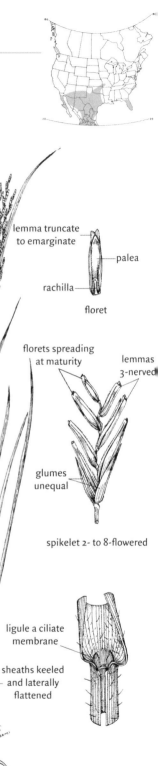

panicle of alternate,
spicate primary
unilateral branches

lemma truncate
to emarginate

palea

rachilla

floret

florets spreading
at maturity

lemmas
3-nerved

glumes
unequal

spikelet 2- to 8-flowered

ligule a ciliate
membrane

sheaths keeled
and laterally
flattened

cleistogamous
spikelets freequently
present within
basal sheaths

Tribe:	ERAGROSTEAE
Species:	*Leptochloa dubia* (Kunth) Nees
Common Name:	Green sprangletop (zacate gigante)
Life Span:	Perennial
Origin:	Native
Season:	Warm

INFLORESCENCE CHARACTERISTICS

type: panicle (5–25 cm long) of 2–15 alternate, spicate primary unilateral branches; branches (4–12 cm long) loosely erect or spreading, widely spaced on the axis

spikelets: 2- to 8-flowered (5–10 mm long), nearly sessile, loosely to closely imbricate, spreading at maturity; lemmas truncate to emarginate (3–5 mm long), 3-nerved, broad, sometimes mucronate, glabrous, nerves with appressed pubescence; cleistogamous spikelets frequently present within basal sheaths

glumes: unequal; second (4–5 mm long) slightly longer than the first (3–4 mm long), lanceolate, translucent with scabrous green nerves

awns: none

VEGETATIVE CHARACTERISTICS

growth habit: cespitose

culms: erect (0.3–1.1 m tall), unbranched above the base; nodes dark brown to black, wiry; base firm

sheaths: keeled, flattened laterally, glabrous or lower sheaths pilose; basal sheaths often purple-tinged; throat pilose

ligules: ciliate membrane (0.3–1.2 mm long)

blades: flat (15–30 cm long, 4–10 mm wide) or involute on drying, glabrous or slightly scabrous, sometimes drooping; midrib prominent adaxially

GROWTH CHARACTERISTICS: starts growth in April, continued growth depends on available moisture, flowers for most of the growing season, reproduces from seeds and tillers, has a relatively short life span, present in mixed stands and is seldom dominant

FORAGE VALUE: good for livestock and fair for wildlife when green, cures relatively well and furnishes forage of fair quality during the dormant season, occasionally cut for hay

HABITAT: rocky hills, canyons, and prairies; most abundant on well-drained rocky or sandy soils, seldom found on clay soils

Mountain muhly
Muhlenbergia montana (Nutt.) A. S. Hitchc.

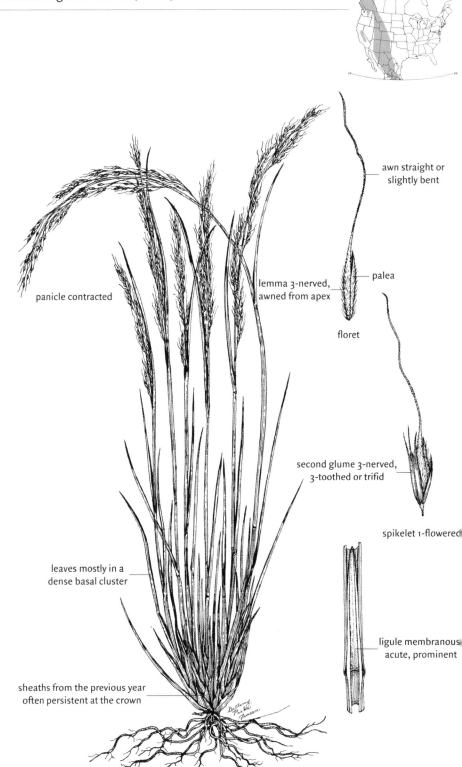

awn straight or
slightly bent

panicle contracted

lemma 3-nerved,
awned from apex

palea

floret

second glume 3-nerved,
3-toothed or trifid

spikelet 1-flowered

leaves mostly in a
dense basal cluster

ligule membranous
acute, prominent

sheaths from the previous year
often persistent at the crown

Tribe:	ERAGROSTEAE
Species:	*Muhlenbergia montana* (Nutt.) A. S. Hitchc.
Common Name:	Mountain muhly (liendrilla de la montaña)
Life Span:	Perennial
Origin:	Native
Season:	Warm

INFLORESCENCE CHARACTERISTICS

type: panicle (5–20 cm long, 1–3 cm wide), contracted, rather loosely flowered, interrupted below; branches ascending, primary panicle branches usually floriferous to near the base (within 1 cm)

spikelets: 1-flowered; lemma (3–4.5 mm long) 3-nerved, greenish to grayish with dark green or purple blotches or bands, scabrous above, with fine hairs on nerves; palea shorter than the lemma

glumes: subequal, slightly keeled, thin, scabrous to nearly glabrous; first acute to obtuse (1.5–3 mm long), ovate, 1-nerved, scabrous; second longer (2–3.3 mm long), 3-nerved, nerves ending as shortly 3-toothed or trifid, scaberulous

awns: lemma awned from apex, flexuous (4–25 mm long), straight or slightly bent, scabrous; glumes awn-tipped

VEGETATIVE CHARACTERISTICS

growth habit: cespitose; leaves mostly in a dense basal cluster, sheaths from the previous year often persistent at the crown

culms: erect (15–80 cm tall), stout, glabrous

sheaths: rounded, lower ones often becoming flat and spreading, longer than the internodes, glabrous

ligules: membranous (4–18 mm long), acute, entire, prominent

blades: flat to involute (5–25 cm long, 1–3 mm wide), arcuate, acuminate, scabrous

GROWTH CHARACTERISTICS: starts growth in late spring, matures August to September, reproduces from seeds and tillers

FORAGE VALUE: good for cattle, horses, and elk; fair for sheep and deer while immature; palatability rapidly declines with maturity

HABITAT: open woodlands, hillsides, canyons, and draws up to 3,000 m in elevation; most abundant in dry loam to clay soils, but will grow in sandy and gravelly soils

Bush muhly
Muhlenbergia porteri Scribn. ex Beal

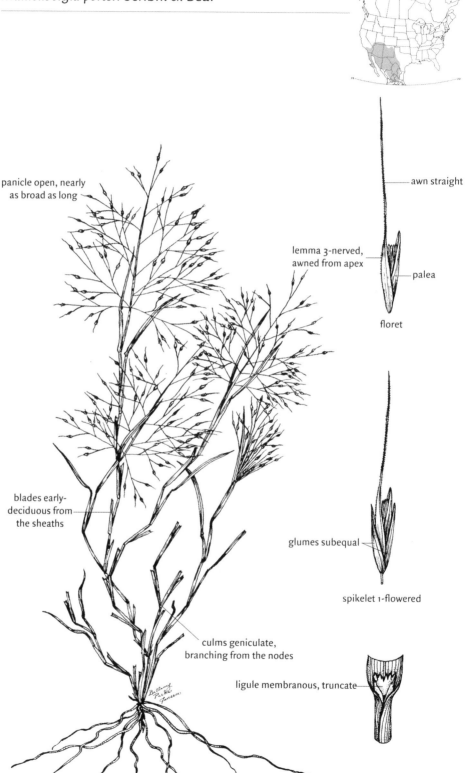

panicle open, nearly
as broad as long

blades early-
deciduous from
the sheaths

culms geniculate,
branching from the nodes

awn straight

lemma 3-nerved,
awned from apex

palea

floret

glumes subequal

spikelet 1-flowered

ligule membranous, truncate

Tribe:	ERAGROSTEAE
Species:	*Muhlenbergia porteri* Scribn. *ex* Beal
Common Name:	Bush muhly (zacate araña, mesquitegrass)
Life Span:	Perennial
Origin:	Native
Season:	Warm

INFLORESCENCE CHARACTERISTICS

type: panicle (5–10 cm long), open, nearly as broad as long, terminal; branches slender, purple, brittle; bearing few spikelets; long-pedicelled (5–20 mm long)

spikelets: 1-flowered, widely spaced; lemma acuminate (3–4 mm long), 3-nerved, sparsely pubescent, often purple; palea nearly the same length as the lemma, acuminate, glabrous

glumes: subequal (1.5–3 mm long), narrow, acuminate, glabrous, keel scabrous

awns: lemma awned (5–13 mm long) from the apex, delicate, straight

VEGETATIVE CHARACTERISTICS

growth habit: loosely cespitose

culms: spreading to ascending (0.3–1 m tall), from a woody, knotty base; wiry, geniculate, branching from the nodes, scaberulous or finely puberulent below and glabrous above

sheaths: round, spreading away from the internodes, mostly shorter than the internodes

ligules: membranous (1–2 mm long), truncate, erose to lacerate

blades: flat, becoming involute (2–8 cm long, 0.5–2 mm wide), thin, acuminate, scabrous, early-deciduous from the sheaths

other: originally existed in extensive stands but now occurs in the protection of shrubs and often ascends through the shrubs

GROWTH CHARACTERISTICS: growth starts at nodes and crown, culms do not die back each year; flowers from June to November, reproduces from seeds and tillers; sensitive to heavy grazing but responds favorably in rotation systems

FORAGE VALUE: excellent for all classes of livestock, deer, and pronghorn; remains green year-long if moisture is adequate, which makes it especially palatable in winter and early spring before other grasses initiate growth

HABITAT: dry mesas, hills, canyons, and rocky deserts; most abundant on calcareous soils

Ring muhly
Muhlenbergia torreyi (Kunth) A. S. Hitchc. ex Bush

SYN = M. *gracillima* Torr.

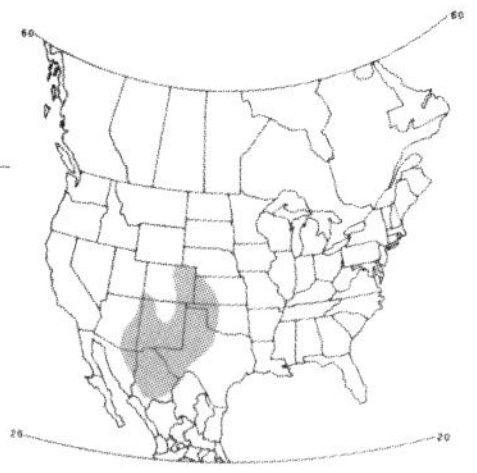

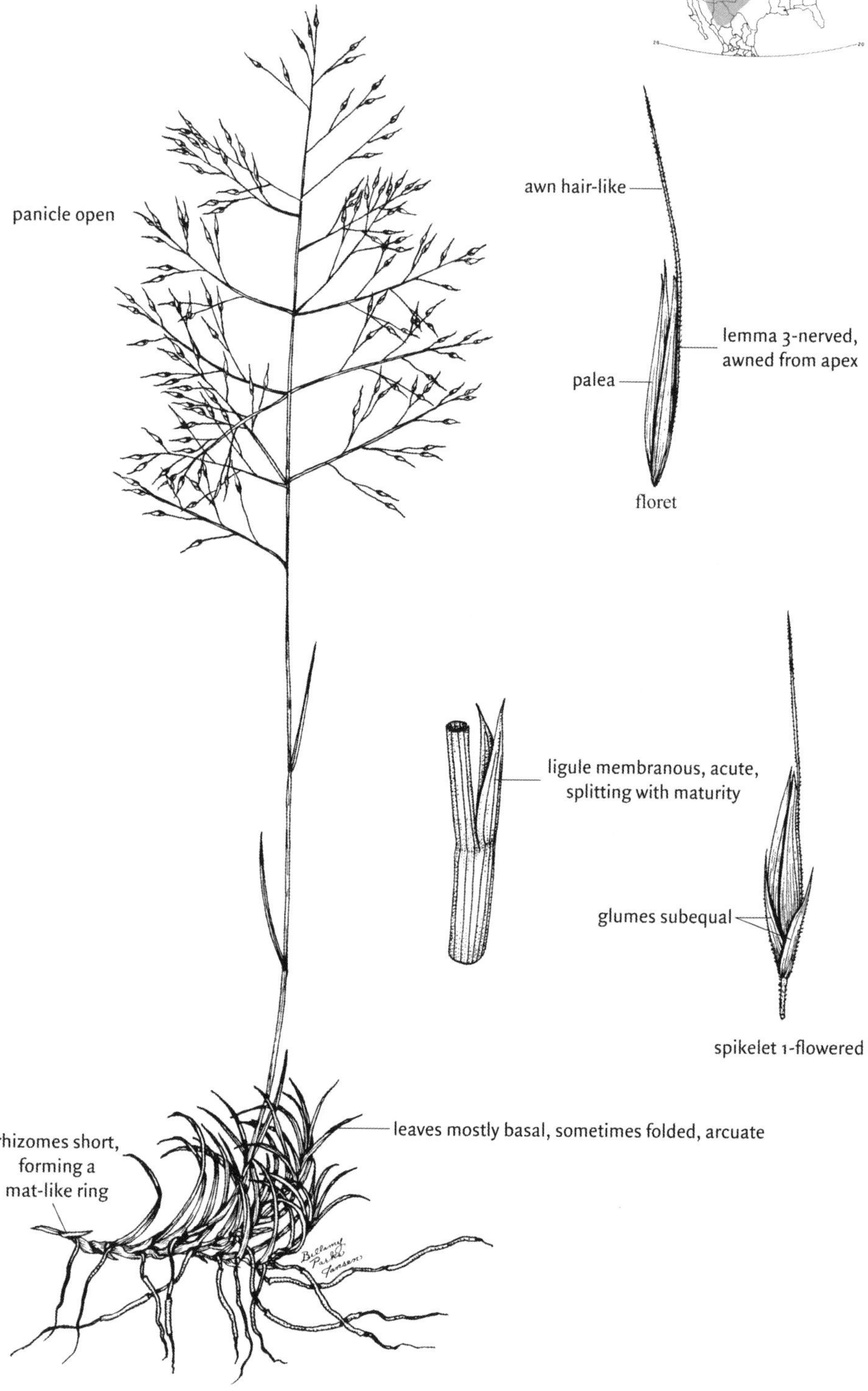

panicle open

awn hair-like

lemma 3-nerved,
awned from apex

palea

floret

ligule membranous, acute,
splitting with maturity

glumes subequal

spikelet 1-flowered

rhizomes short,
forming a
mat-like ring

leaves mostly basal, sometimes folded, arcuate

Tribe:	ERAGROSTEAE
Species:	*Muhlenbergia torreyi* (Kunth) A. S. Hitchc. *ex* Bush
Common Name:	Ring muhly (zacate anillo, ringgrass, ringgrass muhly)
Life Span:	Perennial
Origin:	Native
Season:	Warm

INFLORESCENCE CHARACTERISTICS

type: panicle (7–25 cm long, 4–12 cm wide), open, diffuse, usually red to purplish; branchlets and pedicels appressed or at maturity spreading, pedicels equaling or longer than the spikelets

spikelets: 1-flowered; lemma black (2–3.5 mm long), 3-nerved, minutely bifid, scabrous above; palea equaling or exceeding lemma in length

glumes: subequal (1.5–3 mm long), lanceolate, acute or acuminate to awn-tipped or irregularly toothed, glabrous or scaberulous, 1-nerved

awns: lemma awned from the apex, hair-like (1–4 mm long); glumes awn-tipped

VEGETATIVE CHARACTERISTICS

growth habit: rhizomatous; rhizomes short, forming a mat-like ring as the colony expands outward and dies in the center; leaves mostly basal, crowded

culms: erect (10–30 cm tall) from a decumbent base, somewhat spreading, branching below, nodes not visible

sheaths: rounded, scabrous, puberulent below; margins hyaline

ligules: membranous (2–7 mm long), acute, entire (splitting with maturity)

blades: involute (1–4 cm long, 0.3–1.5 mm wide), sometimes folded, arcuate, sharp-pointed, forming a cushion

GROWTH CHARACTERISTICS: starts growth in late spring or early summer, flowers in midsummer, seeds mature August to September; reproduces from seeds, tillers, and short rhizomes; usually an indicator of abused rangeland or poor sites

FORAGE VALUE: fair to good for cattle when green, poor to fair for wildlife, quality quickly declines with maturity, rated poor for cattle by midsummer, very low forage production

HABITAT: canyons, mesas, rocky slopes, woodlands, and plains; most abundant on sandy to clay loam soils, but it will grow on gravelly soils

Blowoutgrass
Redfieldia flexuosa (Thurb.) Vasey

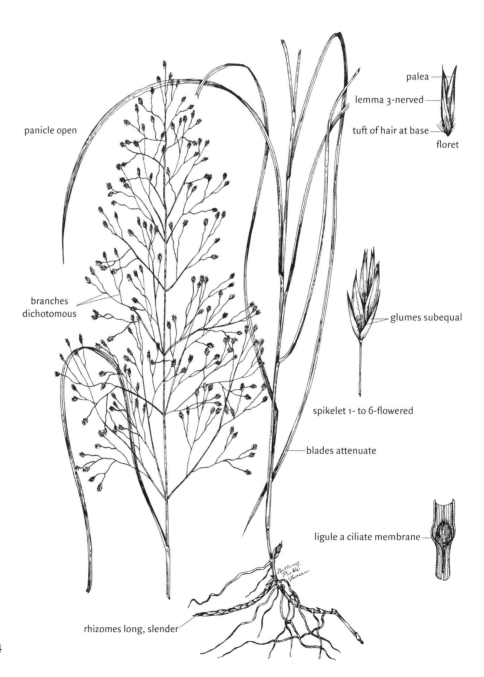

palea

lemma 3-nerved

tuft of hair at base

floret

panicle open

branches
dichotomous

glumes subequal

spikelet 1- to 6-flowered

blades attenuate

ligule a ciliate membrane

rhizomes long, slender

Tribe:	ERAGROSTEAE
Species:	*Redfieldia flexuosa* (Thurb.) Vasey
Common Name:	Blowoutgrass
Life Span:	Perennial
Origin:	Native
Season:	Warm

INFLORESCENCE CHARACTERISTICS

type: panicle (22–60 cm long), open, one-third to one-half as long as the culm, oblong; branches capillary (11–20 cm long), dichotomous

spikelets: 1- to 6-flowered (5–8 mm long, 2.5–5 mm wide), V-shaped, widely spreading at maturity; lemmas acute to acuminate (4–6 mm long), 3-nerved, tuft of hair at base of floret

glumes: subequal, narrow, lanceolate, acuminate, glabrous; first (1.8–4.2 mm long) 1-nerved, second (2.2–5 mm long) 3-nerved

awns: none

VEGETATIVE CHARACTERISTICS

growth habit: rhizomatous; rhizomes long, slender, sharp-pointed, branching

culms: erect to ascending (0.5–1.3 m tall), coarse, glabrous; base usually buried in the sand

sheaths: nearly round, smooth, open, glabrous or lower ones appressed-pubescent near the base, with evenly spaced narrow furrows; lower sheaths becoming fibrous with age; collar slightly expanded

ligules: ciliate membrane (1–3 mm long), membrane minute, truncate to rounded

blades: involute at maturity (15–75 cm long, 1.5–6 mm wide), elongate, attenuate, flexuous, glabrous, evenly spaced narrow furrows on both surfaces

GROWTH CHARACTERISTICS: starts growth in spring, flowers July to October, reproduces from seeds and rhizomes; drought tolerant, resistant to heavy grazing; occurs in large colonies, especially on medium-textured soils and when it is grazed only in winter

FORAGE VALUE: fair for cattle and horses in the summer, but is not readily grazed where other grasses are present, cures rather well and furnishes limited forage in fall and winter

HABITAT: plains, rolling sand hills, and blowouts; most abundant in loose sandy soils that are subject to movement by wind, also occurs in finer textured soils

Burrograss
Scleropogon brevifolius Phil.

SYN = S. *longisetus* Beetle

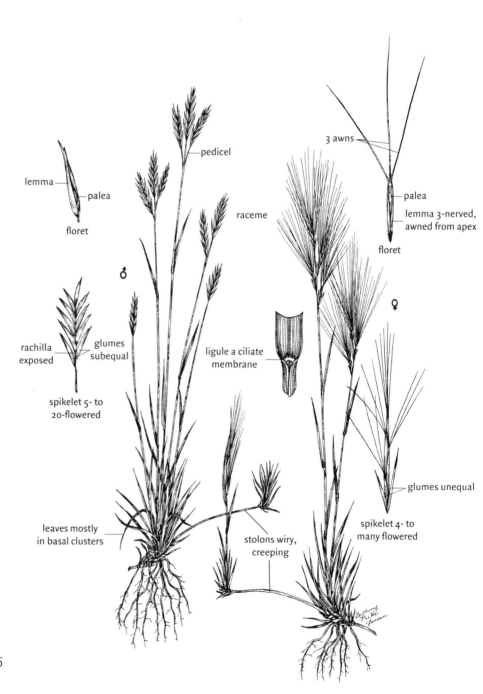

pedicel

lemma
palea
floret

♂

raceme

3 awns

palea

lemma 3-nerved,
awned from apex

floret

♀

rachilla
exposed

glumes
subequal

spikelet 5- to
20-flowered

ligule a ciliate
membrane

glumes unequal

spikelet 4- to
many flowered

leaves mostly
in basal clusters

stolons wiry,
creeping

Tribe:	ERAGROSTEAE
Species:	*Scleropogon brevifolius* Phil.
Common Name:	Burrograss (zacate burro, burrero)
Life Span:	Perennial
Origin:	Native
Season:	Warm

INFLORESCENCE CHARACTERISTICS

type: contracted panicle or raceme (1–6 cm long, excluding awns)

spikelets: unisexual; staminate spikelets 5- to 20-flowered (2–3 cm long), straw-colored, rachilla exposed, persistent lemmas (3.5–7 mm long) similar to glumes; pistillate spikelets with 3–5 fertile florets (2–3 cm long excluding awn), 3-nerved, 1 to many reduced florets, straw-colored to purplish, lemmas firm and rounded

glumes: staminate spikelet glumes subequal (3–10 mm long), thin, lanceolate, first and second glumes separated by a short internode; pistillate spikelet glumes unequal (first 1–2 cm long, second 1.6–3 cm long), thin, lanceolate, subtended by a glume-like bract (1–2 cm long)

awns: pistillate spikelets 3-awned from the apex, lightly twisted (7–14 cm long), divergent, reduced florets usually only represented by an awn; staminate spikelet lemmas awn-tipped

other: plants dioecious (infrequently monoecious)

VEGETATIVE CHARACTERISTICS

growth habit: stoloniferous; stolons wiry, creeping (internodes 5–15 cm long), forming mats; leaves mostly in basal clusters

culms: erect (10–35 cm tall), may be decumbent at the base

sheaths: short, strongly nerved, upper sheaths glabrous, lower sheaths hispid or villous; collar hispid

ligules: ciliate membrane (0.2–0.5 mm long), minute

blades: flat or folded (2–8 cm long, 1–2.5 mm wide), sharp-pointed, twisted, often reflexed

GROWTH CHARACTERISTICS: starts growth in May or June, flowers mostly in late summer and fall but occasionally in the spring; reproduces from seeds and stolons; often develops into large dense stands

FORAGE VALUE: poor for livestock and wildlife; awns may cause eye irritation and contaminate wool

HABITAT: dry, open hills and plains, overstocked ranges; generally on fine-textured, calcareous soils

Alkali sacaton
Sporobolus airoides (Torr.) Torr.

SYN = *Agrostis airoides* Torr.

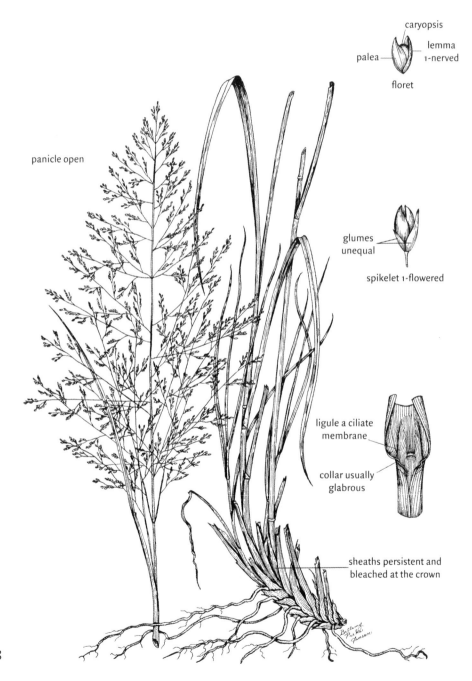

caryopsis

palea

lemma
1-nerved

floret

panicle open

glumes
unequal

spikelet 1-flowered

ligule a ciliate
membrane

collar usually
glabrous

sheaths persistent and
bleached at the crown

Tribe:	ERAGROSTEAE
Species:	*Sporobolus airoides* (Torr.) Torr.
Common Name:	Alkali sacaton (zacatón alcalino)
Life Span:	Perennial
Origin:	Native
Season:	Warm

INFLORESCENCE CHARACTERISTICS

type: panicle, variable in size (20–45 cm long, 15–25 cm wide), subpyramidal to pyramidal, open, exserted or only lower portion enclosed in sheath, often purple; branches without spikelets at the base

spikelets: 1-flowered (1.3–2.8 mm long), mostly on spreading pedicels (0.5–2 mm long), imbricate; lemma equaling the length of the second glume, acute; palea equal to the lemma

glumes: unequal; first (0.5–2 mm long) shorter than the second (1–2.8 mm long), acute, usually 1-nerved

awns: none

VEGETATIVE CHARACTERISTICS

growth habit: densely cespitose; sheaths coarse, persistent at the crown

culms: erect (0.3–1.5 m tall), firm, glabrous, shiny

sheaths: rounded, margins glabrous; collar usually glabrous; lower sheaths usually bleached

ligules: ciliate membrane (0.5 mm long), minute, backed and/or flanked with hairs (1–3 mm long)

blades: flat or becoming involute (5–60 cm long, 2–6 mm wide), firm, pointed, glabrous, rarely long-pilose near the base; prominently ridged adaxially

GROWTH CHARACTERISTICS: starts growth in midspring, flowers June until frost, reproduces from seeds and tillers; withstands flooding and considerable soil deposition, may occur in nearly pure stands

FORAGE VALUE: fair to good for cattle and horses, poor for sheep and wildlife while actively growing, poor for all animals when dry, makes fair hay when cut during or before flowering

HABITAT: alkaline or saline soils in meadows and valleys, sandy soils of desert foothills or roadsides, and dry and gravelly slopes; most abundant on moderately moist alkaline soils of bottomlands where other species are not adapted

Dropseed
Sporobolus compositus (Poir.) Merr.

SYN = S. *asper* (Michx.) Kunth, S. *clandestinus* (Biehler) A. S. Hitchc., S. *drummondii*
(Trin.) Vasey, S. *macer* (Trin.) A. S. Hitchc., S. *pilosa* Vasey

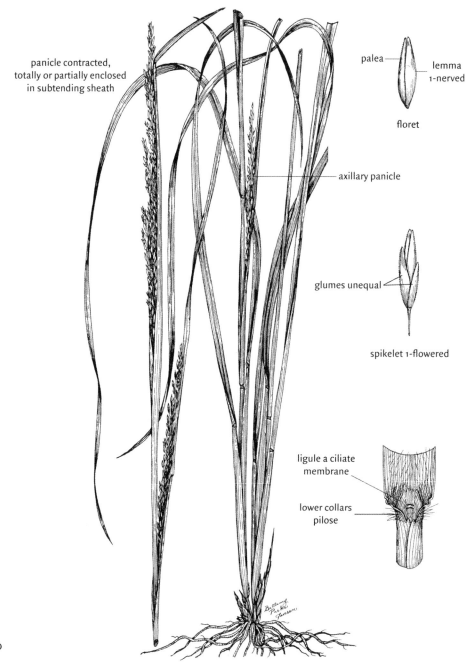

panicle contracted,
totally or partially enclosed
in subtending sheath

palea — — lemma
1-nerved

floret

axillary panicle

glumes unequal

spikelet 1-flowered

ligule a ciliate
membrane

lower collars
pilose

140

Tribe:	ERAGROSTEAE
Species:	*Sporobolus compositus* (Poir.) Merr.
Common Name:	Dropseed (zacatón, tall dropseed, meadow dropseed, rough dropseed, zacate alcalino espigado)
Life Span:	Perennial
Origin:	Native
Season:	Warm

INFLORESCENCE CHARACTERISTICS

type: panicle (5–25 cm long, 4–25 mm wide) contracted, terminal and axillary, solitary at each of the upper culm nodes, completely or partially enclosed in the subtending sheath; panicle branches appressed

spikelets: 1-flowered; lemma flattened, 1-nerved, somewhat rounded at apex (2.5–6.5 mm long), usually longer than the second glume, keeled, glabrous; palea conspicuous, equal to the lemma; cleistogamous spikelets often within axillary panicles

glumes: unequal; first (1.5–4.5 mm long) shorter than the second (2–5 mm long), acute to blunt, keeled, green to purplish with a bright green midnerve

awns: none

VEGETATIVE CHARACTERISTICS

growth habit: usually cespitose, some varieties with short rhizomes

culms: erect (0.6–1.2 m tall), slender, solitary or in small tufts, glabrous

sheaths: open, glabrous or lower ones pilose near the collar and at the throat

ligules: ciliate membrane (about 0.5 mm long), minute, truncate

blades: flat to involute at maturity (10–70 cm long, 1–4.5 mm wide), involute on drying, tapered to filiform apex, scabrous above

other: a variable species with several varieties

GROWTH CHARACTERISTICS: starts growth in late spring, flowers in August, some leaves in dense bunches remain green in winter; reproduces from seeds, tillers, and one variety has rhizomes; drought tolerant, increases during dry periods and with excessive grazing

FORAGE VALUE: fair for livestock and poor for wildlife, most palatable in spring, palatability rapidly declines with maturity

HABITAT: prairies and foothills, on dry clayey to silty soils, most abundant on soils that are intermittently wet and dry, does not grow on deep sandy soils or on soils with a high water table

Sand dropseed
Sporobolus cryptandrus (Torr.) A. Gray

SYN = *Agrostis cryptandra* Torr.

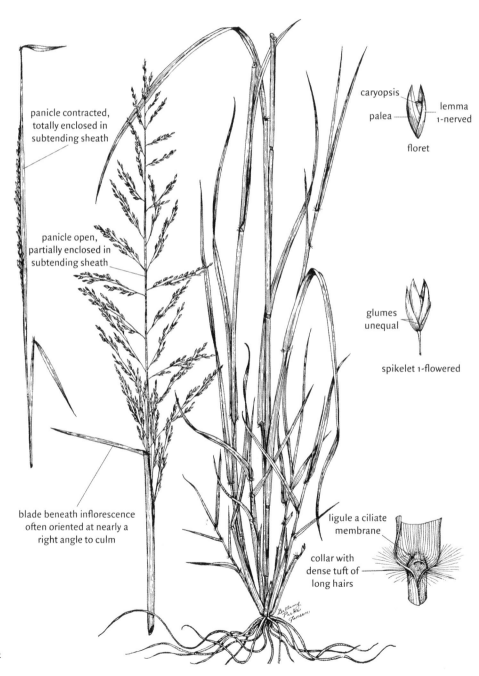

panicle contracted, totally enclosed in subtending sheath

panicle open, partially enclosed in subtending sheath

blade beneath inflorescence often oriented at nearly a right angle to culm

caryopsis

palea

lemma 1-nerved

floret

glumes unequal

spikelet 1-flowered

ligule a ciliate membrane

collar with dense tuft of long hairs

142

Tribe:	ERAGROSTEAE
Species:	*Sporobolus cryptandrus* (Torr.) A. Gray
Common Name:	Sand dropseed (zacatón arenoso, zacate encubierto)
Life Span:	Perennial
Origin:	Native
Season:	Warm

INFLORESCENCE CHARACTERISTICS

type: panicle (15–40 cm long, 2–15 cm wide), terminal, contracted to open above; primary branches distant, occasionally pubescent near panicle axis; pedicels short; inflorescence totally or partially enclosed in the subtending sheath

spikelets: 1-flowered, densely crowded on upper portion of panicle branches, imbricate, lead-gray to purplish; lemma acute (1.4–2.5 mm long), 1-nerved; palea equaling or slightly shorter than the lemma; spikelets within enclosed portion of the inflorescence often cleistogamous

glumes: unequal; first (0.7–1.9 mm long) shorter than the second (1.4–2.7 mm long), 1-nerved or nerveless; second glume equaling or slightly shorter than lemma, thin, acute, 1-nerved

awns: none

VEGETATIVE CHARACTERISTICS

growth habit: cespitose

culms: erect (0.3–1.2 m tall) to geniculate below, flattened to sulcate on one side, glabrous

sheaths: round; margins often ciliate; throat and collar with dense tuft of long hairs (2–4 mm long)

ligules: ciliate membrane (0.3–1 mm long), rounded to truncate

blades: flat to involute (4–35 cm long, 2–8 mm wide), tapering to long and slender apex; margins slightly scabrous; blade beneath inflorescence often oriented at nearly a right angle to the culm

GROWTH CHARACTERISTICS: starts growth in early spring, seeds mature June to August, produces an abundance of seeds, reproduces readily from seeds and tillers; increases with abusive grazing or after drought

FORAGE VALUE: fair to good for livestock and poor for wildlife while actively growing, declines rapidly with maturity

HABITAT: open areas and disturbed sites; most common on sandy soils, but occurs on rocky and silty soils, not tolerant of wet soils

Oniongrass
Melica bulbosa Geyer *ex* Porter & Coult.

SYN = *Bromelica bulbosa* (Geyer ex Porter & Coult.) Weber

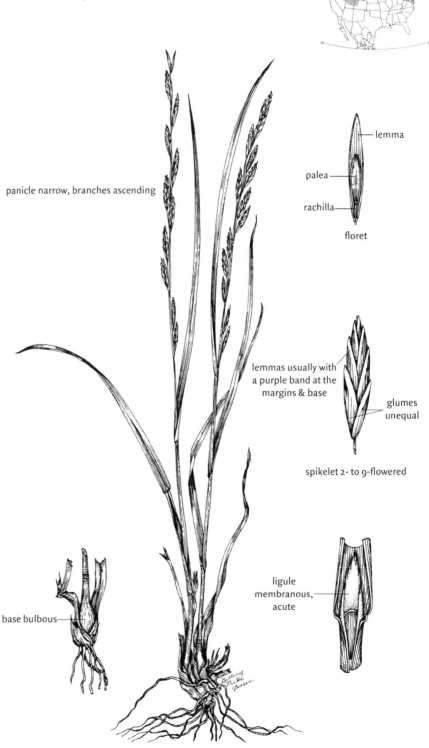

panicle narrow, branches ascending

lemma

palea

rachilla

floret

lemmas usually with a purple band at the margins & base

glumes unequal

spikelet 2- to 9-flowered

base bulbous

ligule membranous, acute

Tribe:	MELICEAE
Species:	*Melica bulbosa* Geyer *ex* Porter & Coult.
Common Name:	Oniongrass (bulbous oniongrass, onion melic)
Life Span:	Perennial
Origin:	Native
Season:	Cool

INFLORESCENCE CHARACTERISTICS

type: panicle (8–20 cm long), narrow, dense, elongate; branches ascending, short and erect

spikelets: 2- to 9-flowered (7–20 mm long), usually with 3 perfect florets, upper florets sterile; lemmas papery (6–11 mm long), broadly acute to obtuse and occasionally emarginate, obscurely nerved, usually with a purple band at the margins and base; palea nerves ciliate

glumes: unequal, ovate, acute to obtuse, scaberulous; first 1- to 5-nerved (6–9 mm long); second 5- to 7-nerved (7–11 mm long); nerves sometimes scabrous

awns: none

VEGETATIVE CHARACTERISTICS

growth habit: cespitose, occasionally with very short rhizomes

culms: erect (0.3–1 m tall), clustered, base bulbous

sheaths: rounded, glabrous to scabrous (infrequently pubescent), margins connate nearly the full length

ligules: membranous (2–6 mm long), acute, erose to deeply lacerate

blades: flat to involute (10–30 cm long, 2–5 mm wide); glabrous to scabrous, occasionally sparsely pubescent adaxially

GROWTH CHARACTERISTICS: starts growth in early spring, flowers in late spring or early summer, seeds mature in July, low seed viability; reproduces from seeds and tillers from bulbous culm bases

FORAGE VALUE: excellent for cattle, sheep, horses, elk, and deer; many species of small animals use the seeds and bulbous culm bases; generally not abundant enough to produce high quantities of forage

HABITAT: meadows, alluvial fans, rocky woods, and hills; occurs on all exposures but is most abundant on north and east exposures; often in mesic sites; most abundant in rich sandy loams or clay loams from mid- to subalpine elevations

Arizona cottontop
Digitaria californica (Benth.) Henr.

SYN = *Trichachne californica* (Benth.) Chase

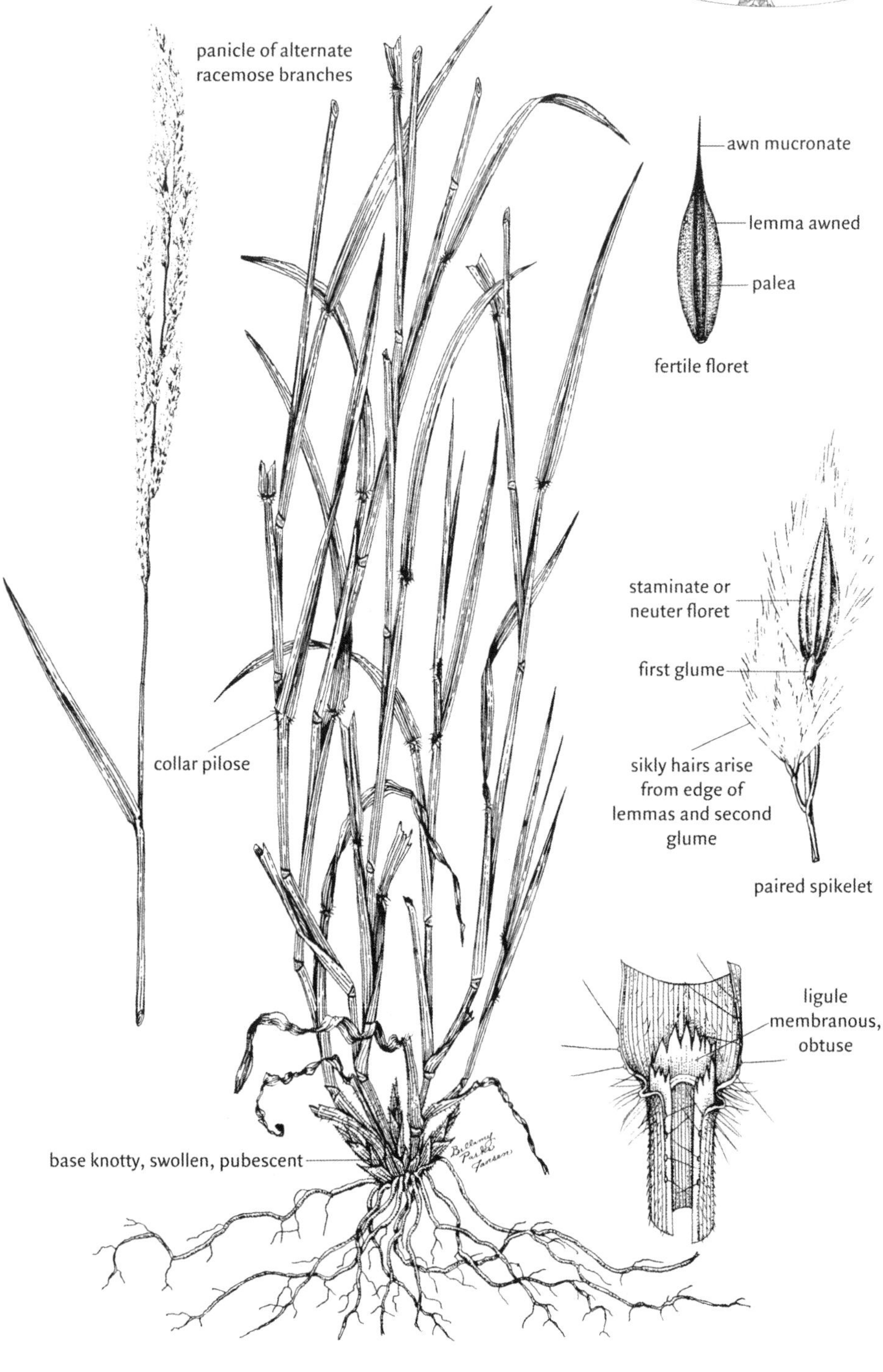

panicle of alternate
racemose branches

awn mucronate

lemma awned

palea

fertile floret

collar pilose

staminate or
neuter floret

first glume

sikly hairs arise
from edge of
lemmas and second
glume

paired spikelet

ligule
membranous,
obtuse

base knotty, swollen, pubescent

Tribe:	PANICEAE
Species:	*Digitaria californica* (Benth.) Henr.
Common Name:	Arizona cottontop (zacate punta blanca, cottongrass, California cottontop)
Life Span:	Perennial
Origin:	Native
Season:	Warm

INFLORESCENCE CHARACTERISTICS

type: panicle (8–20 cm long, 4–20 mm wide) of alternate racemose branches, narrow, densely flowered, erect; branches relatively few (3–5 cm long), appressed, mostly simple

spikelets: in pairs (3–4 mm long), 2-flowered; lower floret staminate or neuter, broad, 3-nerved; upper floret perfect, covered and exceeded by silky hairs that arise from edge of lemmas and second glume; hairs (2–6 mm long) white or silver to purple-tinged

glumes: first minute or absent; second narrow (2–4 mm long), 3-nerved, densely villous

awns: fertile lemma awn-tipped or mucronate

VEGETATIVE CHARACTERISTICS
growth habit: cespitose

culms: erect (0.4–1 m tall), stiff; freely branching at base, often purplish; base knotty, swollen; scale leaves densely pubescent

sheaths: rounded, longer than the internodes, lower sheaths felty pubescent or sparsely hairy, glabrous to sparsely hairy above; collar pilose

ligules: membranous (2–4 mm long), obtuse, erose to lacerate

blades: flat to somewhat folded (2–18 cm long, 2–7 mm wide), involute upon drying, pubescent to glabrous, often bluish-green, occasionally with glandular hairs abaxially, midrib prominent

GROWTH CHARACTERISTICS: starts growth in late spring or early summer, responds quickly to moisture, reproduces primarily from seeds, seed set is good; seldom occurs in pure stands; responds well to summer rest

FORAGE VALUE: good for livestock and fair for wildlife; palatable throughout the year, cures well and some culms remain green in winter, furnishes valuable winter forage; frequently over utilized because of its relatively high winter palatability

HABITAT: plains, dry open ground, steep slopes, chaparral, and semidesert grassland; most abundant in well-drained sandy and gravelly loam soils

Hall panicum
Panicum hallii Vasey

SYN = *P. filipes* Scribn.

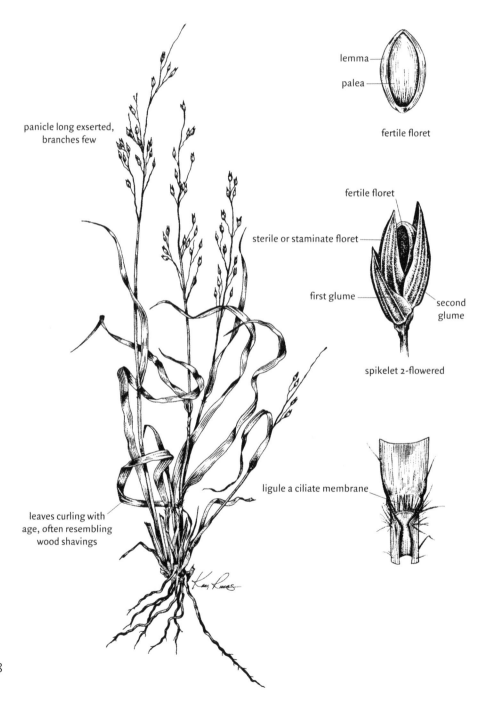

lemma

palea

fertile floret

panicle long exserted,
branches few

fertile floret

sterile or staminate floret

first glume

second
glume

spikelet 2-flowered

ligule a ciliate membrane

leaves curling with
age, often resembling
wood shavings

Tribe:	PANICEAE
Species:	*Panicum hallii* Vasey
Common Name:	Hall panicum (panzio, Hall panicgrass, zacate rizado)
Life Span:	Perennial
Origin:	Native
Season:	Warm

INFLORESCENCE CHARACTERISTICS

type: panicle (6–25 cm long), long-exserted, pyramidal; branches few, spreading; spikelets pedicellate; pedicels short, appressed to slightly spreading

spikelets: 2-flowered (2.2–3.9 mm long, 1–1.5 mm wide); lower floret sterile or staminate, glabrous, acute to acuminate, about as long as the spikelet, strongly 5- to 7-nerved; upper floret fertile; fertile floret lemma elliptical, smooth, dark brown, shiny

glumes: unequal; first one-third to two-thirds length of spikelet, acute; second about as long as spikelet, 5- to 7-nerved

awns: none

VEGETATIVE CHARACTERISTICS

growth habit: cespitose; leaves basal and cauline

culms: erect or ascending (15–90 cm tall), simple or sparingly branched from lower nodes, glaucous; nodes glabrous to appressed-pubescent

sheaths: round, lower overlapping, upper about as long as internodes, glabrous to sparsely papillose-hispid

ligules: ciliate membrane (membrane 0.8–1.5 mm long, hairs about 1 mm long), truncate

blades: flat (4–30 cm long, 2–10 mm wide), stiffly ascending, glabrous to sparsely papillose-hirsute, occasionally glaucous, curling with age, often resembling wood shavings

GROWTH CHARACTERISTICS: starts growth in early spring, flowers April to November; reproduces from seeds and tillers

FORAGE VALUE: fair for livestock and wildlife, palatability rapidly declines with maturity

HABITAT: dry prairies, rocky to gravelly hills, canyons, and bottomlands; sandy to clayey soils, particularly calcareous soils; one variety occurs in moist soils, sometimes found in irrigated fields

Vinemesquite
Panicum obtusum Kunth

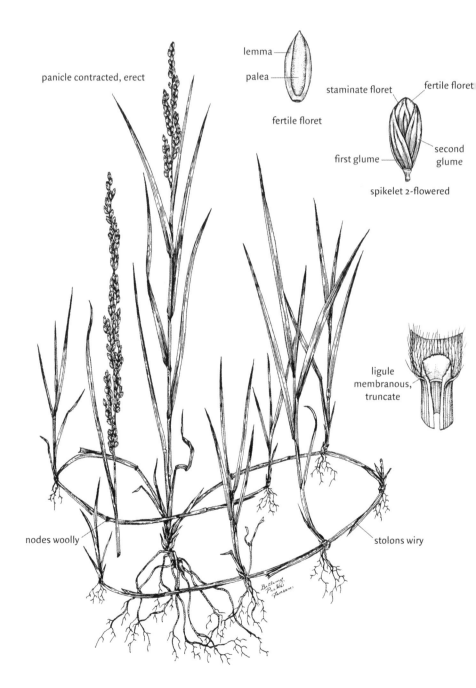

panicle contracted, erect

lemma

palea

fertile floret

staminate floret

fertile floret

first glume

second glume

spikelet 2-flowered

ligule
membranous,
truncate

nodes woolly

stolons wiry

Tribe:	PANICEAE
Species:	*Panicum obtusum* Kunth
Common Name:	Vinemesquite (zacate guía, grapevine mesquite)
Life Span:	Perennial
Origin:	Native
Season:	Warm

INFLORESCENCE CHARACTERISTICS

type: panicle (3–14 cm long, 5–13 mm wide), contracted, erect; primary branches usually unbranched, distant, ascending-appressed

spikelets: 2-flowered (3.2–4.5 mm long), oblong or obovate, blunt, glabrous, hard; lower floret usually staminate, palea often exceeds the lemma; upper floret fertile (2.6–3.6 mm long), elliptic, glabrous, minutely reticulate

glumes: subequal, obtuse to rounded, brown at maturity; first nearly as long as spikelet, 3- to 5-nerved; glume 5- to 9-nerved

awns: none

VEGETATIVE CHARACTERISTICS

growth habit: cespitose; stolons long (to 9 m), wiry, nodes woolly

culms: erect (20–80 cm tall), base knotty, internodes glabrous; nodes pubescent

sheaths: rounded, covers one-half to three-fourths the length of the internode, glabrous to hispid at collar; basal sheaths usually villous

ligules: membranous (1–2 mm long), truncate, erose

blades: flat (5–20 cm long, 2–7 mm wide), firm, elongate; margins scabrous; pilose adaxially at collar; midrib white, prominent abaxially; margins scabrous; stolon leaves may be strongly villous

GROWTH CHARACTERISTICS: starts growth in April to May; flowers May to October; seeds are slow to disseminate; reproduces from seeds, rhizomes, and stolons; may grow in pure stands

FORAGE VALUE: fair to good for livestock and fair for wildlife, withstands heavy grazing; produces fair hay; doves and quail eat the seeds in fall and winter

HABITAT: moist depressions that periodically dry out, banks of rivers and irrigation ditches, and lowland pastures; adapted to a wide range of soil textures, but most abundant on sandy to sandy loam soils

Switchgrass
Panicum virgatum L.

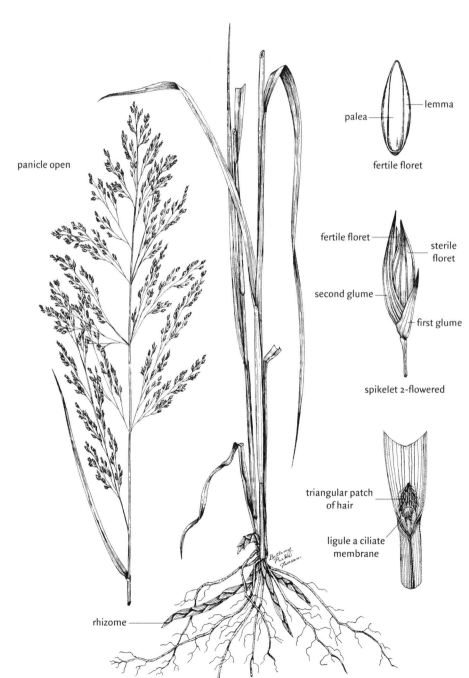

panicle open

lemma

palea

fertile floret

fertile floret

sterile floret

second glume

first glume

spikelet 2-flowered

triangular patch of hair

ligule a ciliate membrane

rhizome

Tribe:	PANICEAE
Species:	*Panicum virgatum* L.
Common Name:	Switchgrass (panzio)
Life Span:	Perennial
Origin:	Native
Season:	Warm

INFLORESCENCE CHARACTERISTICS

type: panicle (15–55 cm long), open; spikelets clustered toward ends of long branches; lower branches in whorls, pairs, or single; inflorescences variable

spikelets: 2-flowered (3–6 mm long); lower floret sterile or staminate, glabrous; upper floret fertile; lemma of fertile floret smooth, shiny, acute, inrolled at base, clasping the palea

glumes: unequal, first (2.3–5 mm long) shorter than the second (3.3–6 mm long), narrowly acute or acuminate; base of the second glume enclosed by the first glume

awns: none

VEGETATIVE CHARACTERISTICS

growth habit: rhizomatous

culms: erect (0.5–3 m tall), glabrous, robust, usually unbranched above base

sheaths: round, glabrous to ciliate at margins, often purplish to reddish at base

ligules: ciliate membrane (1.5–3.5 mm long), mostly hairs, obtuse

blades: flat (10–60 cm long, 3–15 mm wide), elongate, firm; triangular patch of hair adaxially at the base; margins weakly scabrous

GROWTH CHARACTERISTICS: starts growth in April or May, flowers in summer, seeds mature in late summer to early fall; reproduces from seeds, tillers, and rhizomes

LIVESTOCK LOSSES: May contain saponins and cause photosensitization in sheep and horses

FORAGE VALUE: good for all types of livestock and fair for wildlife; as the plants begin to mature in midsummer, nutrient content and palatability decline rapidly; produces good hay if cut before maturity

HABITAT: prairies, open ground, open woods, brackish marshes, and pine woods; adapted to a broad range of soil textures; tolerates flooding for short periods

Knotgrass
Paspalum distichum L.

SYN = *P. vaginatum* Sw.

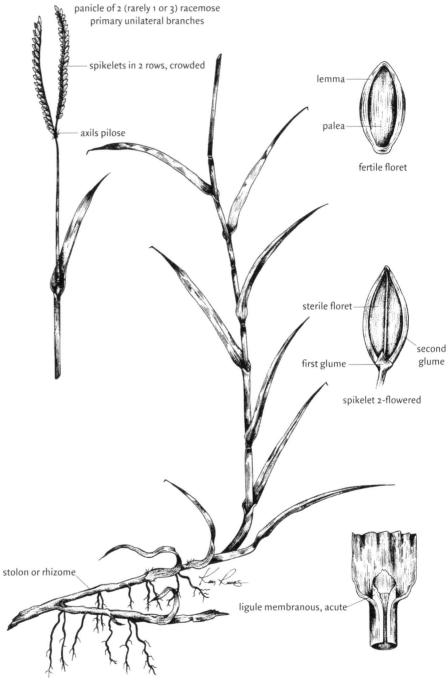

panicle of 2 (rarely 1 or 3) racemose primary unilateral branches

spikelets in 2 rows, crowded

axils pilose

lemma

palea

fertile floret

sterile floret

first glume

second glume

spikelet 2-flowered

stolon or rhizome

ligule membranous, acute

Tribe:	PANICEAE
Species:	*Paspalum distichum* L.
Common Name:	Knotgrass (camalote saladillo, zacate nudoso, zacate de arena, jointgrass)
Life Span:	Perennial
Origin:	Native
Season:	Warm

INFLORESCENCE CHARACTERISTICS

type: panicle of 2 (rarely 1 or 3) racemose primary unilateral branches; branches usually not more than 1 cm apart at the culm apex, erect to somewhat spreading (2–7 cm long), flattened (1–2 mm wide), arcuate; pilose in the axils

spikelets: 2-flowered (2.5–3.3 mm long), dorsally compressed, crowded in 2 rows on lower side of branch; lower floret sterile; lower lemma similar in size, shape, and texture to second glume, glabrous; upper floret perfect, elliptic, blunt to somewhat acute; lemma ovate (2.2–3.1 mm long), greenish, indurate, glabrous, faintly 3-nerved

glumes: unequal; first usually present, minute (to 1.5 mm long), triangular, truncate to lanceolate; second broadly lanceolate (2.5–3.2 mm long), apex obtuse to acute, pubescence appressed

awns: none

VEGETATIVE CHARACTERISTICS

growth habit: stoloniferous or rhizomatous, extensively creeping

culms: ascending to decumbent (0.3–1 m long), often rooting at the lower nodes, compressed; nodes glabrous or pubescent; internodes glabrous

sheaths: somewhat keeled, cross-septate; pilose at base and on upper margins, or all over

ligules: membranous (1–2.5 mm long), acute, erose to lacerate

blades: flat or folded (3–20 cm long, 2.5–6 mm wide), glabrous, occasionally hairs on adaxial surface; margins scabrous

GROWTH CHARACTERISTICS: flowers June to October; reproduces by seed, stolons, rhizomes, and rooting at lower culm nodes

FORAGE VALUE: good for livestock and fair for wildlife while actively growing, becomes less palatable when mature

HABITAT: moist to wet places along streams, canals, and ditches; important species for holding soil along streams, although has been known to restrict the water flow in irrigation ditches

Buffelgrass
Pennisetum ciliare (L.) Link

SYN = *Cenchrus ciliaris* L.

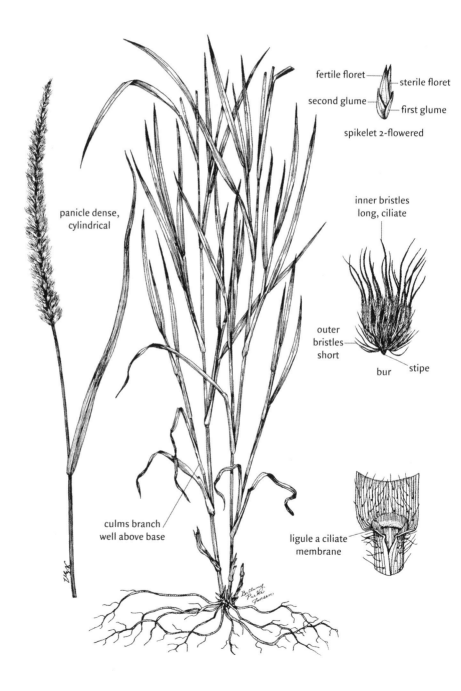

fertile floret — sterile floret

second glume — first glume

spikelet 2-flowered

panicle dense, cylindrical

inner bristles long, ciliate

outer bristles short

bur — stipe

culms branch well above base

ligule a ciliate membrane

Tribe:	PANICEAE
Species:	*Pennisetum ciliare* (L.) Link
Common Name:	Buffelgrass (zacate buffel)
Life Span:	Perennial
Origin:	Introduced (from India and Africa)
Season:	Warm

INFLORESCENCE CHARACTERISTICS

type: panicle (2–13 cm long, 1–2.6 cm wide), dense, cylindrical; inflorescence usually purplish to blackish; burs on a minute stipe; stipe pilose

spikelets: in clusters of 2–4 (2–6 mm long), 2-flowered (lower floret sterile); subtended and enclosed by numerous (about 50) bristles, forming a bur; bristles unequal (4–10 mm long) united at base, erect or spreading, outer ones shorter than inner, long-ciliate on inner bristle margins, purplish, flexuous, terete, connate or free at base

glumes: unequal (1–3.4 mm long)

awns: none

VEGETATIVE CHARACTERISTICS

growth habit: cespitose, from a knotty base

culms: erect or geniculate to spreading (0.5–1.2 m tall), knotty at the base, branched well above the base

sheaths: open, laterally compressed and keeled, glabrous to sparsely pilose

ligules: ciliate membrane (0.5–1.5 mm long), truncate

blades: flat (8–35 cm long, 2.5–8 mm wide), scabrous or slightly pilose, tapering to an acuminate apex

GROWTH CHARACTERISTICS: starts growth in late winter, flowers from spring through fall, reproduces from seeds and rhizomes; will not tolerate extended subfreezing temperatures, commonly seeded on rangeland following mechanical removal of brush

FORAGE VALUE: good for cattle, horses, sheep, and deer; most valuable for spring and early summer grazing, forage quality declines in summer

HABITAT: seeded pastures, road rights-of-way, old fields, and disturbed sites; occurs in all soil textures but is most common in sandy soils, does not tolerate extended flooding

Plains bristlegrass
Setaria leucopila (Scribn. & Merr.) K. Schum.

SYN = *Chaetochloa leucopila* Scribn. & Merr.

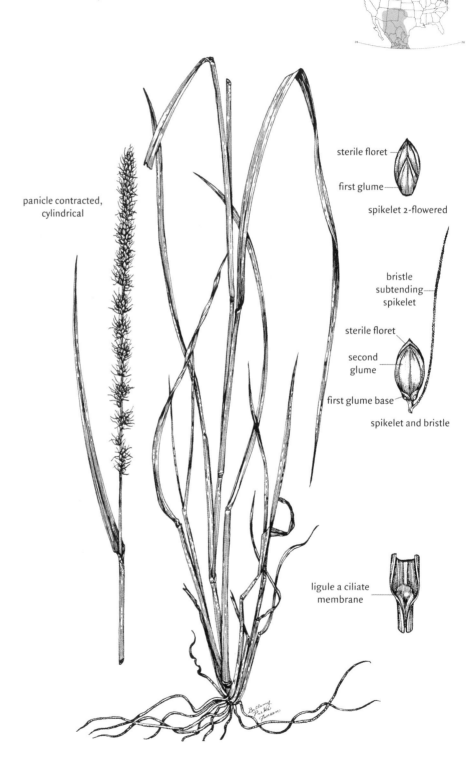

panicle contracted,
cylindrical

sterile floret

first glume

spikelet 2-flowered

bristle
subtending
spikelet

sterile floret

second
glume

first glume base

spikelet and bristle

ligule a ciliate
membrane

Tribe:	PANICEAE
Species:	*Setaria leucopila* (Scribn. & Merr.) K. Schum.
Common Name:	Plains bristlegrass (zacate tempranero)
Life Span:	Perennial
Origin:	Native
Season:	Warm

INFLORESCENCE CHARACTERISTICS

type: panicle (6–25 cm long, 6–15 mm wide), contracted, cylindrical, erect, densely flowered, bristly; bristles representing reduced branches or pedicels

spikelets: 2-flowered (2–3 mm long); lower floret sterile, palea of lower floret one-half to three-fourths as long as the lemma; upper floret perfect, usually subtended by a single bristle (4–15 mm long), lemma and palea rugose

glumes: unequal, first reduced; second as long as spikelet (2–3 mm long)

awns: none

VEGETATIVE CHARACTERISTICS

growth habit: cespitose

culms: stiffly erect to geniculate below (0.2–1.2 m tall), infrequently branched above, scabrous, often pubescent below the nodes

sheaths: keeled, often villous on upper margins

ligules: ciliate membrane (1–2 mm long), obtuse to rounded

blades: flat or folded (8–25 cm long, 2–6 mm wide), glabrous to scabrous or infrequently pubescent, pale or glaucous

GROWTH CHARACTERISTICS: starts growth midspring, flowers from May to September, may produce more than one seed crop depending on available moisture, good seed producer; reproduces from seeds and tillers; cannot withstand heavy grazing; usually does not occur in dense stands

FORAGE VALUE: good for cattle and horses, fair to good for wildlife and sheep

HABITAT: prairies, dry woods, and rocky slopes; in open shade of brush and small trees where it is protected from livestock; most abundant in well-drained alkaline soils along gullies or streams or other areas with occasional abundant moisture

Orchardgrass
Dactylis glomerata L.

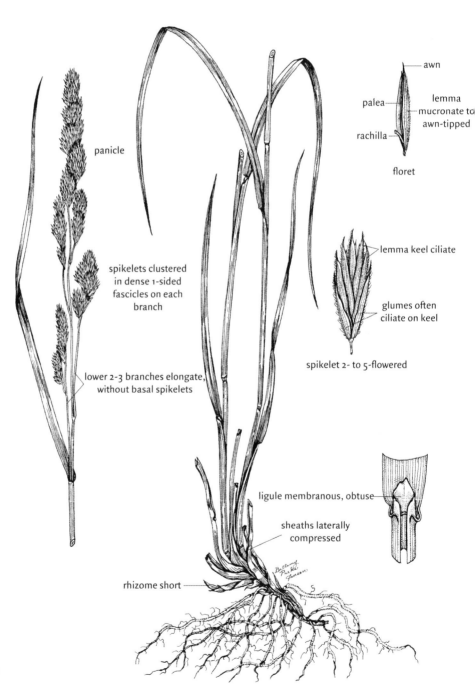

awn

palea — lemma
mucronate to
awn-tipped

rachilla

floret

panicle

lemma keel ciliate

glumes often
ciliate on keel

spikelets clustered
in dense 1-sided
fascicles on each
branch

spikelet 2- to 5-flowered

lower 2-3 branches elongate,
without basal spikelets

ligule membranous, obtuse

sheaths laterally
compressed

rhizome short

Tribe:	POEAE
Species:	*Dactylis glomerata* L.
Common Name:	Orchardgrass (zacate dactilo, zacate ovillo, cocksfoot)
Life Span:	Perennial
Origin:	Introduced (from Europe and Asia)
Season:	Cool

INFLORESCENCE CHARACTERISTICS

type: panicle (3–20 cm long); lower 2–3 branches elongate, without spike-lets basally, occasionally spreading; upper branches short, floriferous to the base, appressed; spikelets clustered in dense 1-sided fascicles on each branch

spikelets: 2- to 5-flowered (5–9 mm long), oblong, nearly sessile; lemmas acute (3–7 mm long), keeled, keel ciliate, margins hyaline

glumes: equal to subequal (2.5–7 mm long), either may be reduced, acute, 1-nerved, keeled; keel ciliate or glabrous

awns: glumes and lemmas mucronate to awn-tipped (up to 2 mm long)

VEGETATIVE CHARACTERISTICS

growth habit: cespitose, rarely with short rhizomes

culms: erect (0.3–1.2 m tall), glabrous

sheaths: laterally compressed, keeled, glabrous to slightly scabrous, most shorter than the internodes

ligules: membranous (2–8 mm long), obtuse, erose to lacerate

blades: flat or folded (10–40 cm long, 2–11 mm wide), elongate, lax; midrib prominent, scabrous; margins scabrous

GROWTH CHARACTERISTICS: starts growth in early spring, reproduces from seeds and tillers; may be injured in areas with dry, cold winters and no snow cover or if subjected to warm temperatures in January or February followed by a period of extremely cold temperatures; does not tolerate extended periods of drought; shade tolerant; responds to nitrogen fertilizer and irrigation

FORAGE VALUE: good to excellent for livestock and wildlife, especially relished by deer, provides early spring forage, cures well as hay, sometimes mixed with alfalfa or other legumes to provide high quality hay; quality of standing forage rapidly declines with maturity

HABITAT: fields, meadows, along ditch banks, and waste places; on fine or coarse soils; only slightly salt tolerant; commonly seeded in mixtures in pastures and on hayland

Rough fescue
Festuca campestris Rydb.

SYN = *F. scabrella* Torr. in part

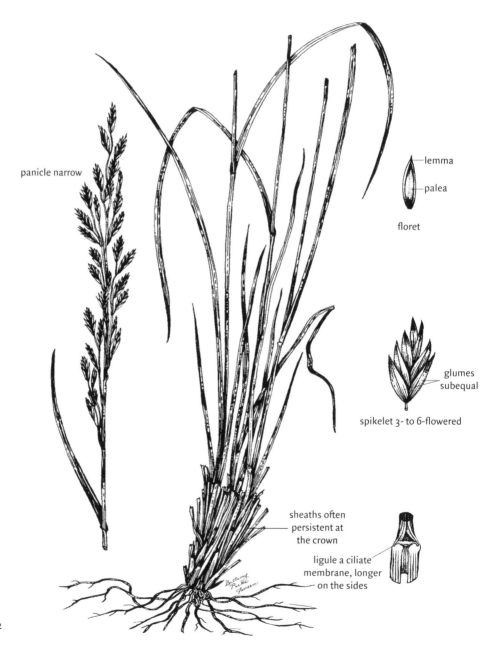

panicle narrow

lemma

palea

floret

glumes
subequal

spikelet 3- to 6-flowered

sheaths often
persistent at
the crown

ligule a ciliate
membrane, longer
on the sides

Tribe:	POEAE
Species:	*Festuca campestris* Rydb.
Common Name:	Rough fescue (buffalo bunchgrass)
Life Span:	Perennial
Origin:	Native
Season:	Cool

INFLORESCENCE CHARACTERISTICS

type: panicle (5–20 cm long), narrow; branches solitary or in pairs above, occasionally in groups of three in the lower portion, appressed, often spreading

spikelets: 3- to 6-flowered (9–15 mm long), often purple; lemmas stout (7–10 mm long), 5-nerved, acute, scabrous

glumes: subequal, acute; first lanceolate (4–7.5 mm long), 1-nerved and slightly shorter than second; second 3-nerved (5.5–8 mm long)

awns: lemma awnless, mucronate, or short-awned (less than 2 mm long)

VEGETATIVE CHARACTERISTICS

growth habit: cespitose in large tufts (to 60 cm in diameter); leaves mostly basal, sheaths often persistent at the crown

culms: erect (0.3–1 m tall), glabrous, scabrous and naked below panicle, stout, purplish at the base

sheaths: enlarged at base, glabrous to scabrous; margins broad, hyaline, open

ligules: ciliate membrane (less than 1 mm long), truncate, longer on the sides

blades: folded or becoming involute (30–70 cm long, 2.5–4.5 mm wide), erect, stiff, pointed, usually scabrous abaxially

GROWTH CHARACTERISTICS: growth period is June to August, reproduces from seeds, tillers, and occasionally from short rhizomes; does not withstand continuous heavy grazing

FORAGE VALUE: excellent for cattle and horses and good for sheep and wildlife during all growth stages; valuable for winter grazing because it retains much of its protein and palatability, produces good hay

HABITAT: foothills, mountains to near timberline, benchland, and valleys; most abundant on dry, deep sandy loam soils

Idaho fescue
Festuca idahoensis Elmer

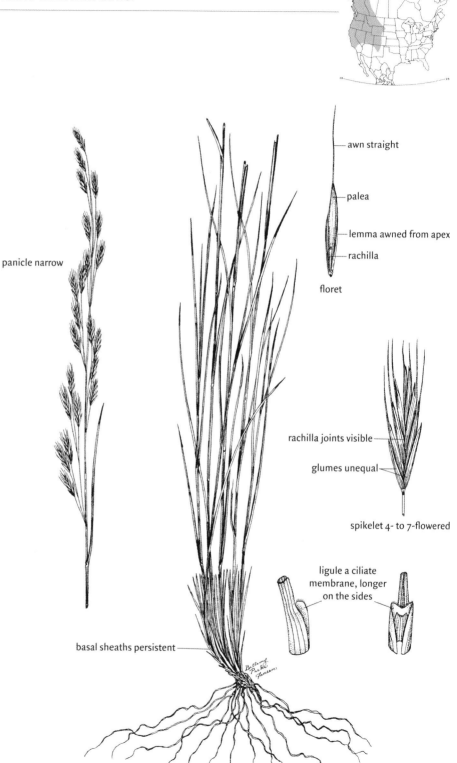

awn straight

palea

lemma awned from apex

rachilla

floret

panicle narrow

rachilla joints visible

glumes unequal

spikelet 4- to 7-flowered

ligule a ciliate membrane, longer on the sides

basal sheaths persistent

Tribe: POEAE
Species: *Festuca idahoensis* Elmer
Common Name: Idaho fescue (blue bunchgrass)
Life Span: Perennial
Origin: Native
Season: Cool

INFLORESCENCE CHARACTERISTICS

type: panicle (7–15 cm long), narrow, loosely contracted; branches ascending, lower branches spreading

spikelets: 4- to 7-flowered (8–14 mm long); lemmas (5–7 mm long) somewhat laterally compressed at maturity, scabrous to glabrous; rachilla joints usually visible

glumes: unequal, lanceolate, acute; first 1-nerved (3–5 mm long), second faintly 5-nerved (4–5.5 mm long)

awns: lemmas awned from apex; awn straight (2–5 mm long)

VEGETATIVE CHARACTERISTICS

growth habit: cespitose; leaves mostly basal

culms: erect (0.3–1 m tall), glabrous to scaberulous, glaucous

sheaths: flattened, keeled, glabrous or scabrous, green or glaucous; basal sheaths short, open, wider than the blade, persistent; collars indistinct; auricles small or absent

ligules: ciliate membrane (less than 2 mm long), slightly longer on the sides

blades: involute (5–25 cm long), filiform, firm, elongate, scabrous, often glaucous, pubescent adaxially, glabrous abaxially

GROWTH CHARACTERISTICS: starts growth in early spring, seeds mature by midsummer, reproduces from seeds and tillers; withstands some excessive grazing

FORAGE VALUE: excellent for livestock and wildlife, especially important late in the growing season because it remains green longer than the associated species; not readily grazed upon drying unless more desirable forage is not available

HABITAT: foothill rangeland, open woods, moist parks, and rocky slopes; grows on all exposures and on many soil types, most abundant on well-drained loams with a neutral to slightly alkaline pH, occurs at elevations from 300 m to nearly 4,000 m

Mutton bluegrass
Poa fendleriana (Steud.) Vasey

SYN = *P. longiligula* Scribn. & Williams

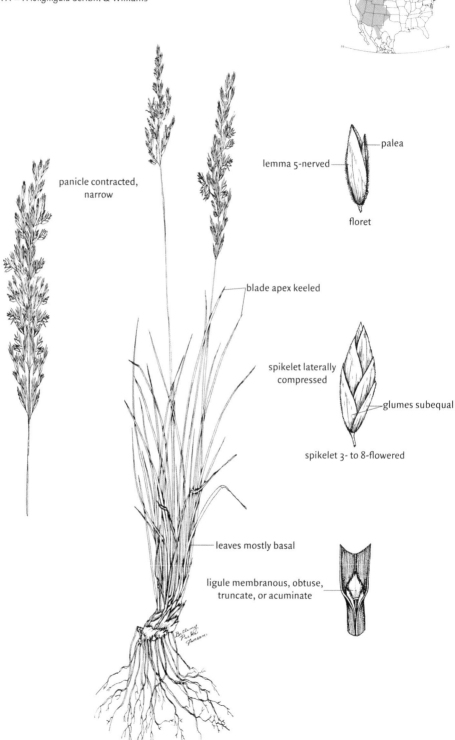

panicle contracted, narrow

palea

lemma 5-nerved

floret

blade apex keeled

spikelet laterally compressed

glumes subequal

spikelet 3- to 8-flowered

leaves mostly basal

ligule membranous, obtuse, truncate, or acuminate

Tribe:	POEAE
Species:	*Poa fendleriana* (Steud.) Vasey
Common Name:	Mutton bluegrass (zacate azúl borreguero, muttongrass)
Life Span:	Perennial
Origin:	Native
Season:	Cool

INFLORESCENCE CHARACTERISTICS

type: panicle (2–10 cm long, 1–2 cm wide), contracted, narrow, densely flowered, exserted above the basal leaves, tan to purple; branches 2–3 per node, short, erect or erect-spreading

spikelets: mostly unisexual; 3- to 8-flowered (6–10 mm long, 1.8–3 mm wide), usually twice as long as wide, laterally compressed, papery; lemma compressed-keeled (4–6 mm long), not webbed, blunt, 5-nerved, marginal nerves pubescent below

glumes: subequal, usually one-half to two-thirds as long as lowest lemma; first 1-nerved (2.8–4.5 mm long); second 3-nerved (3–5.5 mm long), papery, strongly keeled

awns: none

other: incompletely dioecious, most plants pistillate, few staminate, occasional plants with functional perfect flowers

VEGETATIVE CHARACTERISTICS

growth habit: cespitose, rarely with rhizomes; rhizomes short, slender; leaves mostly basal

culms: erect (15 to 80 cm tall) to decumbent at base, glabrous, scabrous below inflorescence

sheaths: rounded, short, glabrous to scabrous; sheath bases bleached, expanded, and persistent for several years; margins hyaline, open

ligules: membranous, highly variable (usually 0.5 mm long, occasionally up to 3–7 mm long); obtuse, truncate, or acuminate; erose

blades: folded or involute (10–20 cm long, 1–4 mm wide), stiff, erect, scabrous, glaucous, double midrib, often remains green after drying; apex keeled

GROWTH CHARACTERISTICS: starts growth in early spring and matures in June or July, reproduces from seeds and tillers and rarely from rhizomes

FORAGE VALUE: excellent for cattle and horses, good for sheep, elk, and deer; values decline rapidly with maturity

HABITAT: mesas, foothills, mountains, dry open woods, cold deserts, and rocky hills; grows on a broad range of soils

Kentucky bluegrass
Poa pratensis L.

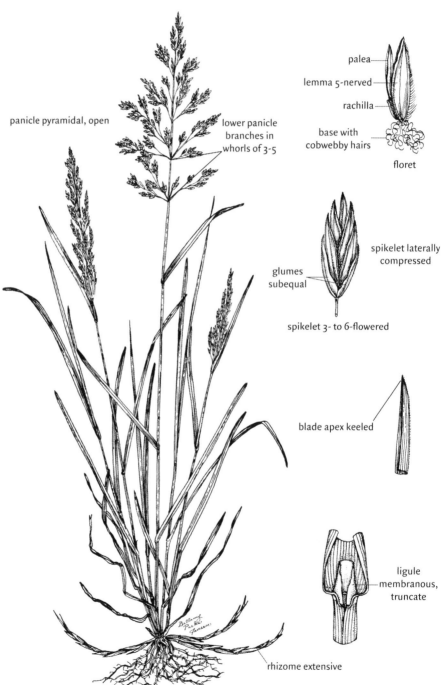

panicle pyramidal, open

lower panicle branches in whorls of 3-5

palea

lemma 5-nerved

rachilla

base with cobwebby hairs

floret

spikelet laterally compressed

glumes subequal

spikelet 3- to 6-flowered

blade apex keeled

ligule membranous, truncate

rhizome extensive

Tribe:	POEAE
Species:	*Poa pratensis* L.
Common Name:	Kentucky bluegrass (zacate azúl de Kentucky)
Life Span:	Perennial
Origin:	Introduced (from Europe)
Season:	Cool

INFLORESCENCE CHARACTERISTICS

type: panicle (3–13 cm long, 3–8 cm wide), pyramidal, open; branches long, flexuous, lower branches in whorls of 3–5 (commonly 5)

spikelets: 3- to 6-flowered (3–6 mm long), laterally compressed, nearly as wide as long, ovate; lemmas keeled (2.5–4 mm long), acute or obtuse; 5-nerved; nerves pubescent with a tuft of long, cobwebby hairs at the base; palea slightly shorter than the lemma

glumes: subequal (2–3.5 mm long), acute, keeled, scabrous on keel; first 1-nerved (1.7–3.3 mm long); second 3-nerved (2–3.7 mm long)

awns: none

VEGETATIVE CHARACTERISTICS

growth habit: rhizomatous; rhizomes extensive, forming a dense sod

culms: erect (0.2–1 m tall), occasionally decumbent at base, slender, wiry, somewhat flattened

sheaths: rounded, glabrous to scabrous, distinctly veined; margins connate about one-half of the length

ligules: membranous (1–2 mm long), truncate, entire

blades: flat or folded (5–40 cm long, 1–5 mm wide), elongated, glabrous to slightly pubescent, double midrib; apex keeled

GROWTH CHARACTERISTICS: growth from rhizomes initiated in spring or fall; initiates aerial culms in early spring and summer, becomes dormant during summer if moisture is limiting; reproduces from seeds, tillers, and rhizomes; not tolerant of drought; able to withstand continuous heavy grazing

FORAGE VALUE: good for livestock and wildlife in early spring when other plants are not growing; undesirable in hay meadows and tallgrass prairies because of its low growth form, poor yield, and early maturity

HABITAT: meadows, open woods, open ground, and disturbed sites; commonly planted for lawns and some pastures, adapted to a broad range of soil textures; most common on sites with abundant soil moisture

Sandberg bluegrass
Poa secunda Presl

SYN = *P. sandbergii* Vasey

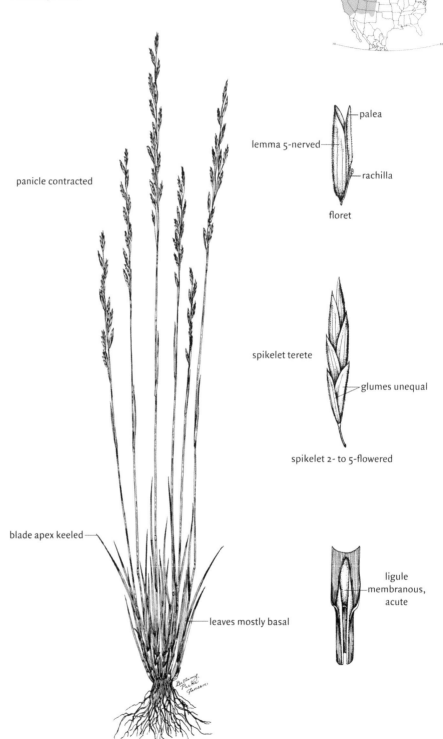

panicle contracted

palea

lemma 5-nerved

rachilla

floret

spikelet terete

glumes unequal

spikelet 2- to 5-flowered

blade apex keeled

leaves mostly basal

ligule membranous, acute

Tribe:	POEAE
Species:	*Poa secunda* Presl
Common Name:	Sandberg bluegrass
Life Span:	Perennial
Origin:	Native
Season:	Cool

INFLORESCENCE CHARACTERISTICS

type: panicle (2–8 cm long, usually less than 1 cm wide), contracted, not densely flowered, yellowish-green to purplish; branches per node 2–3, short, appressed, rarely spreading below

spikelets: 2- to 5-flowered (4–7 mm long, 0.9–2 mm wide), terete, acute, longer than wide; lemmas convex on back (3–5.5 mm long), purplish, apex acute, base short-pubescent to scabrous, not webbed, 5-nerved

glumes: unequal, papery, acute; first 1- to 3-nerved (2.2–5 mm long); second 3-nerved (3–5 mm long), shorter than lowermost lemma

awns: none

VEGETATIVE CHARACTERISTICS

growth habit: strongly cespitose; leaves mostly basal

culms: erect (10–45 cm tall), decumbent at base, wiry, glabrous; nodes occasionally reddish; leaves 1–3 per culm

sheaths: rounded, glabrous to scabrous, veins prominent, persistent; margins hyaline, basal portion connate

ligules: membranous (1–4 mm long), acute, usually entire

blades: flat, folded, or involute (3–16 cm long, 1–3 mm wide), glabrous, double midrib, apex keeled; margins slightly barbed

GROWTH CHARACTERISTICS: one of the first plants to start growth in early spring, seeds mature in early summer, reproduces from seeds and tillers; drought tolerant; withstands moderate to heavy grazing; forage yield is generally low

FORAGE VALUE: good for cattle and fair for sheep, deer, and pronghorn in spring and early summer; with adequate moisture, remains green and furnishes good forage throughout the summer

HABITAT: plains, dry woods, and rocky slopes; adapted to a wide variety of soils, most abundant on deep sandy to silt loam soils

Sixweeks fescue
Vulpia octoflora (Walt.) Rydb.

SYN = *Festuca octoflora* Walt.

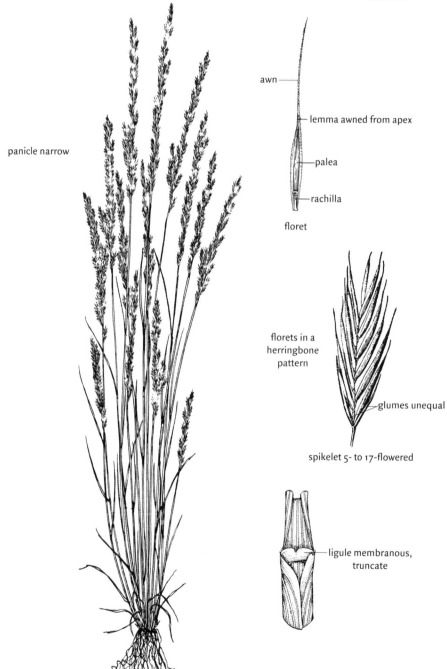

panicle narrow

awn

lemma awned from apex

palea

rachilla

floret

florets in a
herringbone
pattern

glumes unequal

spikelet 5- to 17-flowered

ligule membranous,
truncate

172

Tribe:	POEAE
Species:	*Vulpia octoflora* (Walt.) Rydb.
Common Name:	Sixweeks fescue (cañuela anual, vulpia, sixweeksgrass)
Life Span:	Annual
Origin:	Native
Season:	Cool

INFLORESCENCE CHARACTERISTICS

type: panicle (1–10 cm long), narrow; branches short, appressed to slightly spreading

spikelets: 5- to 17-flowered (4–10 mm long, excluding the awns), glabrous, scabrous, or pubescent; florets closely arranged in a herringbone pattern; lemmas (3–7 mm long) lanceolate, convex, rounded on the back, obscurely nerved

glumes: unequal, subulate-lanceolate; first 1-nerved (1.7–4.5 mm long), second 3-nerved (2.7–6.7 mm long)

awns: lemma awned (3–7 mm long) from apex; glumes awnless or occasionally with a short awn

VEGETATIVE CHARACTERISTICS

growth habit: loosely cespitose or solitary

culms: erect (5–60 cm tall), occasionally decumbent, slender and weak, glabrous or pubescent

sheaths: rounded, ridged, smooth, glabrous or sparingly pubescent; margins open

ligules: membranous (0.5–2 mm long), truncate, erose or ciliate

blades: involute (2–8 cm long, 0.5–2 mm wide), glabrous to finely pubescent adaxially

GROWTH CHARACTERISTICS: starts growth in early spring and matures about 6 weeks later, reproduces from seeds; extreme variation in plant height is directly related to available moisture; often an indicator of improper grazing

FORAGE VALUE: little forage value for livestock or wildlife except during a 2- to 3-week period in early spring; livestock commonly pull the roots from the soil when grazing

HABITAT: prairies, plains, mesas, open ground, waste areas, and disturbed sites; adapted to a broad range of soils, most common on coarse-textured soils

Needleandthread
Hesperostipa comata (Trin. & Rupr.) Barkw.

SYN = *Stipa comata* Trin. & Rupr.

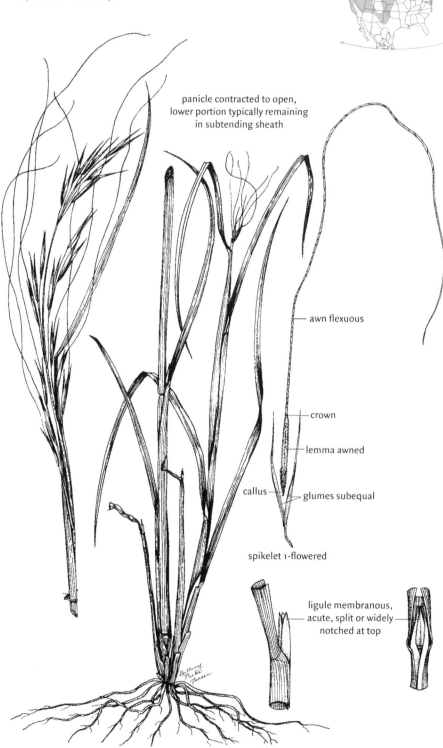

panicle contracted to open, lower portion typically remaining in subtending sheath

awn flexuous

crown

lemma awned

callus

glumes subequal

spikelet 1-flowered

ligule membranous, acute, split or widely notched at top

Tribe: STIPEAE
Species: *Hesperostipa comata* (Trin. & Rupr.) Barkw.
Common Name: Needleandthread
Life Span: Perennial
Origin: Native
Season: Cool

INFLORESCENCE CHARACTERISTICS

type: panicle (10–40 cm long), contracted to open, lower portion typically remaining in subtending sheath; branches slender and ascending

spikelets: 1-flowered, large, few, drooping at anthesis; lemma (1–1.5 cm long) slender, pale to brown at maturity, 5-nerved, lightly pubescent; crown hairs minute; callus (3 mm long) bearded with stiff hairs (0.2–0.5 mm long), sharp

glumes: subequal (1.5–3.5 cm long); first longer, narrow, acuminate, 3- to 5-nerved, margins hyaline

awns: lemma awned (10–20 cm long), flexuous, twisted; lower segment short-pubescent; terminal segment glabrous or scabrous, not twisted

VEGETATIVE CHARACTERISTICS

growth habit: densely cespitose

culms: erect (0.3–1.1 m tall); nodes glabrous to puberulent

sheaths: round, usually longer than the internodes, glabrous to scabrous, prominently veined, open

ligules: membranous (2–6 mm long), acute, split or widely notched at top

blades: flat or involute (5–40 cm long, 1–3 mm wide), scabrous adaxially, glabrous to scaberulous abaxially

GROWTH CHARACTERISTICS: starts growth in early spring or when moisture is available, seeds mature in early summer, reproduces from seeds and tillers

LIVESTOCK LOSSES: sharp-pointed callus and long awns may cause injury by working into the eyes, tongue, and ears; sheep are especially susceptible to injury, awns may contaminate the fleece and carcass

FORAGE VALUE: fair to good for livestock and poor to fair for wildlife, extensively utilized by elk in winter and deer in spring; cures well to provide fall and winter forage for livestock

HABITAT: prairies, plains, dry hills, foothills, alluvial fans, and sandy benches; most abundant on excessively drained soils

Texas wintergrass
Nassella leucotricha (Trin. & Rupr.) Pohl

SYN = *Stipa leucotricha* Trin. & Rupr.

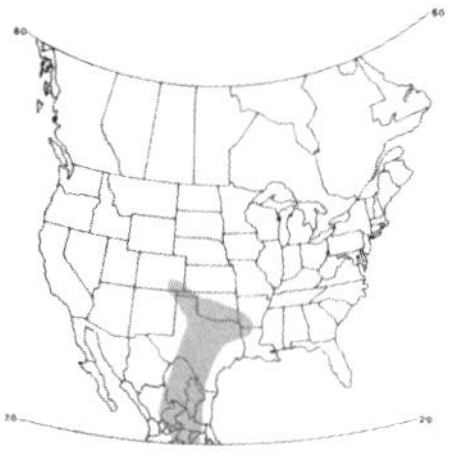

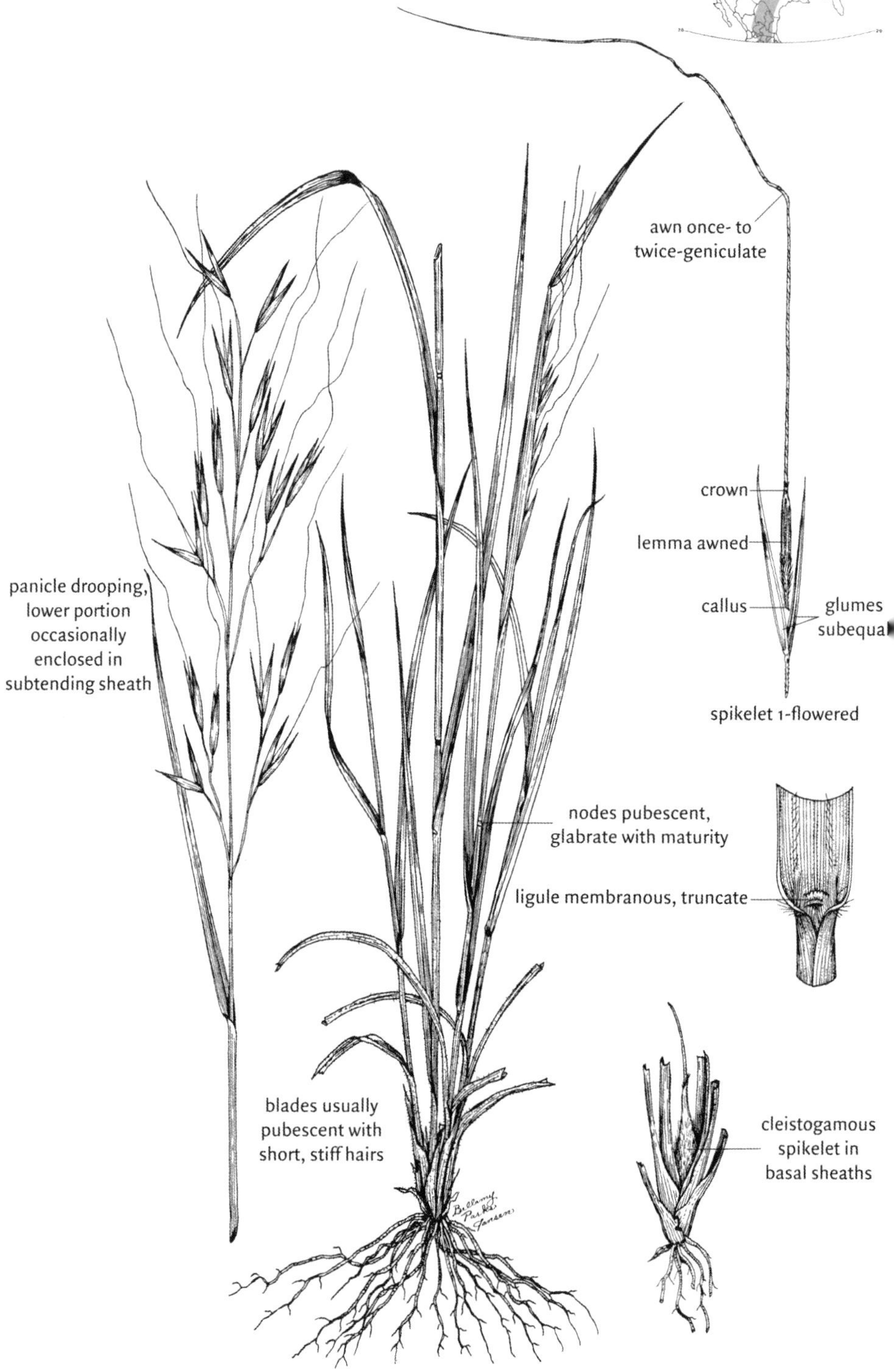

awn once- to twice-geniculate
crown
lemma awned
callus
glumes subequal
spikelet 1-flowered
panicle drooping, lower portion occasionally enclosed in subtending sheath
nodes pubescent, glabrate with maturity
ligule membranous, truncate
blades usually pubescent with short, stiff hairs
cleistogamous spikelet in basal sheaths
Bellamy, Parks, Jansen

Tribe:	STIPEAE
Species:	*Nassella leucotricha* (Trin. & Rupr.) Pohl
Common Name:	Texas wintergrass (flechilla, Texas needlegrass, speargrass)
Life Span:	Perennial
Origin:	Native
Season:	Cool

INFLORESCENCE CHARACTERISTICS

type: panicle (6–25 cm long, usually less than 10 cm wide), lower portion occasionally enclosed in subtending sheath; lower branches flexuous, slender, spreading; upper branches ascending

spikelets: 1-flowered; lemma light brown (9–12 mm long), appressed pubescence below, rugose on body above base, rounded, summit of the neck white; crown hairs stiff, minute (0.5–1 mm long); callus sharp (4 mm long) with silky hairs (to 3 mm long)

glumes: subequal (1.2–1.8 cm long); first longest, thin, glabrous, 5- to 7-nerved, acuminate to awn-tipped, somewhat hyaline

awns: lemma awned (4–10 cm long), stout, once- to twice-geniculate; lower portion scabrous to pubescent, twisted (2–3.5 cm long); upper portions essentially glabrous, loosely twisted

other: produces cleistogamous spikelets in basal sheaths

VEGETATIVE CHARACTERISTICS

growth habit: cespitose

culms: erect to ascending (30–90 cm tall), spreading at base; nodes pubescent, glabrate with maturity

sheaths: rounded, pubescent to nearly glabrous; collar with long hairs on sides

ligules: membranous (absent to 1 mm long), variable, truncate, entire to erose

blades: flat, becoming involute (5–40 cm long, 1–5 mm wide), usually pubescent with short and stiff hairs on one or both surfaces

GROWTH CHARACTERISTICS: starts growth in late fall, remains green through winter and spring; reproduces from seeds, cleistogamous spikelets, and tillers

LIVESTOCK LOSSES: sharp-pointed callus and awns may cause injury

FORAGE VALUE: fair for both livestock and wildlife, most valuable for early spring forage

HABITAT: dry grasslands, disturbed sites, and heavily grazed areas

Purple needlegrass
Nassella pulchra (A. S. Hitchc.) Barkw.

SYN = *Stipa pulchra* A. S. Hitchc.

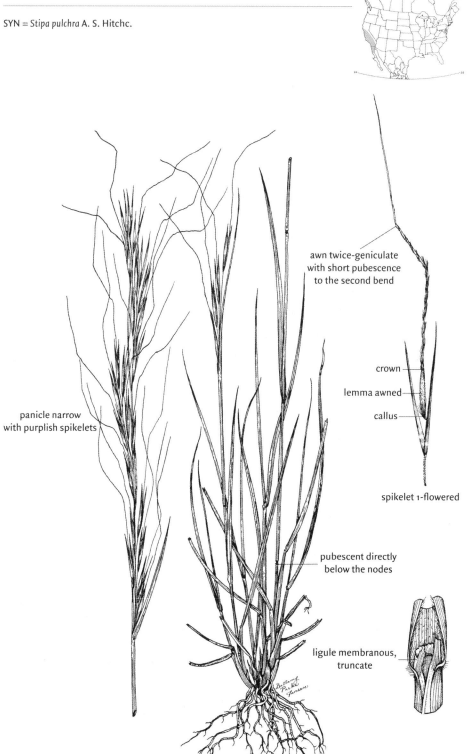

awn twice-geniculate
with short pubescence
to the second bend

crown

lemma awned

callus

spikelet 1-flowered

panicle narrow
with purplish spikelets

pubescent directly
below the nodes

ligule membranous,
truncate

Tribe:	STIPEAE
Species:	*Nassella pulchra* (A. S. Hitchc.) Barkw.
Common Name:	Purple needlegrass (flechilla púrpura, flechilla, purple tussockgrass)
Life Span:	Perennial
Origin:	Native
Season:	Cool

INFLORESCENCE CHARACTERISTICS

type: panicle (15–20 cm long), nodding, narrow, loose; lower branches spreading (2.5–5 cm long)

spikelets: 1-flowered, often strongly purplish; lemma narrow (7–13 mm long), fusiform, sparingly pilose, summit with smooth neck; crown ciliate; callus sharp (2–3 mm long), with white hairs (1 mm long)

glumes: subequal (2 cm long); first longest, narrow, long acuminate, 3-nerved

awns: lemma awned (5–9 cm long), twice-geniculate, short pubescence to the second bend; first segment twisted (1.5–2 cm long), second segment shorter and loosely twisted, third segment straight (4–6 cm long)

VEGETATIVE CHARACTERISTICS

growth habit: densely cespitose; leaves mostly basal

culms: erect (0.6–1 m tall), pubescent only directly below the nodes

sheaths: rounded, essentially glabrous; margins ciliate and open; collar pilose

ligules: membranous (1 mm long), truncate, erose

blades: flat or involute (10–30 cm long, 3–5 mm wide), lightly pubescent

GROWTH CHARACTERISTICS: starts growth in late fall or early spring, flowers April to June, reproduces from seeds and tillers; productivity reduced if grazed during period of maximum growth

LIVESTOCK LOSSES: callus of floret may cause livestock injury similar to those caused by *Hesperostipa comata*

FORAGE VALUE: good for cattle, sheep, and horses; fair for deer; basal leaves remain green for 9 or 10 months of the year

HABITAT: warm slopes and well-drained flats of prairies, open timbered areas of foothills and valleys, waste places, and disturbed sites; adapted to a broad range of soils, most abundant on sandy loams

Indian ricegrass
Stipa hymenoides Roemer & Schultes

SYN = *Oryzopsis hymenoides* (Roemer & Schultes) Ricker

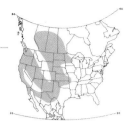

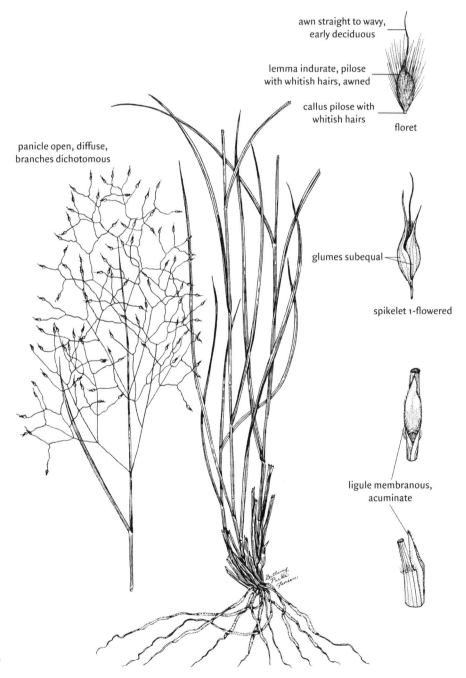

awn straight to wavy,
early deciduous

lemma indurate, pilose
with whitish hairs, awned

callus pilose with
whitish hairs

floret

panicle open, diffuse,
branches dichotomous

glumes subequal

spikelet 1-flowered

ligule membranous,
acuminate

Tribe:	STIPEAE
Species:	*Stipa hymenoides* Roemer & Schultes
Common Name:	Indian ricegrass (arrocillo, zacate arroz de indio, sand bunchgrass)
Life Span:	Perennial
Origin:	Native
Season:	Cool

INFLORESCENCE CHARACTERISTICS

type: panicle (7–25 cm long), diffuse; branches spreading, dichotomous, flexuous; pedicels curved (5–30 mm long)

spikelets: 1-flowered (5–8 mm long, excluding awn); lemma indurate (3–5 mm long), brown to black at maturity, usually covered with whitish hairs (2.5–4.5 mm long); palea slightly shorter than the lemma

glumes: subequal, first (5–8.2 mm long) longer than the second (4.2–7.5 mm long), glabrous to puberulent, ovate to acuminate, 3- to 5-nerved, nerves prominent, margins papery

awns: lemma awned (3–8 mm long), stout, straight to wavy, early deciduous; glumes sometimes short-awned

VEGETATIVE CHARACTERISTICS

growth habit: cespitose

culms: stiffly erect (30–80 cm tall), densely tufted, slender, glabrous

sheaths: rounded, glabrous to ciliate on the overlapping margin, papery, shorter than the internodes, open, persistent; collar pubescent

ligules: membranous (3–9 mm long), acuminate, may be deeply notched to erose

blades: involute (5–40 cm long, 1–3 mm wide), filiform, scabrous to hirsute adaxially, glabrous abaxially with a prominent midrib

GROWTH CHARACTERISTICS: starts growth in early spring, flowers in late spring, reproduces from seeds and tillers; drought tolerant

FORAGE VALUE: good for cattle, sheep, horses, and wildlife; especially valuable for winter grazing because the plants cure well and lower plant parts remain somewhat green; seeds are high in protein; some Native Americans used the seeds to make flour

HABITAT: prairies, plains, deserts, foothills, mesas, and disturbed sites; most abundant on sandy soils, but will occur on well-drained silty and limy soils; moderately salt and alkali tolerant

Columbia needlegrass
Stipa nelsonii Scribn.

SYN = *Stipa columbiana* Macoun

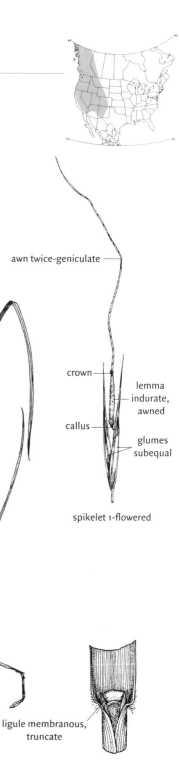

awn twice-geniculate

panicle narrow,
branches short,
appressed

crown

lemma
indurate,
awned

callus

glumes
subequal

spikelet 1-flowered

ligule membranous,
truncate

Tribe:	STIPEAE
Species:	*Stipa nelsonii* Scribn.
Common Name:	Columbia needlegrass (subalpine needlegrass)
Life Span:	Perennial
Origin:	Native
Season:	Cool

INFLORESCENCE CHARACTERISTICS

type: panicle (7–30 cm long), narrow, dense, sometimes purplish; branches short, appressed

spikelets: 1-flowered; lemma slender (5–7 mm long), indurate, slightly pubescent; crown with tuft of hairs (longest 0.7–1 mm); callus sharp (0.7–1 mm long)

glumes: subequal (7–11 mm long), exceed lemma, acuminate to mucronate, 3- or 5-nerved, glabrous, papery

awns: lemma awned (1.8–3.3 cm long), twice-geniculate; lower segment minutely scabrous and twisted; upper segment not twisted (often 1.5 cm or longer)

VEGETATIVE CHARACTERISTICS

growth habit: cespitose; leaves mostly basal

culms: erect (0.3–1 m tall), stout, straight; nodes may be purple, glabrous

sheaths: rounded to slightly keeled, prominently veined, open, glabrous to scabrous; margins occasionally pubescent at apex; throat naked

ligules: membranous (0.5–2 mm long), usually longer on the sides, truncate, erose

blades: flat to mostly involute (10–30 cm long, 1–5 mm wide), glabrous; margins barbed

GROWTH CHARACTERISTICS: starts growth in midspring, matures by September, reproduces from seeds and tillers; may regrow in the fall if moisture is adequate

LIVESTOCK LOSSES: sharp-pointed callus may work into the ears, eyes, nostrils, and tongues of grazing animals; sheep are especially susceptible to injury; awns may contaminate the fleece

FORAGE VALUE: fair to good for cattle and horses; fair for sheep, deer, and elk; becomes nearly unpalatable at maturity

HABITAT: dry plains, meadows, and open woods in foothills and mountains; adapted to soils ranging from sandy loams to clays, most abundant in fine-textured soils

Crested wheatgrass
Agropyron cristatum (L.) Gaertn.

SYN = *A. desertorum* (Fisch.) J. A. Schultes, *A. sibericum* (Willd.) Beauv.

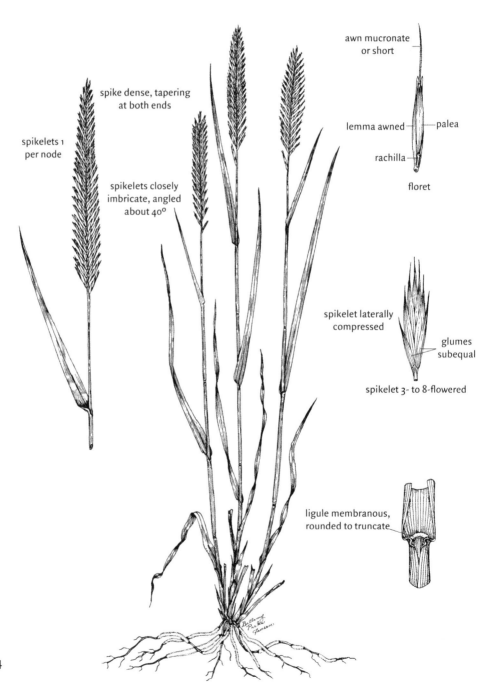

spikelets 1
per node

spike dense, tapering
at both ends

spikelets closely
imbricate, angled
about 40°

awn mucronate
or short

lemma awned — palea

rachilla

floret

spikelet laterally
compressed

glumes
subequal

spikelet 3- to 8-flowered

ligule membranous,
rounded to truncate

Tribe:	TRITICEAE
Species:	*Agropyron cristatum* (L.) Gaertn.
Common Name:	Crested wheatgrass
Life Span:	Perennial
Origin:	Introduced (from eastern Europe and Asia)
Season:	Cool

INFLORESCENCE CHARACTERISTICS

type: spike (2–9 cm long, 7–20 mm wide), dense, tapering at both ends; spikelets 1 per node, closely imbricate, several times longer than rachis internodes, spreading to ascending, angled about 40°; rachis scabrous to pilose, occasionally wavy

spikelets: 3- to 8-flowered (5–15 mm long), laterally compressed; lemmas firm (4–7 mm long), acute, margins ciliate; upper florets sterile

glumes: subequal (3–6 mm long); second longest, lanceolate, glabrous to pilose, midnerve prominent and ciliate

awns: glumes and lemmas mucronate or tapering to short awns (1–5 mm long)

VEGETATIVE CHARACTERISTICS

growth habit: cespitose, rarely with rhizomes

culms: erect to ascending (0.2–1 m tall), glabrous; base occasionally geniculate

sheaths: rounded, glabrous (sometimes pubescent below); margins overlapping, open; slender auricles (1 mm long)

ligules: membranous (0.5–1.5 mm long), rounded to truncate, erose

blades: flat (5–20 cm long, 2–8 mm wide), glabrous to puberulent, nerves raised adaxially, smooth abaxially; margins weakly scabrous

GROWTH CHARACTERISTICS: starts growth in early spring, flowers in late spring, reproduces from seeds and tillers; drought and cold tolerant, may regrow in the fall if moisture is sufficient

FORAGE VALUE: good for livestock and fair for wildlife; cures well for use as winter forage, produces good hay if cut early, most valuable for early spring grazing

HABITAT: planted in pastures, hay meadows, and roadsides; most abundant on dry, medium-textured soils, less adapted to heavy clays and sands; relatively salt tolerant; a major species for reseeding in sagebrush areas

Canada wildrye
Elymus canadensis L.

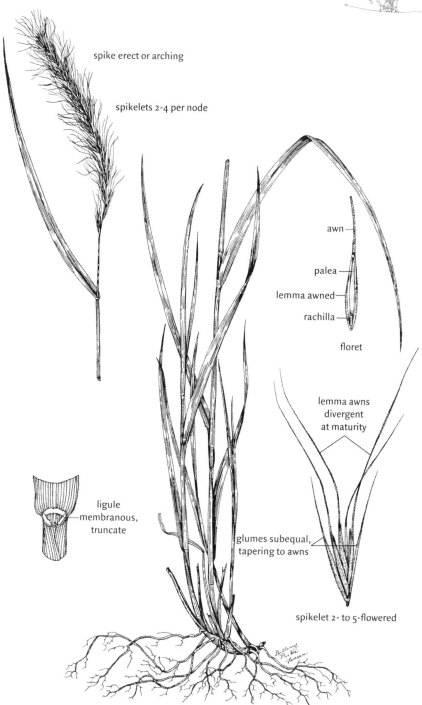

spike erect or arching

spikelets 2-4 per node

awn

palea

lemma awned

rachilla

floret

lemma awns divergent at maturity

ligule membranous, truncate

glumes subequal, tapering to awns

spikelet 2- to 5-flowered

Tribe:	TRITICEAE
Species:	*Elymus canadensis* L.
Common Name:	Canada wildrye (centeno silvestre)
Life Span:	Perennial
Origin:	Native
Season:	Cool

INFLORESCENCE CHARACTERISTICS

type: spike (8–25 cm long), erect or arching, totally exserted, somewhat elliptical, thick, bristly, occasionally interrupted below; spikelets 2–4 per node, imbricate, slightly spreading

spikelets: 2- to 5-flowered (1.2–1.5 cm long); lemmas broad at base (8–15 mm long), scabrous to hirsute

glumes: subequal (1–2.5 cm long, 0.8–1.5 mm wide), subulate, nerves glabrous, scabrous, or ciliate

awns: lemma awned (1.5–4 cm long), flexuous, divergent at maturity; glumes tapering to awns (1–3 cm long), somewhat straight

VEGETATIVE CHARACTERISTICS

growth habit: cespitose, rarely with rhizomes

culms: erect to decumbent at base (1–1.5 m tall), coarse

sheaths: rounded, glabrous, rarely pubescent; auricles well developed (1–2 mm long), clasping, finger-like

ligules: membranous (0.5–2 mm long), truncate, erose or rarely ciliate

blades: flat or folded (5–40 cm long, 7–20 mm wide), elongate, ascending, tapering to a fine point, scabrous adaxially, midrib prominent abaxially; margins finely toothed

GROWTH CHARACTERISTICS: starts growth in fall and makes some growth in winter in the southern portion of its range, seeds mature by late spring to early summer, reproduces from seeds and tillers

FORAGE VALUE: good for cattle and horses and fair for sheep and wildlife during the spring when green and growing, value decreases sharply when the plants mature; inflorescences may become infested with ergot and be potentially dangerous to livestock

HABITAT: prairies, stream banks, disturbed areas, and ditches; usually in open areas, adapted to a broad range of dry and moist soils

Squirreltail
Elymus elymoides (Raf.) Swezey

SYN = *Sitanion hystrix* (Nutt.) J. G. Smith

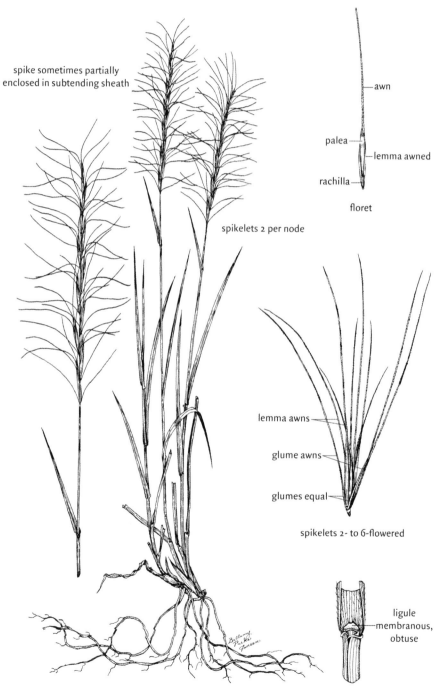

spike sometimes partially
enclosed in subtending sheath

awn

palea

lemma awned

rachilla

floret

spikelets 2 per node

lemma awns

glume awns

glumes equal

spikelets 2- to 6-flowered

ligule
membranous,
obtuse

Tribe:	TRITICEAE
Species:	*Elymus elymoides* (Raf.) Swezey
Common Name:	Squirreltail (zacate triguillo, cola de zorra, bottlebrush squirreltail)
Life Span:	Perennial
Origin:	Native
Season:	Cool

INFLORESCENCE CHARACTERISTICS

type: spike (2–15 cm long excluding awns), sometimes partially enclosed in subtending sheath, erect; spikelets usually 2 per node; rachis disarticulating readily

spikelets: 2- to 6-flowered; lower 1–2 florets may be reduced to awn-like structures; lemmas convex (8–10 mm long), glabrous to lightly pubescent

glumes: equal (3.5–8.5 cm long), narrow, subulate to lanceolate, entire or bifid

awns: lemma awned (5–15 mm long); glumes awned (2–10 cm long), widely spreading, stiff, scabrous, green to purplish

VEGETATIVE CHARACTERISTICS

growth habit: cespitose

culms: erect to spreading (10–60 cm tall), stiff, glabrous to puberulent

sheaths: round, pubescent or glabrous; margins open, translucent; auricles small (1 mm long), often purplish

ligules: membranous (0.6–1 mm long), obtuse, erose to ciliolate

blades: flat to involute (5–20 cm long, 1–5 mm wide), stiffly ascending, tapering to a fine point, glabrous to lightly pubescent, prominently veined

GROWTH CHARACTERISTICS: starts growth in early spring, flowers in late spring, may regrow and flower a second time with favorable moisture, reproduces from seeds and tillers

LIVESTOCK LOSSES: sharp-pointed callus and awns may cause injury to soft tissue, awns may contaminate fleece

FORAGE VALUE: fair for cattle and horses and poor for sheep before inflorescences develop, may be consumed in late summer and early fall after inflorescences have broken and fallen, unpalatable during winter

HABITAT: dry hills, plains, open woods, and rocky slopes of deserts; most abundant on disturbed sites on either deep or shallow soils; may grow in saline or alkaline soils

Intermediate wheatgrass
Elymus hispidus (P. Opiz) Melderis

SYN = *Agropyron intermedium* (Host) Beauv., *Elytrigia intermedia* (Host) Nevski,
Thinopyrum intermedium (Host) Barkw. & D. R. Dewey

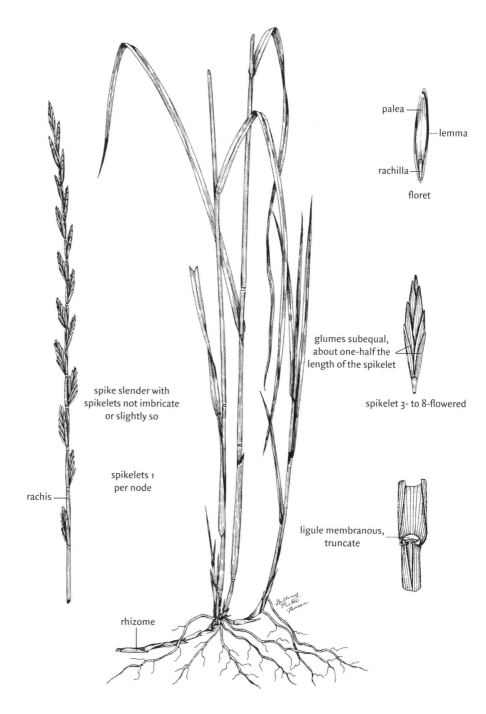

palea

lemma

rachilla

floret

glumes subequal,
about one-half the
length of the spikelet

spikelet 3- to 8-flowered

spike slender with
spikelets not imbricate
or slightly so

spikelets 1
per node

rachis

ligule membranous,
truncate

rhizome

Tribe:	TRITICEAE
Species:	*Elymus hispidus* (P. Opiz) Melderis
Common Name:	Intermediate wheatgrass
Life Span:	Perennial
Origin:	Introduced (from Europe and Russia)
Season:	Cool

INFLORESCENCE CHARACTERISTICS

type: spike (8–25 cm long), slender; spikelets 1 per node, not imbricate or slightly so, may be curved away from the rachis at maturity; rachis segments scabrous (6–15 mm long)

spikelets: 3- to 8-flowered (1–2 cm long); lemmas lanceolate (7–11 mm long), broad, blunt, glabrous to hirsute

glumes: subequal (4–9 mm long); second longest, about one-half the length of the spikelet, lanceolate, blunt to acuminate, glabrous to hirsute, distinctly nerved

awns: lemma may be awn-tipped

VEGETATIVE CHARACTERISTICS

growth habit: cespitose or solitary, with rhizomes

culms: erect (0.4–1.2 m tall), robust, glabrous, glaucous

sheaths: round, mostly glabrous; margins ciliate, open; auricles acute (1–2 mm long)

ligules: membranous (1–2 mm long), truncate, erose to entire

blades: flat to loosely involute (10–40 cm long, 5–10 mm wide), stiff, broad at base and tapering to a point, strongly veined, green or glaucous, glabrous to scabrous to pilose adaxially

GROWTH CHARACTERISTICS: starts growth in early spring; matures June to August; little growth during the summer, even with adequate moisture; reproduces from seeds, tillers, and rhizomes; responds to nitrogen fertilization

FORAGE VALUE: good to excellent for all classes of livestock, fair for wildlife; produces good to excellent hay if cut early; forage cures relatively well and remains palatable

HABITAT: meadows, hills, roadsides, and disturbed sites; seeded in both dryland and irrigated pastures and hay meadows, adapted to a broad range of soil textures and soil moisture conditions

Western wheatgrass
Elymus smithii (Rydb.) Gould

SYN = *Agropyron smithii* Rydb., *Pascopyrum smithii* (Rydb.) A. Löve

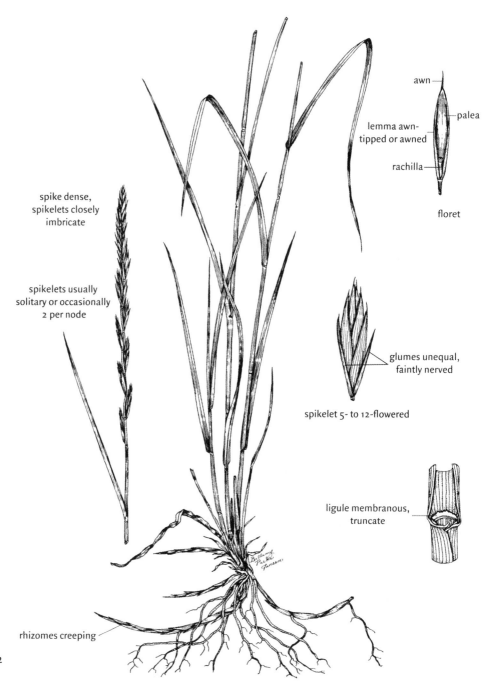

awn

palea

lemma awn-
tipped or awned

rachilla

floret

spike dense,
spikelets closely
imbricate

spikelets usually
solitary or occasionally
2 per node

glumes unequal,
faintly nerved

spikelet 5- to 12-flowered

ligule membranous,
truncate

rhizomes creeping

Tribe:	TRITICEAE
Species:	*Elymus smithii* (Rydb.) Gould
Common Name:	Western wheatgrass (agropiro del oeste, bluestem)
Life Span:	Perennial
Origin:	Native
Season:	Cool

INFLORESCENCE CHARACTERISTICS

type: spike (6–20 cm long), often dense, stiff, erect; spikelets usually solitary or occasionally 2 per node, closely imbricate (about one-half of each spikelet overlaps); rachis internodes scabrous (5–6 mm long)

spikelets: 5- to 12-flowered (1.5–2.5 cm long), glaucous, occasionally densely; lemmas acute (8–14 mm long), glabrous to pubescent on margins, 5-nerved

glumes: unequal, first shorter (6–12 mm long) than the second (7–15 mm long), asymmetrical, acute, narrow, rigid, faintly 3- to 5-nerved, glabrous to scabrous

awns: glumes occasionally awn-tipped; lemmas occasionally awn-tipped or awned (to 5 mm long)

VEGETATIVE CHARACTERISTICS

growth habit: rhizomatous (sometimes appearing solitary); rhizomes creeping and may form a loose sod

culms: erect (30–90 cm tall), single or in small clusters, glabrous to glaucous

sheaths: rounded, glabrous or scabrous, open; auricles short (1–2 mm long) or absent

ligules: membranous (to 1 mm long), truncate, erose to minutely ciliate

blades: flat to involute (10–25 cm long, 2–6 mm wide), rigid, tapering to a short point, strongly veined, glaucous, scabrous to rarely pilose adaxially

GROWTH CHARACTERISTICS: growth starts in early spring, dormant in summer, begins growth again in fall if soil moisture is adequate; reproduces from seeds and rhizomes

FORAGE VALUE: good for all classes of livestock, fair for pronghorn and other wildlife; cures well, making good winter forage

HABITAT: prairies, foothills, sagebrush deserts, swales, alkaline meadows, ditch banks, and road rights-of-way; occurs in all soil textures, but most abundant in fine-textured soils

Bluebunch wheatgrass
Elymus spicatus (Pursh) Gould

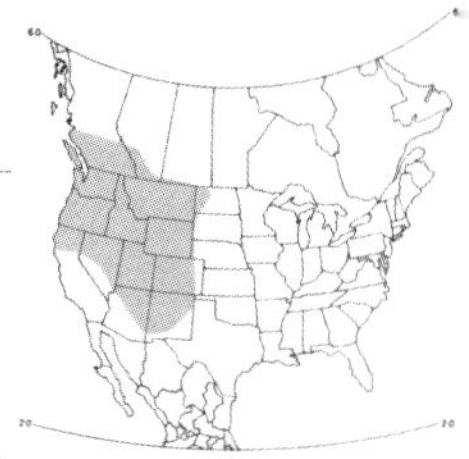

SYN = *Agropyron spicatum* (Pursh) Scribn. & J. G. Smith, *Pseudoroegneria spicata* (Pursh) A. Löve

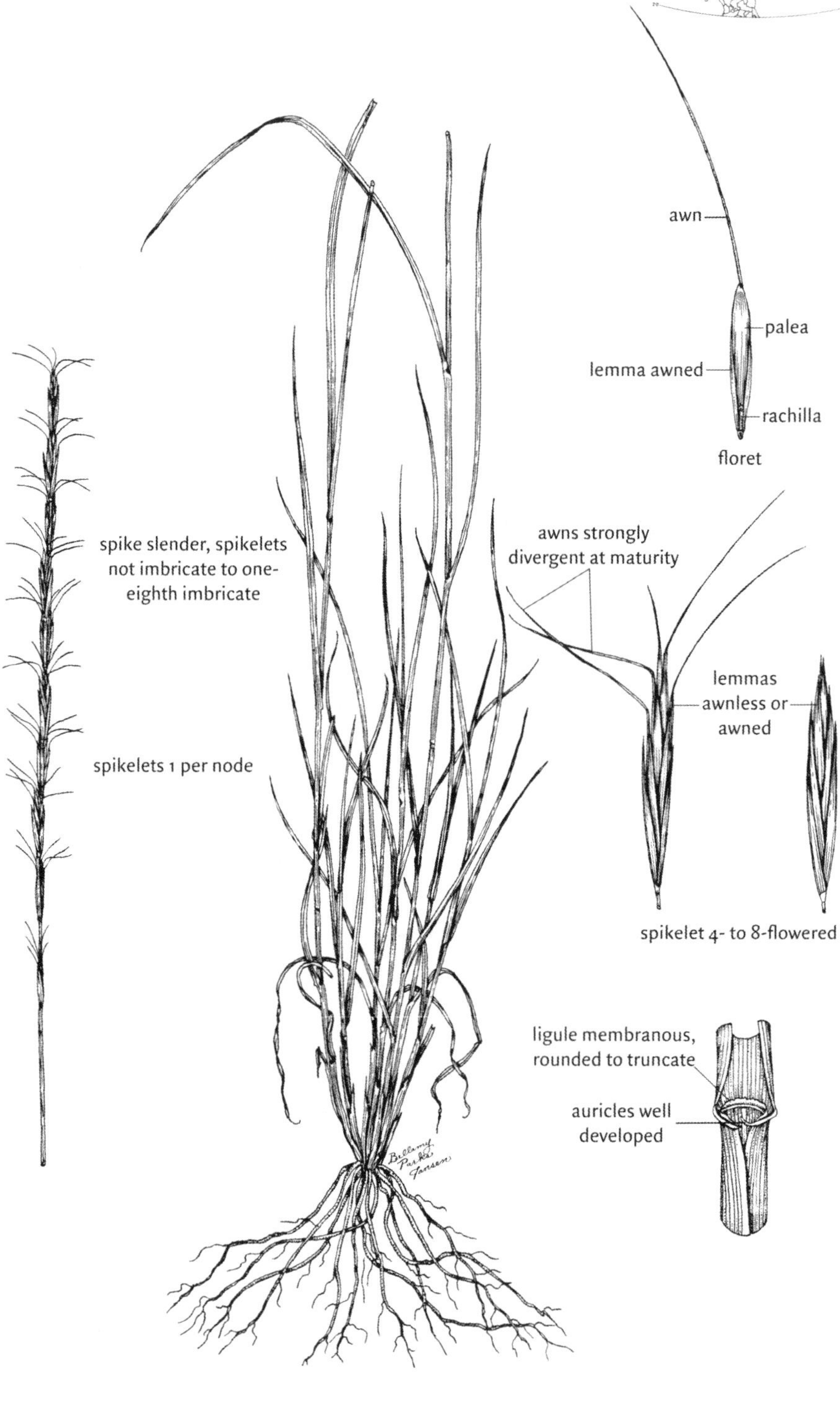

awn

palea

lemma awned

rachilla

floret

spike slender, spikelets not imbricate to one-eighth imbricate

awns strongly divergent at maturity

lemmas awnless or awned

spikelets 1 per node

spikelet 4- to 8-flowered

ligule membranous, rounded to truncate

auricles well developed

Tribe:	TRITICEAE
Species:	*Elymus spicatus* (Pursh) Gould
Common Name:	Bluebunch wheatgrass (beardless wheatgrass, big bunchgrass)
Life Span:	Perennial
Origin:	Native
Season:	Cool

INFLORESCENCE CHARACTERISTICS

type: spike (6–15 cm long), slender; spikelets 1 per node, usually only 5–14, not imbricate to one-eighth imbricate; rachis internodes 1–2 cm long

spikelets: 4- to 8-flowered (1–2 cm long); lemmas acute (7–11 mm long), glabrous, faintly 5-nerved

glumes: subequal (4.5–11 mm long); second longest, narrow, obtuse to acute, about one-half as long as spikelet, 4- to 5-nerved; nerves glabrous or scabrous

awns: lemmas awnless or awned; awns strongly divergent at maturity (1–2 cm long); glumes rarely awn-tipped

VEGETATIVE CHARACTERISTICS

growth habit: cespitose or rarely with short rhizomes, bunches up to 15 cm in diameter at the base

culms: erect (0.2–1 m tall), slender, glaucous, glabrous or puberulent below the nodes

sheaths: rounded, glabrous to appressed-puberulent, prominently veined; margins imbricate, open; old sheaths strongly persisting; auricles well-developed (0.5–1 mm long), acute, clasping, reddish

ligules: membranous (0.5–2 mm long), rounded or truncate, erose to ciliate

blades: loosely involute to flat (5–25 cm long, 1.5–4.5 mm wide), elongate, pubescent adaxially; margins white and weakly barbed

GROWTH CHARACTERISTICS: growth begins in April, plants stay green well into the summer; regrowth occurs following fall rains; reproduces from seeds, tillers, and rarely by rhizomes

FORAGE VALUE: excellent for cattle and horses; good for sheep, elk, and deer; cures well and makes good standing winter feed

HABITAT: plains, mountain slopes, canyons, open woods, and stream banks; will tolerate moist soils, but most abundant in dry soils; sometimes seeded alone or in mixtures for grazing or hay

Slender wheatgrass
Elymus trachycaulus (Link) Shinners

SYN = *Agropyron caninum* (L.) Beauv. subsp. *majus* (Vasey) C. L. Hitchc., *A. pauciflorum* A. S. Hitchc., *A. subsecundum* (Link) A. S. Hitchc., *A. trachycaulum* (Link) A. S. Hitchc.

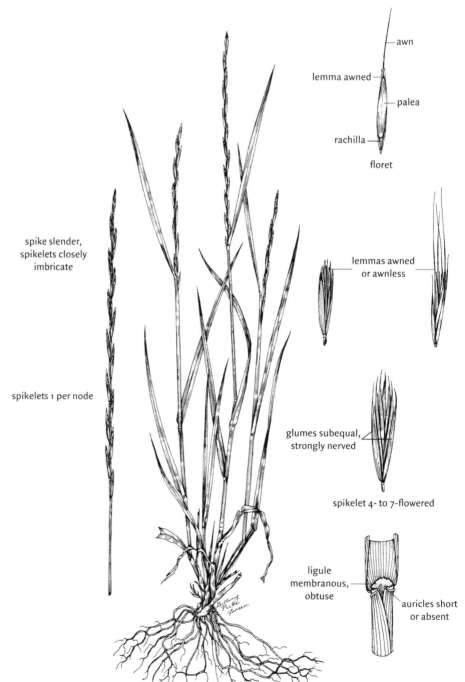

awn

lemma awned

palea

rachilla

floret

spike slender, spikelets closely imbricate

lemmas awned or awnless

spikelets 1 per node

glumes subequal, strongly nerved

spikelet 4- to 7-flowered

ligule membranous, obtuse

auricles short or absent

196

Tribe:	TRITICEAE
Species:	*Elymus trachycaulus* (Link) Shinners
Common Name:	Slender wheatgrass (agropiro delgado, bearded wheatgrass)
Life Span:	Perennial
Origin:	Native
Season:	Cool

INFLORESCENCE CHARACTERISTICS

type: spike (5–25 cm long), slender, sometimes violet-tinged; spikelets 1 per node, closely imbricate (usually about one-half of each spikelet overlaps); rachis segments (4–11 mm long) scabrous

spikelets: 4- to 7-flowered (1–1.8 cm long); lemmas acute (8–13 mm long), glabrous to scabrous to occasionally hirsute on margins; rachilla pubescent

glumes: subequal (6–14 mm long); second longest, broad, lanceolate, acute, nearly enclosing the florets, strongly 5- to 7-nerved, nerves dark green, margins hyaline, scabrous

awns: lemmas awned (1–30 mm long) or awnless; glumes may taper to short awns

VEGETATIVE CHARACTERISTICS

growth habit: cespitose

culms: erect to decumbent at base (0.5–1.5 m tall), slender, green or glaucous, glabrous

sheaths: round, glabrous to pilose, open; auricles short (0.3–1 mm long) or absent, 1 often rudimentary

ligules: membranous (0.4–0.8 mm long), obtuse, erose to ciliate

blades: flat or folded (5–25 cm long, 2–8 mm wide), slender with pointed apex, elongate, scabrous to glabrous; margins with narrow white band

GROWTH CHARACTERISTICS: starts growth in midspring, seeds mature by August to September, reproduces from seeds and tillers

FORAGE VALUE: excellent for both sheep and cattle when green, good when mature; good to excellent for wildlife

HABITAT: riverbanks, open forests, mountain slopes, and rolling hills; most abundant in well-drained medium- to fine-textured soils; tolerant of moderate drought as well as long, wet periods

Foxtail barley
Hordeum jubatum L.

spicate raceme sometimes partially enclosed in subtending sheath

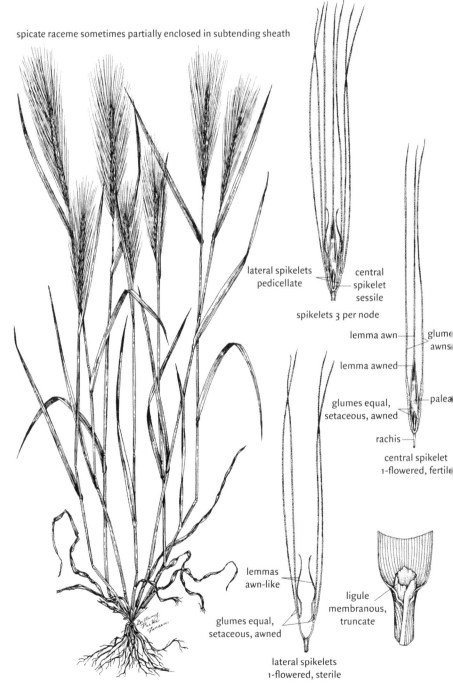

lateral spikelets
pedicellate

central
spikelet
sessile

spikelets 3 per node

lemma awn

glume
awns

lemma awned

glumes equal,
setaceous, awned

palea

rachis

central spikelet
1-flowered, fertile

lemmas
awn-like

glumes equal,
setaceous, awned

lateral spikelets
1-flowered, sterile

ligule
membranous,
truncate

Tribe:	TRITICEAE
Species:	*Hordeum jubatum* L.
Common Name:	Foxtail barley (cola de zorra)
Life Span:	Perennial
Origin:	Native
Season:	Cool

INFLORESCENCE CHARACTERISTICS

type: spicate raceme (4–10 cm long, 4–6 cm wide excluding awns), some-times partially enclosed in the subtending sheath, nodding, purplish, spike-lets 3 per node; central spikelet sessile; 2 lateral spikelets pedicellate, reduced; rachis readily disarticulating at maturity

spikelets: central spikelet 1-flowered, fertile, lemma body narrow (4–8 mm long); lateral spikelets 1-flowered, sterile, of similar length (0.7–1.2 mm long), reduced to little more than a series of 3 awns

glumes: equal, setaceous; central spikelet glumes awn-like (2.5–7 cm long), narrow, scabrous; lateral spikelet glumes slightly shorter, similar in shape

awns: lemma and glumes awned (2–6 cm long), thin; glume awns scabrous

VEGETATIVE CHARACTERISTICS

growth habit: cespitose

culms: erect (30–75 cm tall), decumbent below, slender; nodes dark, gla-brous to soft-pubescent

sheaths: round, glabrous to lightly pilose; auricles small (to 0.5 mm long) or absent, sometimes present on some leaves of a plant but not on others

ligules: membranous (0.2–1 mm long), truncate, erose to ciliolate

blades: flat (5–15 cm long, 2–5 mm wide), tapering to fine points, scabrous, glabrous to occasionally pilose abaxially

GROWTH CHARACTERISTICS: starts growth in late April to May, matures June to August, reproduces from seeds and tillers, short-lived

FORAGE VALUE: poor for all classes of livestock and wildlife, it may be lightly grazed before inflorescence development, presence in hay greatly reduces hay value; awns may cause sores in and around the nose, eyes, and mouth and contaminate fleece

HABITAT: open ground, meadows, waste places, and alkaline and saline sites; adapted to a broad range of soil types, most abundant where water occasionally accumulates

Little barley
Hordeum pusillum Nutt.

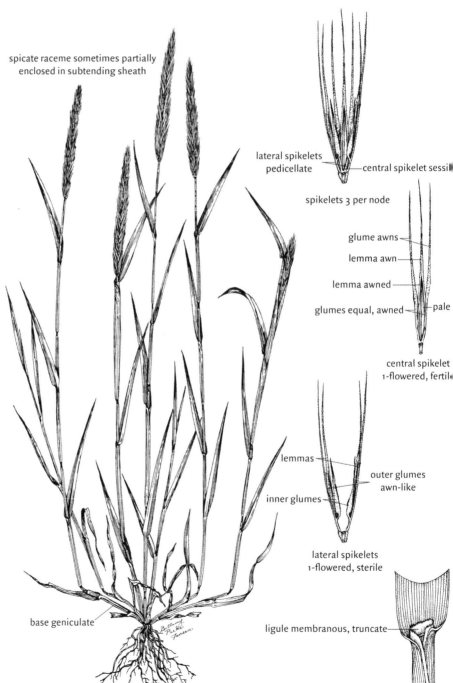

spicate raceme sometimes partially
enclosed in subtending sheath

lateral spikelets
pedicellate

central spikelet sessi

spikelets 3 per node

glume awns

lemma awn

lemma awned

glumes equal, awned — pale

central spikelet
1-flowered, fertil

lemmas

outer glumes
awn-like

inner glumes

lateral spikelets
1-flowered, sterile

base geniculate

ligule membranous, truncate

Tribe:	TRITICEAE
Species:	*Hordeum pusillum* Nutt.
Common Name:	Little barley (cebadilla, cola de ardilla)
Life Span:	Annual
Origin:	Native
Season:	Cool

INFLORESCENCE CHARACTERISTICS

type: spicate raceme (4–8 cm long, 3–8 mm wide, excluding awns), erect, stiff, sometimes partially enclosed in subtending sheath; spikelets 3 per node, central spikelet sessile, 2 lateral spikelets pedicellate (pedicels 0.3–0.7 mm long), reduced; rachis readily disarticulating at maturity

spikelets: central spikelet 1-flowered, fertile, lemma body narrow (5–7 mm long), scabrous; lateral spikelets 1-flowered, sterile, smaller, lemma body acuminate (1.5–3.5 mm long)

glumes: equal; central spikelet glumes lanceolate (3.4–5.5 mm long), broadened above base, faintly 3-nerved, scabrous; lateral glumes shorter, inner 2 glumes similar to central spikelet glumes, outer 2 glumes awn-like, scabrous

awns: lemma of central spikelet awned (2–7 mm long); lemmas of lateral spikelets short-awned; outer glumes of lateral spikelets awn-like; other glumes awned (7–15 mm long)

VEGETATIVE CHARACTERISTICS

growth habit: weakly cespitose or solitary

culms: erect (10–40 cm tall), base geniculate, glabrous, nodes dark

sheaths: round, glabrous or with short spreading pubescence, inflated; auricles absent

ligules: membranous (0.4–0.7 mm long), truncate, erose to short ciliate

blades: flat (1–12 cm long, 2–5 mm wide), erect; margin weakly scabrous, glabrous to pubescent

GROWTH CHARACTERISTICS: starts growth in early spring, matures by May to June, reproduces from seeds; especially frequent during years with abundant winter and spring moisture

FORAGE VALUE: essentially no value for either livestock or wildlife, although it is sometimes lightly grazed in the spring

HABITAT: plains and open ground; most abundant on dry or alkaline soils of formerly cultivated land and deteriorated rangeland

Basin wildrye
Leymus cinereus (Scribn. & Merr.) A. Löve

SYN = *Elymus cinereus* Scribn. & Merr.

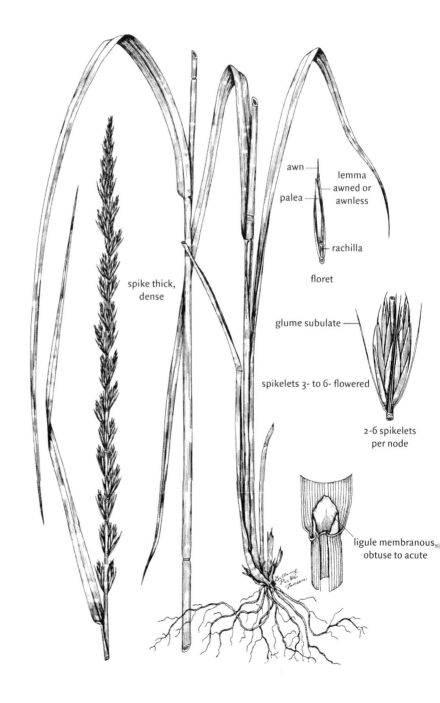

awn
lemma
awned or
awnless
palea
rachilla
floret

spike thick,
dense

glume subulate

spikelets 3- to 6- flowered

2-6 spikelets
per node

ligule membranous,
obtuse to acute

Tribe:	TRITICEAE
Species:	*Leymus cinereus* (Scribn. & Merr.) A. Löve
Common Name:	Basin wildrye (giant wildrye, Great Basin wildrye)
Life Span:	Perennial
Origin:	Native
Season:	Cool

INFLORESCENCE CHARACTERISTICS

type: spike or rarely spicate raceme (10–25 cm long), thick, dense, generally erect, stiff, occasionally interrupted below, rarely develops short branches with up to 6 nodes; spikelets 2–6 per node (1 may be pedicellate), highly imbricate

spikelets: 3- to 6-flowered (9–20 mm long); lemmas acute to blunt (7–12 mm long), glabrous to sparsely strigose, margins hyaline

glumes: unequal (7–16 mm long, 0.2–0.4 mm wide), subulate, awn-pointed, scabrous, nerves usually obscure

awns: lemma awned (1–5 mm long) or awnless; glumes awn-pointed

VEGETATIVE CHARACTERISTICS

growth habit: cespitose, rarely with short rhizomes

culms: erect (1–2.5 m tall), coarse, robust, glabrous to harshly puberulent; nodes glabrous or puberulent

sheaths: rounded, glabrous to puberulent, margins open; auricles well developed (1–2 mm long) to lacking, finger-like

ligules: membranous (3–7 mm long), obtuse to acute, entire to erose

blades: flat to involute (20–60 cm long, 5–15 mm wide), firm, narrowing to an acute tip, glabrous to harshly puberulent, strongly nerved

GROWTH CHARACTERISTICS: starts growth in early spring, seeds mature by August, reproduces from seeds and tillers; not tolerant of heavy grazing because of the relatively high position (10–15 cm above the ground) of the growing points

FORAGE VALUE: good for cattle and fair for sheep and wildlife, provides abundant forage in early spring, relatively unpalatable in the summer, furnishes important winter feed for most classes of livestock but usually requires a protein supplement

HABITAT: riverbanks, ravines, moist or dry slopes, and plains; adapted to a broad range of soils, including moderately saline soils

Medusahead rye
Taeniatherum caput-medusae (L.) Nevski

SYN - *Elymus caput-medusae* L., *T. asperum* (Simonkai) Nevski

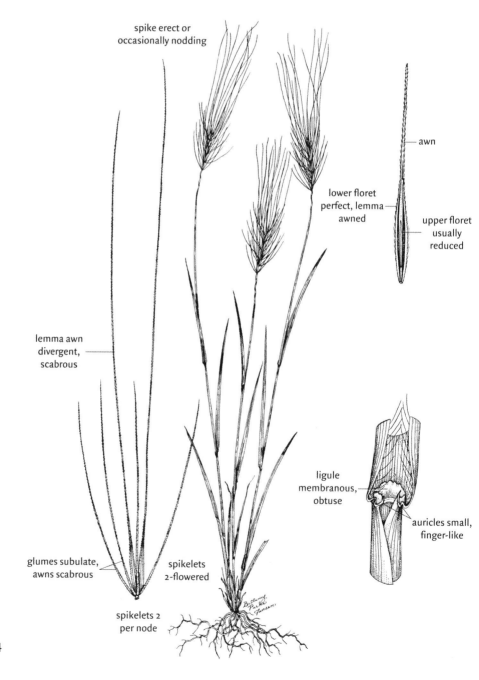

spike erect or occasionally nodding

awn

lower floret perfect, lemma awned

upper floret usually reduced

lemma awn divergent, scabrous

ligule membranous, obtuse

auricles small, finger-like

glumes subulate, awns scabrous

spikelets 2-flowered

spikelets 2 per node

Tribe:	TRITICEAE
Species:	*Taeniatherum caput-medusae* (L.) Nevski
Common Name:	Medusahead rye
Life Span:	Annual
Origin:	Introduced (from Europe)
Season:	Cool

INFLORESCENCE CHARACTERISTICS

type: spike (2–5 cm long, excluding awns), erect or occasionally nodding, dense, bristly; spikelets 2 per node

spikelets: 2-flowered (7–12 mm long, excluding awns); lower floret perfect, upper floret usually reduced; fertile lemma body lanceolate (5–8 mm long), narrow, indistinctly 3-nerved, scabrous

glumes: equal (1.5–3.5 cm long), subulate, stiff, indurate below, nerveless, glabrous

awns: lemma awned (3–8 cm long), flat, divergent, scabrous; glumes awned (1–2.5 cm long), scabrous

VEGETATIVE CHARACTERISTICS

growth habit: cespitose or solitary

culms: ascending from a decumbent and geniculate base (20–60 cm tall), slender, weak, often branching at the base

sheaths: round, slightly inflated; margins open, strigose-puberulent to glabrous; auricles small, finger-like

ligules: membranous (0.2–0.5 mm long), obtuse, erose

blades: involute (3–10 cm long, 1–3 mm wide), glabrous to puberulent; margins sometimes ciliate

GROWTH CHARACTERISTICS: starts growth in March or April, matures May to June, abundant seed production, reproduces from seeds; an aggressive, rapidly spreading weed

LIVESTOCK LOSSES: stiff awns may caused injury to grazing animals by working into the ears, eyes, nose, and tongue and may contaminate fleece

FORAGE VALUE: poor for livestock early in the spring, becomes worthless with production of inflorescences; worthless for wildlife at all times

HABITAT: open ground, disturbed areas, waste places, and deteriorated range-land; adapted to a broad range of soil types

Grass-like Plants

Threadleaf sedge
Carex filifolia Nutt.

SYN = *C. elyniformis* Porsild

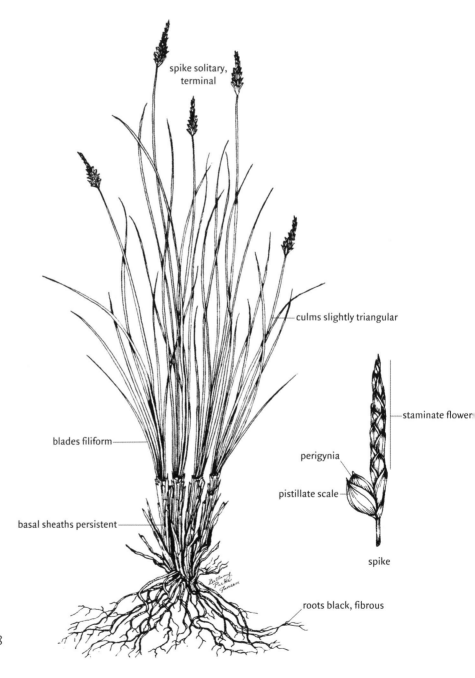

spike solitary, terminal

culms slightly triangular

staminate flower

blades filiform

perigynia

pistillate scale

basal sheaths persistent

spike

roots black, fibrous

208

Family:	CYPERACEAE
Species:	*Carex filifolia* Nutt.
Common Name:	Threadleaf sedge (blackroot)
Life Span:	Perennial
Origin:	Native
Season:	Cool

GROWTH FORM

monoecious, grass-like herb (5–30 cm tall), erect; cespitose with old basal sheaths persistent; starts growth in early spring, flowers April to May, reproduces from seeds and tillers

FLORAL AND FRUIT CHARACTERISTICS

inflorescences: spikes (5–20 mm long, 2–6 mm wide); solitary, terminal, compactly cylindrical first, broader at maturity, generally tapering to both ends, comprised of staminate flowers above and pistillate flowers below

flowers: unisexual; staminate flowers 3–25, stamens 3; pistillate scales broadly obovate, reddish-brown, apex obtuse; margins hyaline, broad, conspicuous, white; perigynia 1–15, erect or ascending, obovoid to obovoid-orbicular (3–4.5 mm long, 2–2.5 mm wide), obscurely 2-ribbed, straw-colored; abruptly contracted into a short, cylindric, obliquely cut beak (0.2–0.5 mm long); stigmas 3

fruits: achenes; obovoid (2.3–3 mm long), triangular in cross-section, brown to black

VEGETATIVE CHARACTERISTICS

leaves: grass-like; blades filiform (3–20 cm long, 0.3–1 mm wide), stiff, mostly basal, generally 2–3 per culm, involute, light green, glabrous, margins scabrous; sheaths brown near base, glabrous

stems: culms slightly triangular, filiform, stiff, wiry

other: roots fibrous, stout, black

HISTORIC, FOOD AND MEDICINAL USES: some Native Americans used culm bases as famine food

LIVESTOCK LOSSES: none

FORAGE VALUE: excellent for sheep, horses, and wildlife; good for cattle; extremely valuable early spring forage, maintains high palatability throughout the growing season; retains good forage quality during winter

HABITAT: prairies, hills, ridges, and valleys; adapted to a broad range of soil types; most abundant on dry soils

Elk sedge
Carex geyeri Boott

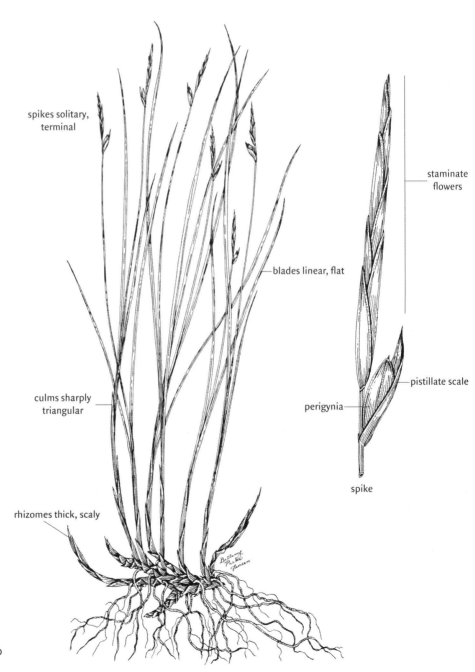

spikes solitary, terminal

blades linear, flat

culms sharply triangular

rhizomes thick, scaly

staminate flowers

pistillate scale

perigynia

spike

Family:	CYPERACEAE
Species:	*Carex geyeri* Boott
Common Name:	Elk sedge (Geyer sedge, pine sedge)
Life Span:	Perennial
Origin:	Native
Season:	Cool

GROWTH FORM
monoecious, grass-like herb (10–50 cm tall), erect; clustered or loosely cespitose, with thick, scaly rhizomes; starts growth in early spring, flowers April to June, reproduces from seeds and rhizomes

FLORAL AND FRUIT CHARACTERISTICS
inflorescences: spikes (5–25 mm long, 1.5–3 mm wide); solitary, terminal, cylindric above, light brown; staminate portion elevated on a short peduncle

flowers: unisexual; staminate flowers 2–15, stamens 3; pistillate scales usually longer and wider than the perigynia, brownish to greenish, often with a lighter midrib, margins hyaline, lower scales short-awned and surpassing the perigynium; perigynia oblong-obovoid (5.5–7 mm long, 2.5–3.5 mm wide), ribs 2, rather abruptly contracted above to the scarcely beaked tip; stigmas 3

fruits: achenes; obovoid (4–5 mm long), sharply triangular in cross-section, sides concave, completely filling the perigynia

VEGETATIVE CHARACTERISTICS
leaves: grass-like; blades linear (10–50 cm long, 1.5–4 mm wide), short at flowering and then elongating, commonly 2 per culm, veins parallel, flat, margins scabrous, tips often withered; sheaths tight, truncate at the mouth

stems: culms sharply triangular, stiffly erect; numerous sterile culms present

HISTORIC, FOOD AND MEDICINAL USES: some Native Americans boiled and ate the culm bases

LIVESTOCK LOSSES: none

FORAGE VALUE: fair to good for cattle, fair for sheep and deer; grazed in early spring and fall since it starts growth earlier than most forages and remains green late into the fall; good for elk in winter

HABITAT: slopes, open woodlands, and dry meadows; adapted to a broad range of soil types

Nebraska sedge
Carex nebrascensis D. R. Dewey

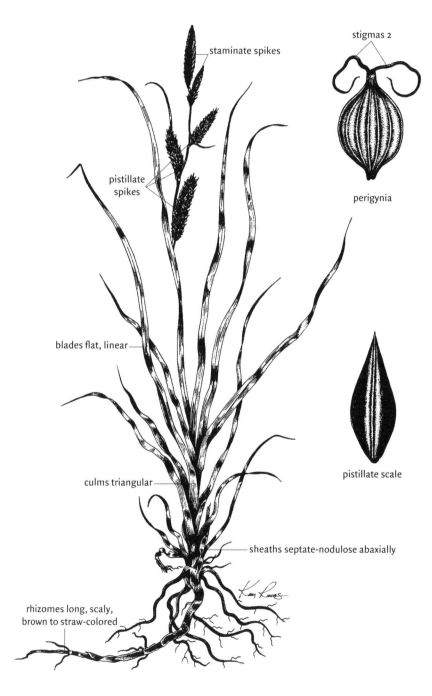

staminate spikes

stigmas 2

perigynia

pistillate spikes

blades flat, linear

pistillate scale

culms triangular

sheaths septate-nodulose abaxially

rhizomes long, scaly, brown to straw-colored

Family:	CYPERACEAE
Species:	*Carex nebrascensis* D. R. Dewey
Common Name:	Nebraska sedge
Life Span:	Perennial
Origin:	Native
Season:	Cool

GROWTH FORM
monoecious, grass-like herb (25–110 cm tall), erect; rhizomes long, scaly, brown to straw-colored; flowers May to July, reproduces from seeds and rhizomes

FLORAL AND FRUIT CHARACTERISTICS
inflorescences: spikes (3–6) unisexual or 1–2 androgynous; upper 1–2 spikes staminate (1.5–7 cm long, 3–9 mm wide), broadly linear, sessile; basal 2–5 spikes pistillate (15–60 mm long, 5–9 mm wide), occasionally androgynous, cylindrical; lowest bract leaf-like, usually exceeding inflorescence, not sheathing

flowers: unisexual; staminate flowers several per spike; scales brownish to purplish-black, margins hyaline; pistillate flowers 30–150 per spike; scales brown to black, lanceolate, obtuse to acuminate, about the same size as the perigynia, midrib white, excurrent; perigynia stramineous with reddish dots, oblong to obovate (3–4 mm long, 2 mm wide), flattened planoconvex or biconvex, strongly nerved, beak bidentate (0.5–1 mm long); stigmas 2

fruits: achenes; lenticular (1.5–2.2 mm long, about 1.2 mm wide)

VEGETATIVE CHARACTERISTICS
leaves: grass-like; blades flat, linear (10–40 cm long, 3–12 mm wide), 8–15 per culm, mostly on the lower one-third of the culm, flat or channeled near base, glaucous, thick, firm; sheaths septate-nodulose abaxially, hyaline and yellowish-brown adaxially

stems: culms triangular, stout, erect

HISTORIC, FOOD AND MEDICINAL USES: some Native Americans ate raw stem bases as a famine food

LIVESTOCK LOSSES: none

FORAGE VALUE: poor to fair for sheep, fair to good for cattle and wildlife; although not as palatable as some sedge species, it is a valuable late season forage; makes good hay

HABITAT: wet meadows, swamps, streams, marshes, edges of lakes, ponds, and ditches; adapted to a broad range of soil textures

Beaked sedge
Carex utriculata Boott

SYN = *C. inflata* Huds., *C. rostrata* Stokes

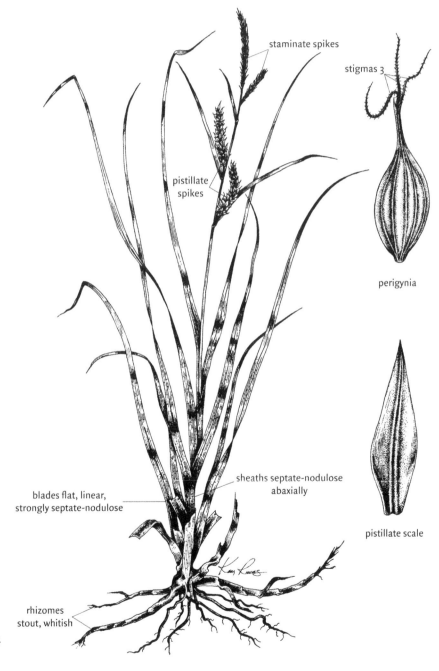

staminate spikes

stigmas 3

pistillate spikes

perigynia

sheaths septate-nodulose abaxially

blades flat, linear, strongly septate-nodulose

pistillate scale

rhizomes stout, whitish

Family:	CYPERACEAE
Species:	*Carex utriculata* Boott
Common Name:	Beaked sedge (Northwest Territory sedge)
Life Span:	Perennial
Origin:	Native
Season:	Cool

GROWTH FORM

monoecious, grass-like herb (40–120 cm tall), erect; cespitose with short root stalks producing stout, whitish rhizomes; flowers June to August; reproduces by seeds and rhizomes

FLORAL AND FRUIT CHARACTERISTICS

inflorescences: spikes unisexual or 1–2 androgynous; upper 2–4 spikes staminate (1–6 cm long, 3–4 mm wide), narrowly linear, upper one peduncled, lower ones sessile; basal 2–5 spikes pistillate (1–10 cm long, 8–12 mm wide), occasionally androgynous, densely 40- to 150-flowered, cylindrical, sessile above to short-peduncled below; bracts leaf-like, little if at all sheathing, shorter than to slightly exceeding inflorescence

flowers: unisexual; pistillate scales 3-nerved, margins hyaline; perigynia oval to ovoid (3.5–8 mm long, 2.5–3.5 mm wide), yellowish-green to brown, inflated, subterete, strongly 8- to 16-nerved; beak bidentate (1–2 mm long); stigmas 3, style S-curved toward base

fruits: achenes; trigonous (1.7–2 mm long, 1.2–1.4 mm wide), style persistent

VEGETATIVE CHARACTERISTICS

leaves: grass-like; blades flat, linear (20–60 cm long, 4–12 mm wide), 4–10 per culm, strongly septate-nodulose, thick, stiff; margins revolute, more or less channeled at base; sheaths septate-nodulose abaxially, white hyaline adaxially

stems: culms, stout, erect; bluntly triangular below the inflorescence; base thickened, spongy, reddish

HISTORIC, FOOD AND MEDICINAL USES: some Native Americans used stem bases for food

LIVESTOCK LOSSES: none

FORAGE VALUE: low in palatability, rarely utilized by domestic livestock in North America, however reported to be excellent forage in Iceland and Siberia; good for elk and moose

HABITAT: wet meadows, marshes, swamps, fens, lake and pond edges, springs

Hardstem bulrush
Schoenoplectus acutus (Muhl. ex Bigel.) A. & D. Löve

SYN = *Scirpus acutus* Muhl. *ex* Bigel.

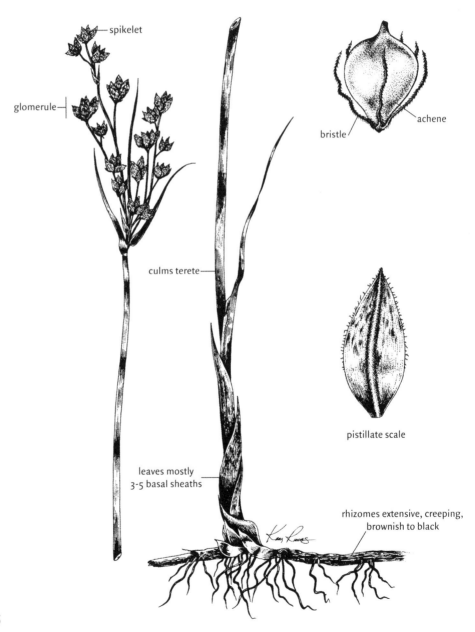

spikelet

glomerule

achene

bristle

culms terete

pistillate scale

leaves mostly
3-5 basal sheaths

rhizomes extensive, creeping,
brownish to black

Family:	CYPERACEAE
Species:	*Schoenoplectus acutus* (Muhl. *ex* Bigel.) A. & D. Löve
Common Name:	Hardstem bulrush (tule)
Life Span:	Perennial
Origin:	Native
Season:	Cool

GROWTH FORM

grass-like herb (1–4 m tall, 0.8–2.3 cm wide at base), forming dense colonies from extensive, creeping, brownish to black rhizomes; flowers June to mid-August, reproduces from seeds and rhizomes

FLORAL AND FRUIT CHARACTERISTICS

inflorescences: panicle-like (3–10 cm long), appearing lateral with up to 60 spikelets, compact to open; spikelets in glomerules of 2–15 (rarely solitary), pendulous with short to long pedicels; principal involucral bract appearing as a continuation of the culm (1.5–10 cm long), solitary or rarely 2–3

spikelets: perfect; 20- to 50-flowered, ovoid to linear (5–15 mm long, 3–5 mm wide), grayish-brown, acute; scales pale-brown with reddish spots (2–8 mm long, 2–2.6 mm wide), scarious, lower ones puberulent abaxially, apex acute to cleft, mucronate; perianth of 4 to 6 bristles, retrorsely barbed, shorter than or equalling achene; stamens 3, often persisting; styles 2- to 3-cleft

fruits: achenes; trigonous (1.8–2.9 mm long, 1.2–1.9 mm wide), light green to dark brown, planoconvex or unequally trigonous, beak to 0.5 mm long

VEGETATIVE CHARACTERISTICS

leaves: consisting mostly of 3–5 basal sheaths, upper sheaths occasionally with tapering blades to 25 cm long

stems: culms erect, terete, stout, dark green, firm, unyielding to crushing when green

HISTORIC, FOOD AND MEDICINAL USES: some Native Americans ate young shoots raw or boiled, pollen was used in bread, starchy root was boiled or ground into meal; stems were woven into mats

LIVESTOCK LOSSES: none

FORAGE VALUE: usually unpalatable and poor for livestock, poor to fair for wildlife

HABITAT: emergent in shallow to deep waters of sloughs, ponds, lakes, ditches, and roadsides, especially in brackish water

Wire rush
Juncus balticus Willd.

SYN = *J. arcticus* Willd.

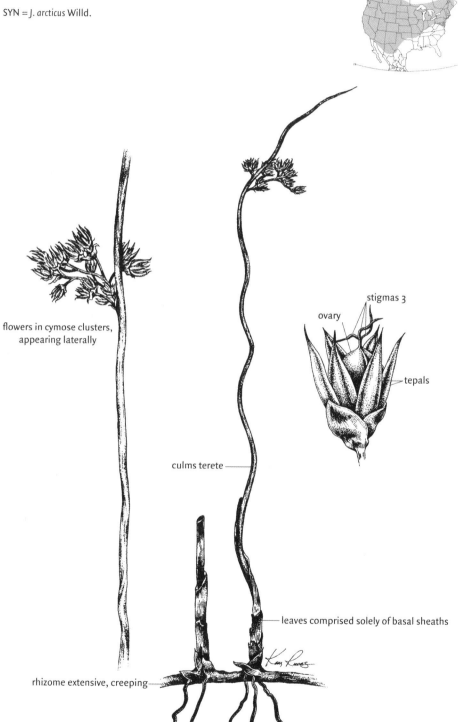

flowers in cymose clusters, appearing laterally

stigmas 3

ovary

tepals

culms terete

leaves comprised solely of basal sheaths

rhizome extensive, creeping

Family:	JUNCACEAE
Species:	*Juncus balticus* Willd.
Common Name:	Wire rush (baltic rush)
Life Span:	Perennial
Origin:	Native
Season:	Cool

GROWTH FORM

grass-like herb (to 110 cm tall), erect, wiry; from extensive creeping rhizomes; flowers May to August, reproduces from seeds and rhizomes

FLORAL AND FRUIT CHARACTERISTICS

inflorescences: cymose clusters (1–7 cm long and wide); appearing laterally near the top of the plant, dense to spreading; involucral bract terete to somewhat flattened (to 20 cm long), erect, appearing as a continuation of the stem, obtuse to mucronate; flowers borne singly, subsessile or on brownish pedicels (3–4 mm long)

flowers: perfect; tepals 6, greenish becoming straw colored or dark brown (3–6 mm long), lanceolate, acute to ovate, margins hyaline, inner ones shorter than outer ones; stamens 6; stigmas 3, ovary 3-locular

fruits: capsule; narrowly ovoid, ranging from shorter than to longer than perianth, acute, mucronate, 3-locular; seeds ovoid-ellipsoid (0.3–0.8 mm long), grayish brown, ends apiculate

VEGETATIVE CHARACTERISTICS

leaves: with the exception of the involucre, comprised solely of basal sheaths, blades and auricles absent; sheaths rounded, clustered near base of plant (2–15 cm long), light brown to reddish-brown

stems: culms erect, terete, stout (1.5–4 mm thick), crowded, arising from thick, dark brown, creeping rhizomes, often forming a sod

other: variable species with about 6 varieties recognized in North America

HISTORIC, FOOD AND MEDICINAL USES: some Native Americans made mats and baskets from the stems

LIVESTOCK LOSSES: none

FORAGE VALUE: fair to good for cattle and sheep when actively growing, often more palatable in hay than when growing, worthless when mature

HABITAT: lake and stream marshes, wet meadows, ditches, seeps, lake and pond shores; often associated with alkaline sites; frequently abundant

Forbs and Woody Plants

Skunkbrush sumac
Rhus aromatica Aiton

SYN = R. *trilobata* Nutt.

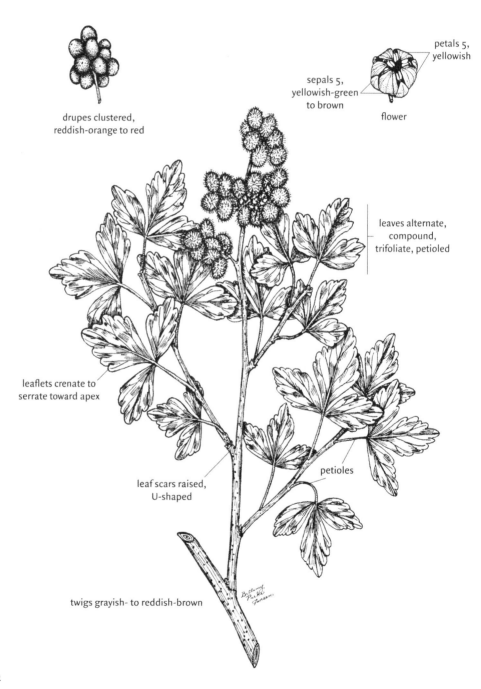

drupes clustered,
reddish-orange to red

sepals 5,
yellowish-green
to brown

petals 5,
yellowish

flower

leaves alternate,
compound,
trifoliate, petioled

leaflets crenate to
serrate toward apex

leaf scars raised,
U-shaped

petioles

twigs grayish- to reddish-brown

Family: ANACARDIACEAE
Species: *Rhus aromatica* Aiton
Common Name: Skunkbrush sumac (lemita, skunkbush, polecat bush, fragrant sumac, landrisco)
Life Span: Perennial
Origin: Native
Season: Cool

GROWTH FORM

dioecious (polygamo-dioecious) shrub (to 2.5 m tall); dense, ascending branches from several trunks; flowers April to May, reproduces from seeds

FLORAL AND FRUIT CHARACTERISTICS

inflorescence: cluster; spike-like (6 cm long, 3 cm wide), terminal

flowers: regular, unisexual and perfect; sepals 5, obtuse (1 mm long), united at base, yellowish-green to brown; petals 5, obtuse (1–2 mm long), spreading at maturity, yellowish; stamens 5

fruits: drupe (4–7 mm in diameter), globose, reddish-orange to red, lightly to densely hirsute, hairs simple to glandular; occurring in clusters; persistent in winter

other: flowers formed late in the previous season and before or as leaves appear in early spring

VEGETATIVE CHARACTERISTICS

leaves: alternate, compound, trifoliate; terminal leaflet ovate (2–9 cm long, 2–8 cm wide), lateral leaflets ovate (3.5–5 cm long, 2–4 cm wide), crenate to serrate toward apex, entire at base; adaxially dull, glabrous or pubescent; abaxially pale, densely to lightly pubescent; aromatic; leaflets sessile to subsessile; petiole (1–1.5 cm long)

stems: twigs grayish- to reddish-brown, slender, glabrous or pubescent; leaf scars elongate, raised, U-shaped; bark dark brown, smooth to fissured; fragrant when bruised

other: highly variable species consisting of many varieties

HISTORIC, FOOD, AND MEDICINAL USES: some tribes of Native Americans made use of the berry-like fruits as food, medicine, and lemonade-like drinks; slender shoots were used for basket weaving

LIVESTOCK LOSSES: none

FORAGE VALUE: poor to fair for cattle, horses, and sheep; good browse for wildlife and goats; birds and small mammals eat the fruits

HABITAT: hillsides, ravines, thickets, woodland, and roadsides; adapted to a wide range of soils; planted for wildlife habitat and erosion control

Poison hemlock
Conium maculatum L.

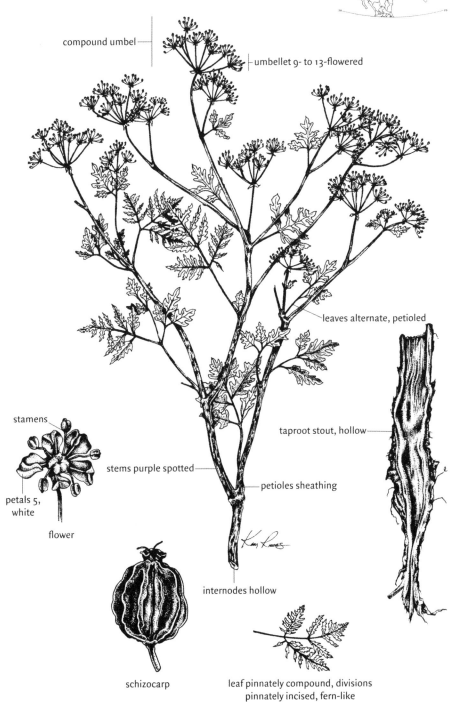

compound umbel

umbellet 9- to 13-flowered

leaves alternate, petioled

stamens

petals 5, white

flower

stems purple spotted

petioles sheathing

taproot stout, hollow

internodes hollow

schizocarp

leaf pinnately compound, divisions pinnately incised, fern-like

Family:	APIACEAE
Species:	*Conium maculatum* L.
Common Name:	Poison hemlock (hemlock, spotted hemlock)
Life Span:	Perennial (biennial)
Origin:	Introduced (from Europe)
Season:	Cool

GROWTH FORM

forb (0.5–3 m tall); erect, highly branched; flowers May to August; reproduces by seed

FLORAL AND FRUIT CHARACTERISTICS

inflorescence: compound umbels (4–7 cm wide), convex; involucre bracts short, ovate, acuminate; 8–17 unequal rays (1.5–2.5 cm long), spreading to ascending; involucre of numerous bractlets similar to bracts, midrib conspicuous, shorter than pedicels; umbellets 9- to 13-flowered; pedicels spreading (4–6 mm long)

flowers: perfect, regular; sepals absent; petals 5 (1–1.5 mm long), notched, clawed, white

fruits: schizocarp (2–2.5 mm long), broadly ovoid, laterally flattened, glabrous, grayish-brown; ribs prominent, undulate, crenate; 2 seeded; seed flattened to concave, deeply and narrowly sulcate

VEGETATIVE CHARACTERISTICS

leaves: alternate, pinnately compound, divisions pinnately incised, broadly ovate (15–30 cm long, 5–30 cm wide), fern-like; lobes oblong to lanceolate; glabrous; petioles sheathing

stems: erect from a stout and hollow taproot, highly branched, glabrous, glaucous, purple spotted, internodes hollow

other: plant is a biennial producing a rosette of leaves the first year and flowering the second year; plant has a sour, musty or mousy odor; may be fatal if eaten by humans

HISTORIC, FOOD, AND MEDICINAL USES: an extract from this species is believed to have been used to put Socrates to death

LIVESTOCK LOSSES: highly poisonous to livestock, contains several alkaloids; alkaloid poisons in the roots, leaves, and fruits are the most toxic

FORAGE VALUE: usually unpalatable

HABITAT: moist soils of pastures, rangelands, roadsides, stream banks, and disturbed sites

Western yarrow
Achillea millefolium L.

SYN = *A. lanulosa* Nutt.

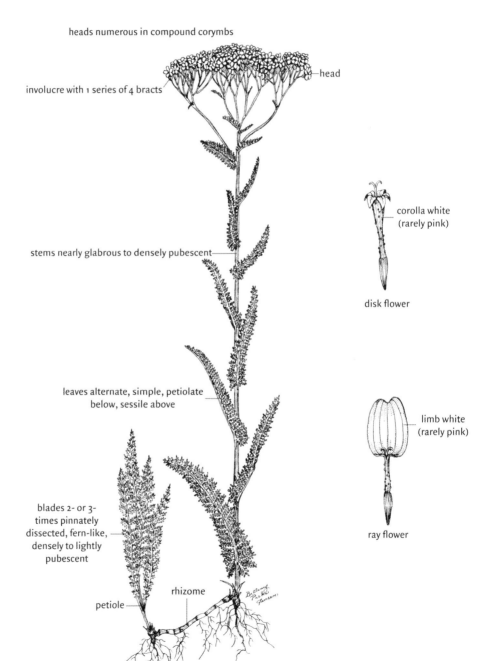

heads numerous in compound corymbs

head

involucre with 1 series of 4 bracts

corolla white
(rarely pink)

disk flower

stems nearly glabrous to densely pubescent

leaves alternate, simple, petiolate
below, sessile above

limb white
(rarely pink)

ray flower

blades 2- or 3-
times pinnately
dissected, fern-like,
densely to lightly
pubescent

petiole

rhizome

Family:	ASTERACEAE
Tribe:	ANTHEMIDEAE
Species:	*Achillea millefolium* L.
Common Name:	Western yarrow (hierba del oro, yarrow, milfoil, milenrama)
Life Span:	Perennial
Origin:	Native
Season:	Cool

GROWTH FORM
forb (0.2–1 m tall); erect, arising singly or as a loose cluster; flowers April to July, reproduces from seeds and rhizomes

FLORAL AND FRUIT CHARACTERISTICS
inflorescence: heads numerous (5–7 mm tall) in compound corymbs; flat to round-topped

flowers: perfect, heads radiate; involucre with 1 series of 4 bracts, bract margins scarious; ray flowers 5–20, typically about 5; limb ovate (2–5 mm long), white (rarely pink or pinkish-white), 5-toothed; disk flowers 10–40, corolla tube somewhat flattened, white (rarely pink or pinkish-white)

fruits: achene, oblong, flattened, glabrous; pappus absent

VEGETATIVE CHARACTERISTICS
leaves: alternate, simple; blades 2- or 3-times pinnately dissected (3–15 cm long, 5–30 mm wide), fern-like, lanceolate in outline, densely to lightly pubescent; petiolate below, sessile above; aromatic

stems: erect, simple to sparingly branched above, nearly glabrous to densely pubescent; aromatic

HISTORIC, FOOD, AND MEDICINAL USES: a decoction of leaves and flowers was used by the Blackfoot as an eyewash; the Winnebago steeped whole plants and poured the liquid into aching ears; green leaves were used to relieve itching, chewed for toothaches, and used as a mild laxative; leaves boiled for tea were used as a cold remedy; during the Civil War, powdered leaves were applied to wounds to stop bleeding, and its common name was soldiers' woundwort

LIVESTOCK LOSSES: not generally considered to be poisonous, but may contain toxic alkaloids and glycosides

FORAGE VALUE: poor to fair for cattle and fair to good for sheep; usually grazed only when green; heads may be eaten by pronghorn, deer, and sheep

HABITAT: prairies, sagebrush plains, pastures, roadsides, and disturbed sites; adapted to a wide range of soils; frequently planted in flower gardens

Silver sagebrush
Artemisia cana Pursh

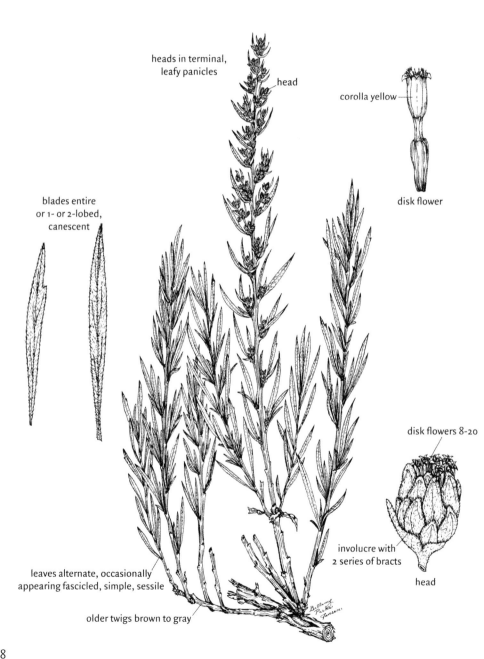

heads in terminal, leafy panicles

head

corolla yellow

disk flower

blades entire or 1- or 2-lobed, canescent

disk flowers 8-20

involucre with 2 series of bracts

head

leaves alternate, occasionally appearing fascicled, simple, sessile

older twigs brown to gray

Family:	ASTERACEAE
Tribe:	ANTHEMIDEAE
Species:	*Artemisia cana* Pursh
Common Name:	Silver sagebrush (istafiate, white sagebrush, hoary sagebrush)
Life Span:	Perennial
Origin:	Native
Season:	Evergreen

GROWTH FORM

shrub (to 1.5 m tall); densely branched, crown rounded; often forming colonies from extensive rhizomes; flowers August to September, reproduces from seeds and rhizomes

FLORAL AND FRUIT CHARACTERISTICS

inflorescence: heads in terminal, leafy panicles (15–30 cm long, 2–6 cm wide); heads in groups of 3–10, subtended by leaf-like bract often surpassing the heads

flowers: perfect, heads discoid; involucre campanulate (3.5–5 mm long, 2.3–4.5 mm wide) with 2 series of bracts; outer series involucral bracts ovate, acute, densely canescent; inner series bracts elliptic, canescent to glabrous, margins scarious; disk flowers 8–20, corolla yellow and tubular (2.5 mm long)

fruits: achene (2.5 mm long, 1 mm wide), cylindrical, angular, ribs 5–6, light brown; pappus absent

VEGETATIVE CHARACTERISTICS

leaves: alternate, occasionally appearing fascicled, simple; blades linear (2–9 cm long, 1–7 mm wide), acute; margins entire to 1- or 2-lobed; canescent; sessile

stems: twigs green to straw colored, rigid, finely canescent; older twigs brown to gray, glabrous; trunk bark tan to grayish-brown, exfoliating into long fibrous strips

other: plants often dry to a goldish hue with a yellowish stem, mildly aromatic

HISTORIC, FOOD, AND MEDICINAL USES: decoction used by some Native Americans to stop coughing, an extract was thought to restore hair

LIVESTOCK LOSSES: none

FORAGE VALUE: good to excellent in fall and winter for cattle and sheep, increases under browsing by cattle and decreases under browsing by sheep; some varieties are not palatable

HABITAT: river valleys, terraces, and uplands in moist to moderately dry soils; most abundant in deep loamy to sandy soils, alkali tolerant

Sand sagebrush
Artemisia filifolia Torr.

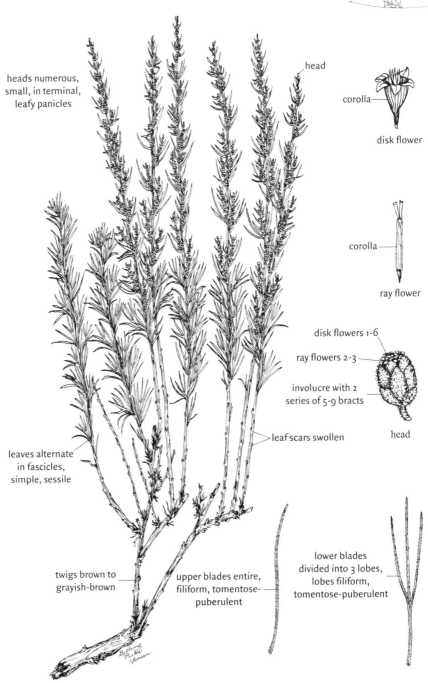

heads numerous, small, in terminal, leafy panicles

head

corolla

disk flower

corolla

ray flower

disk flowers 1-6

ray flowers 2-3

involucre with 2 series of 5-9 bracts

head

leaf scars swollen

leaves alternate in fascicles, simple, sessile

twigs brown to grayish-brown

upper blades entire, filiform, tomentose-puberulent

lower blades divided into 3 lobes, lobes filiform, tomentose-puberulent

Family:	ASTERACEAE
Tribe:	ANTHEMIDEAE
Species:	*Artemisia filifolia* Torr.
Common Name:	Sand sagebrush (artemisa, sandhill sage)
Life Span:	Perennial
Origin:	Native
Season:	Warm

GROWTH FORM
shrub (to 1.2 m tall); freely branching, rounded crown, aromatic; flowers August to October, reproduces from seeds

FLORAL AND FRUIT CHARACTERISTICS
inflorescence: heads numerous and small, in terminal, leafy panicles (15–20 cm long, 5–20 mm wide), plume-like, subtended by leaf-like bracts often surpassing the heads; peduncles short and stiff

flowers: unisexual, heads inconspicuously radiate; involucre with 2 series of 5–9 bracts (1.8 mm long), outer series short and indurate; inner series elliptic and thin, obtuse, woolly; ray flowers 2–3 (0.6–0.7 mm long), pistillate; disk flowers 1–6 (1.5 mm long), staminate

fruits: achene (0.7–1 mm long, 0.4–0.5 mm wide), obovoid, lobes 5, brown, lightly 4- to 5-ribbed, glabrous; pappus absent

VEGETATIVE CHARACTERISTICS
leaves: alternate in fascicles, simple (3.5–7.5 cm long); upper blades simple, entire, filiform; lower blades divided into 3 lobes, lobes filiform; margins entire; tomentose-puberulent; sessile

stems: twigs brown to grayish-brown, erect, slender, freely branching, striate, pubescent to glabrous, eventually exfoliating into thin shreds

other: entire plant is aromatic

HISTORIC, FOOD, AND MEDICINAL USES: in Mexico, a decoction of leaves was taken for intestinal worms and stomach problems

LIVESTOCK LOSSES: may cause sage sickness in horses

FORAGE VALUE: poor to worthless for cattle, poor to fair for horses and sheep; furnishes fair forage for pronghorn and deer; rarely grazed if other forages are present; achenes are an important food for grouse and other birds

HABITAT: plains and pastures in well drained sandy soil, generally considered an indicator of sandy soil; often abundant on sand

Fringed sagebrush
Artemisia frigida Willd.

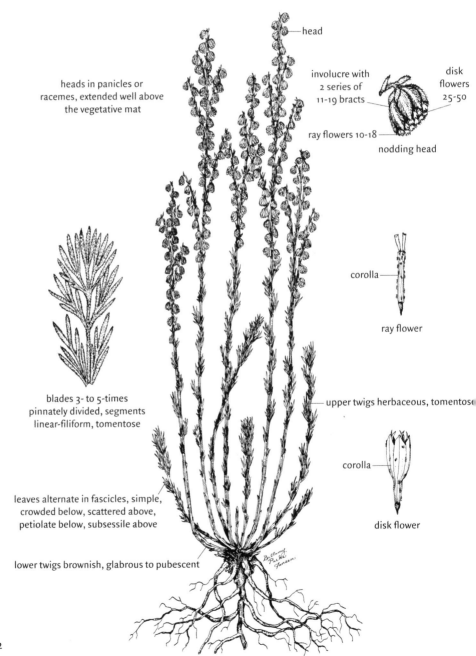

head

heads in panicles or racemes, extended well above the vegetative mat

involucre with 2 series of 11-19 bracts

disk flowers 25-50

ray flowers 10-18

nodding head

corolla

ray flower

blades 3- to 5-times pinnately divided, segments linear-filiform, tomentose

upper twigs herbaceous, tomentose

corolla

disk flower

leaves alternate in fascicles, simple, crowded below, scattered above, petiolate below, subsessile above

lower twigs brownish, glabrous to pubescent

Family:	ASTERACEAE
Tribe:	ANTHEMIDEAE
Species:	*Artemisia frigida* Willd.
Common Name:	Fringed sagebrush (artemisa, prairie sagewort, wormwood)
Life Span:	Perennial
Origin:	Native
Season:	Cool

GROWTH FORM

subshrub (to 40 cm tall); semi-erect to decumbent, mat-forming; flowers August to September, reproduces from seeds

FLORAL AND FRUIT CHARACTERISTICS

inflorescence: heads in panicles or racemes (5–30 cm long, 1–10 cm wide), leafy, extended well above the vegetative mat; leafy, heads sessile and nodding

flowers: pistillate and perfect, heads inconspicuously radiate; involucre loosely tomentose (2–3 mm tall); with 2 series of 11–19 bracts, lanceolate to ovate; ray flowers 10–18, pistillate; disk flowers 25–50, perfect, tubular

fruits: achene, subcylindrical, narrowing to base, ribbed, glabrous; pappus absent

VEGETATIVE CHARACTERISTICS

leaves: alternate in fascicles, simple; crowded below and scattered above; blade outline rounded (5–12 mm long), 3- to 5-times pinnately divided, segments linear-filiform; tomentose; petiolate below, subsessile above

stems: twigs branching at base; upper twigs herbaceous, tomentose; lower twigs brownish, glabrous to pubescent

other: entire plant tomentose and aromatic

HISTORIC, FOOD, AND MEDICINAL USES: some Native Americans called it "woman sage" and used it to eliminate the greasy smell from dried meat, to bandage cuts after it was chewed, to make mats, fans, menstrual pads, and as toilet paper

LIVESTOCK LOSSES: none

FORAGE VALUE: fair to good for sheep, goats, and pronghorn; poor to fair for cattle; important winter feed for elk and deer; increases rapidly under heavy grazing; drought tolerant

HABITAT: prairies, plains, and foothills; most abundant in well-drained sandy or rocky soils; occasionally planted as an ornamental

Cudweed sagewort
Artemisia ludoviciana Nutt.

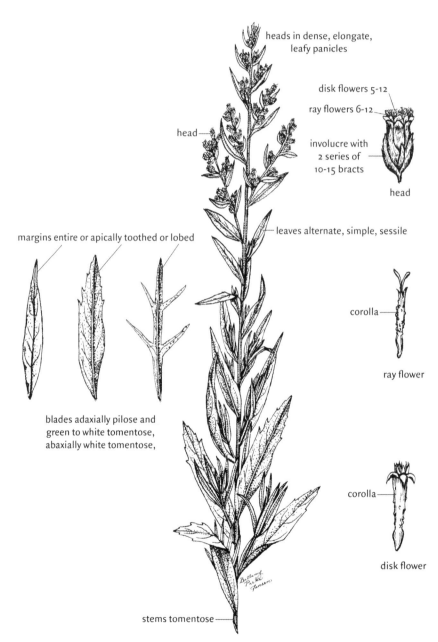

heads in dense, elongate, leafy panicles

disk flowers 5-12

ray flowers 6-12

head

involucre with 2 series of 10-15 bracts

head

margins entire or apically toothed or lobed

leaves alternate, simple, sessile

corolla

ray flower

blades adaxially pilose and green to white tomentose, abaxially white tomentose,

corolla

disk flower

stems tomentose

Family:	ASTERACEAE

Family: ASTERACEAE
Tribe: ANTHEMIDEAE
Species: *Artemisia ludoviciana* Nutt.
Common Name: Cudweed sagewort (iztafiate, Louisiana wormwood, western mugwort)
Life Span: Perennial
Origin: Native
Season: Warm

GROWTH FORM

forb (0.3–1.5 m tall); erect, 1 to few stems; flowers August to September, reproduces from seeds and rhizomes, often forming colonies

FLORAL AND FRUIT CHARACTERISTICS

inflorescence: heads in dense, elongate, leafy panicles (20–50 cm long); heads fascicled or on spike-like branches

flowers: pistillate and perfect, heads inconspicuously radiate; involucre (1.2 mm tall, 3–4 mm wide) with 2 series of 10–15 bracts; outer series small, tomentose; inner series elliptic and obtuse, tomentose; ray flowers 6–12, pistillate (1 mm long); disk flowers 5–12, perfect (1.4–2 mm long)

fruits: achene (1–1.2 mm long), elliptical, brownish, smooth, glabrous, obscurely nerved; pappus absent

other: inflorescence can vary from compact and dense to loose and relatively open

VEGETATIVE CHARACTERISTICS

leaves: alternate, simple, mostly cauline; blades linear to lanceolate or elliptic (2–12 cm long, to 1 cm wide, exclusive of the lobes), reduced above; margins entire or apically toothed or lobed; adaxially pilose and green to white tomentose, abaxially white tomentose; sessile

stems: simple or short branched above, slender, tomentose

other: variable in size, leaves, and pubescence; all plant parts are aromatic

HISTORIC, FOOD, AND MEDICINAL USES: some Native Americans called it "man sage" and used it for ceremonial and purification purposes; to deodorize feet; cure headaches, treat coughs, hemorrhoids, stomach troubles, and wounds on horses; made into pillows and saddle pads; burned to drive mosquitoes away

LIVESTOCK LOSSES: none

FORAGE VALUE: fair to poor for cattle and sheep; somewhat palatable to elk, deer, and pronghorn; decreasing forage value from south to north

HABITAT: prairies, plains, foothills, open woods, disturbed sites, and roadsides

Black sagebrush
Artemisia nova A. Nels.

SYN = *A. arbuscula* Nutt. subsp. *nova* (A. Nels.) G. H. Ward

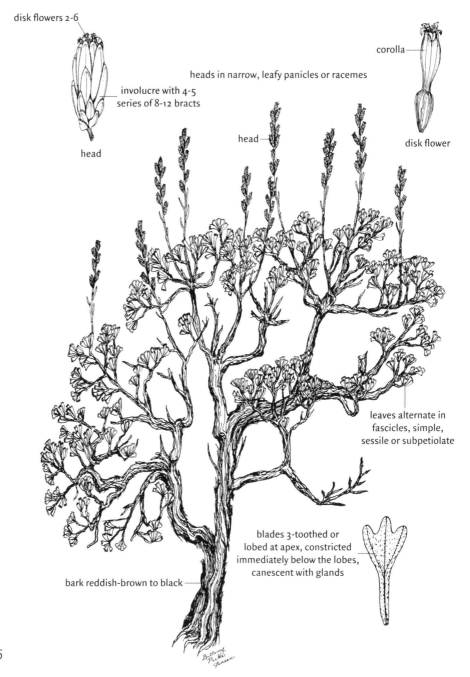

disk flowers 2-6

involucre with 4-5
series of 8-12 bracts

head

heads in narrow, leafy panicles or racemes

head

corolla

disk flower

leaves alternate in
fascicles, simple,
sessile or subpetiolate

blades 3-toothed or
lobed at apex, constricted
immediately below the lobes,
canescent with glands

bark reddish-brown to black

Family:	ASTERACEAE
Tribe:	ANTHEMIDEAE
Species:	*Artemisia nova* A. Nels.
Common Name:	Black sagebrush (black sage)
Life Span:	Perennial
Origin:	Native
Season:	Evergreen

GROWTH FORM

shrub (to 40 cm tall); branches numerous, decumbent, spreading, crown rounded; flowers August to September, reproduces from seeds

FLORAL AND FRUIT CHARACTERISTICS

inflorescence: heads in narrow, leafy panicles or racemes (5–15 cm long, 1–2 cm wide); heads few, small, subsessile, erect to drooping, occasionally 2–3 in a cluster

flowers: perfect, heads discoid; involucre (2–5 mm long, 1 mm wide) with 4–5 series of 8–12 bracts; outer bracts short, ovate, canescent; inner bracts elliptic, lightly canescent to glabrous; disk flowers 2–6, tubular (2.5–3.5 mm long)

fruits: achene (1.5–2 mm long), obovoid, brownish, flattened, glabrous, resinous; pappus absent

other: reddish-brown flower stalks, persistent through winter

VEGETATIVE CHARACTERISTICS

leaves: alternate in fascicles, simple; blades obdeltoid (3–20 mm long, 2–8 mm wide), spatulate to linear above; apex 3-toothed or lobed, constricted immediately below the lobes; dark to pale green, often canescent, glandular; blades maybe reduced and entire above; sessile or subpetiolate

stems: twigs round, short, rigid, light to dark reddish-brown becoming black with age, occasionally canescent; trunk dark, reddish-brown to black

other: appearing darker in color than the other species of *Artemisia*, entire plant aromatic, many leaves are persistent throughout the winter

HISTORIC, FOOD, AND MEDICINAL USES: some Native Americans drank a decoction of the boiled stems, leaves, and twigs for bronchitis; leaves were crushed and the vapor inhaled to relieve nasal congestion

LIVESTOCK LOSSES: none

FORAGE VALUE: good for livestock but poorly consumed in summer; good winter browse for livestock and wildlife

HABITAT: rocky slopes and windswept ridges in dry, shallow soils at elevations between 1500 and 2900 m

Budsage
Artemisia spinescens (DC.) Eaton

SYN = *Picrothamnus desertorum* Nutt.

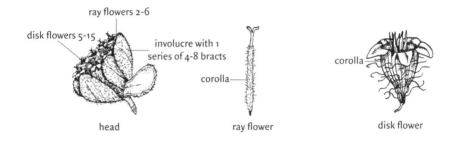

ray flowers 2-6

disk flowers 5-15

involucre with 1 series of 4-8 bracts

corolla

head

ray flower

corolla

disk flower

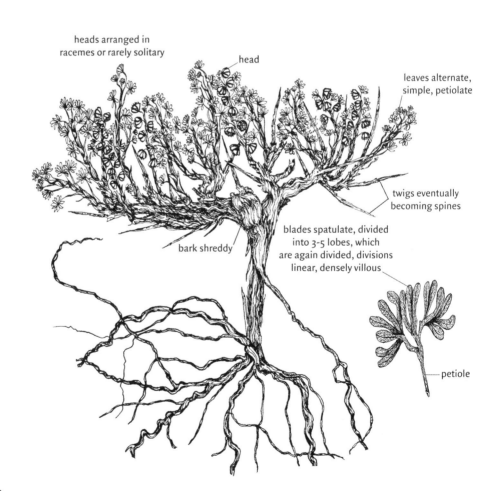

heads arranged in racemes or rarely solitary

head

leaves alternate, simple, petiolate

twigs eventually becoming spines

blades spatulate, divided into 3-5 lobes, which are again divided, divisions linear, densely villous

bark shreddy

petiole

238

Family:	ASTERACEAE
Tribe:	ANTHEMIDEAE
Species:	*Artemisia spinescens* (DC.) Eaton
Common Name:	Budsage (bud sagewort, spiny sagebrush, bud sagebrush)
Life Span:	Perennial
Origin:	Native
Season:	Cool

GROWTH FORM
shrub (to 60 cm tall); much-branched, crown rounded, rigid, spinescent; flowers March to June, reproduces from seeds

FLORAL AND FRUIT CHARACTERISTICS
inflorescence: heads arranged in racemes (1–5 cm long) or rarely solitary; axillary and/or terminal, leafy; heads solitary or fascicled, small, nodding, subsessile

flowers: unisexual, heads inconspicuously radiate; involucre (2–3.5 mm tall, 3–4.5 mm wide) with 1 series of 4–8 bracts; bracts ovate, obtuse, densely villous; ray flowers 2–6, pistillate, corolla hairy; disk flowers 5–15, staminate, corolla hairy

fruits: achene, oblong or ellipsoid, densely villous; pappus absent

VEGETATIVE CHARACTERISTICS
leaves: alternate, simple; blades spatulate (5–15 mm long, less than 1 cm wide), divided into 3–5 lobes, which are again divided, divisions linear; densely villous; petiolate

stems: older twigs rigid, thickened, gray to dark brown, shreddy; newer twigs ascending, tomentose to hairy, eventually becoming spines; trunk short, gray to brown, shreddy; leaves usually absent most of the year leaving tawny-colored stems

other: leaves usually deciduous by midsummer

HISTORIC, FOOD, AND MEDICINAL USES: pollen commonly causes hay fever

LIVESTOCK LOSSES: reported to be poisonous to cattle when consumed alone, not poisonous to sheep

FORAGE VALUE: good for sheep in Utah, Nevada, and California where it is especially valuable for early browse; poor to worthless for cattle and horses; good to fair for wildlife

HABITAT: desert mesas, hills, and plains; adapted to a broad range of dry or well-drained soils, an indicator of alkaline soils

Big sagebrush
Artemisia tridentata Nutt.

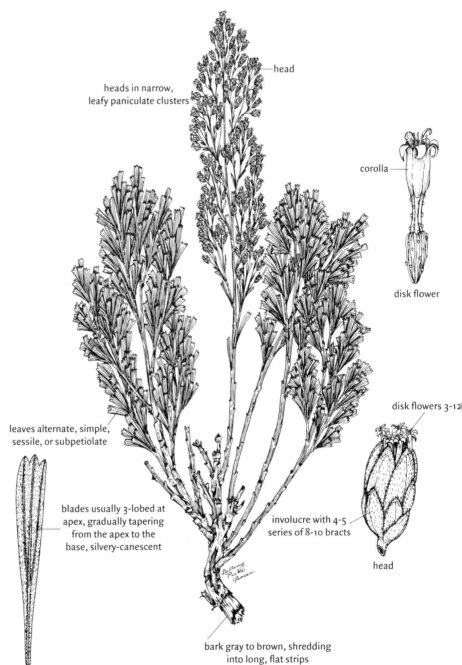

head

heads in narrow,
leafy paniculate clusters

corolla

disk flower

leaves alternate, simple,
sessile, or subpetiolate

disk flowers 3-12

blades usually 3-lobed at
apex, gradually tapering
from the apex to the
base, silvery-canescent

involucre with 4-5
series of 8-10 bracts

head

bark gray to brown, shredding
into long, flat strips

Family:	ASTERACEAE
Tribe:	ANTHEMIDEAE
Species:	*Artemisia tridentata* Nutt.
Common Name:	Big sagebrush (chamiso hediondo, sagebrush)
Life Span:	Perennial
Origin:	Native
Season:	Evergreen

GROWTH FORM
shrub (to 6 m tall); erect, highly branched, crown rounded or spreading, trunk short; flowers August to September, reproduces from seeds

FLORAL AND FRUIT CHARACTERISTICS
inflorescence: heads in narrow, leafy paniculate clusters (5–15 cm long); heads numerous, erect to drooping, subsessile

flowers: perfect, heads discoid; involucre with 4–5 series (2–4 mm long, 2–2.8 mm wide) of 8–10 bracts; outer bracts short, ovate, canescent; inner bracts oblong, canescent to almost glabrous; disk flowers 3–12 (2.5–3.5 mm long)

fruits: achene (2–3 mm long, 1 mm wide), obovoid, brownish, flattened, pubescent, resinous; pappus absent

VEGETATIVE CHARACTERISTICS
leaves: alternate, simple; blades narrowly obdeltoid (1–5 cm long, 3–14 mm wide), 3-lobed at truncate apex, sometimes entire; gradually tapering from the apex to the base; silvery-canescent; sessile or subpetiolate; elongate

stems: twigs round, rigid, green above, brown below, new growth pubescent; bark gray-brown, shredding into long, flat strips

other: entire plant aromatic, highly variable in growth with several recognized varieties

HISTORIC, FOOD, AND MEDICINAL USES: wood once used for thatch and firewood, decoctions used by some Native Americans as a laxative, pollen causes hay fever

LIVESTOCK LOSSES: volatile oils may cause rumen stasis

FORAGE VALUE: good for sheep and wildlife on winter range, fair for cattle; high in protein but also high in volatile oils; food source and cover for sage grouse and many other kinds of wildlife

HABITAT: valleys, plains, basins, and mountain slopes; most abundant in dry, well-drained, gravelly or rocky soils

Annual broomweed
Amphiachyris dracunculoides (DC.) Nutt. *ex* Rydb.

SYN = *Gutierrezia dracunculoides* (DC.) Blake, *Xanthocephalum dracunculoides*
(DC.) Shinners

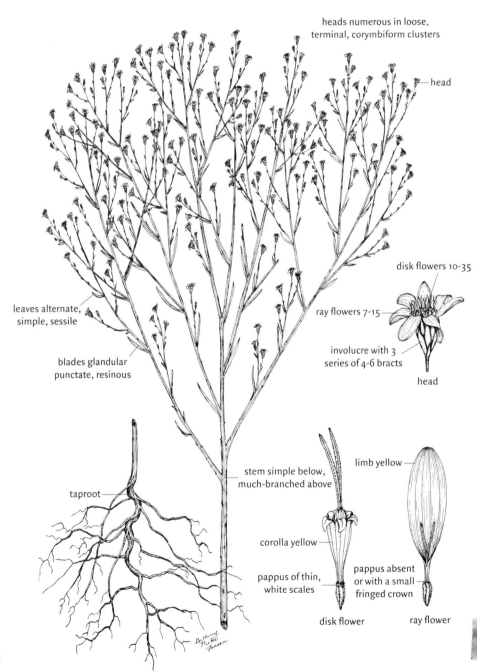

heads numerous in loose, terminal, corymbiform clusters

head

disk flowers 10-35

ray flowers 7-15

involucre with 3 series of 4-6 bracts

head

leaves alternate, simple, sessile

blades glandular punctate, resinous

taproot

stem simple below, much-branched above

limb yellow

corolla yellow

pappus of thin, white scales

pappus absent or with a small fringed crown

disk flower

ray flower

Family:	ASTERACEAE
Tribe:	ASTEREAE
Species:	*Amphiachyris dracunculoides* (DC.) Nutt. *ex* Rydb.
Common Name:	Annual broomweed (escobilla anual, common broomweed, prairie broomweed)
Life Span:	Annual
Origin:	Native
Season:	Warm

GROWTH FORM

forb (0.1–1.2 m tall); erect, single-stemmed at base with a taproot, freely branching above, appearing bushy; flowers August to October, reproduces from seeds

FLORAL AND FRUIT CHARACTERISTICS

inflorescences: heads in numerous clusters; corymbiform, loose, terminal

flowers: unisexual, heads radiate; involucre turbinate (3–5 mm long), with 3 series of 4–6 bracts; bracts imbricate, indurate, straw-colored, with green herbaceous tips, shiny; ray flowers 7–15, pistillate, limb yellow (4–6 mm long); disk flowers 10–35, staminate, corolla yellow

fruits: achene (1.5–2.2 mm long), dark, pubescent; ray flower achenes with pappus absent or with a small fringed crown; disk flower achenes with a pappus of 5 or more basally united scales; scales thin, white awn-like

VEGETATIVE CHARACTERISTICS

leaves: alternate, simple; blades lanceolate to linear or filiform (1–5 cm long, 0.6–3 mm wide); margins entire; glandular punctate, resinous, glabrous; sessile

stems: simple below, much-branched above, branches erect or ascending, slender, glabrous or slightly resinous

HISTORIC, FOOD, AND MEDICINAL USES: none

LIVESTOCK LOSSES: herbage and pollen can cause dermatitis in livestock (and humans) and a condition similar to pinkeye

FORAGE VALUE: worthless to cattle, will be consumed if no other forage is present; increases in abundance under heavy grazing pressure and during dry periods

HABITAT: dry upland prairies, limestone barrens, roadsides, railroad rights-of-way, and other disturbed sites

Hairy goldaster
Chrysopsis villosa (Pursh) Nutt.

SYN = *Heterotheca villosa* (Pursh) Shinners

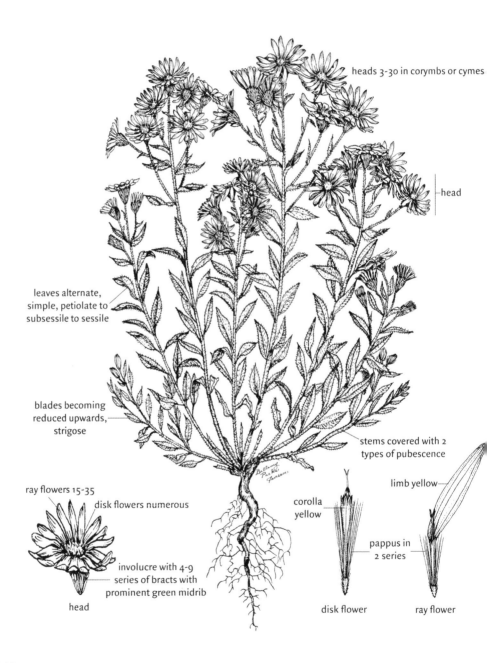

heads 3-30 in corymbs or cymes

head

leaves alternate, simple, petiolate to subsessile to sessile

blades becoming reduced upwards, strigose

stems covered with 2 types of pubescence

limb yellow

corolla yellow

pappus in 2 series

ray flowers 15-35

disk flowers numerous

involucre with 4-9 series of bracts with prominent green midrib

head

disk flower

ray flower

Family:	ASTERACEAE
Tribe:	ASTEREAE
Species:	*Chrysopsis villosa* (Pursh) Nutt.
Common Name:	Hairy goldaster (hierba velluda, telegraph plant, false goldenaster, golden aster)
Life Span:	Perennial
Origin:	Native
Season:	Warm

GROWTH FORM

forb (20–60 cm tall); stems arising singularly or in a small group from a woody caudex, erect or ascending; flowers July to September, reproduces from seeds and rarely from rhizomes

FLORAL AND FRUIT CHARACTERISTICS

inflorescences: heads 3–30 in corymbs or cymes

flowers: pistillate and perfect, heads radiate; involucre (7–12 mm long, 7–15 mm wide) with 4–9 series of bracts; bracts imbricate, narrow, linear, inner series long, prominent green midrib, pubescent; ray flowers 15–35, pistillate, yellow, revolute, limb yellow (8–14 mm long), apex 5-toothed; disk flowers (5–8 mm long) numerous, perfect, corolla yellow

fruits: achene, small, oblong, villous, somewhat flattened; pappus in 2 series, outer pappus of uneven scales, inner pappus of uneven bristles

VEGETATIVE CHARACTERISTICS

leaves: alternate, simple; blades (1–4 cm long, 3–10 mm wide) becoming reduced upwards, lower blades oblanceolate, and petiolate, mid-cauline blades oblanceolate and subsessile, upper blades linear to oblanceolate and sessile; margins entire; strigose, hairs appressed to ascending

stems: simple or branched above; covered with 2 types of pubescence; first pustulate hispid, elongate, and persistent; second divergent to appressed

other: a variable species with several varieties

HISTORIC, FOOD, AND MEDICINAL USES: some Native Americans consumed a decoction from the tops and stems as a soothing, quieting medicine to aid sleep

LIVESTOCK LOSSES: none

FORAGE VALUE: fair for sheep in poor semidesert areas of the West, otherwise considered worthless

HABITAT: prairies, plains, and semidesert areas; most abundant in dry open areas in sandy and rocky or calcareous soils

Rubber rabbitbrush
Chrysothamnus nauseosus (Pall.) Britt.

SYN = *Ericameria nauseosa* (Pall. *ex* Pursh) Nesom & Baird

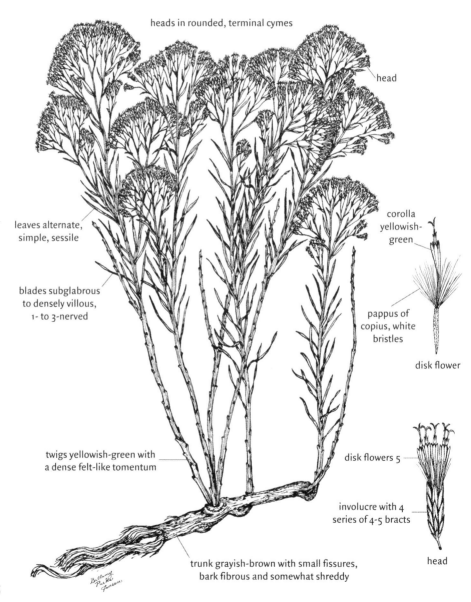

heads in rounded, terminal cymes

head

leaves alternate,
simple, sessile

blades subglabrous
to densely villous,
1- to 3-nerved

corolla
yellowish-
green

pappus of
copius, white
bristles

disk flower

twigs yellowish-green with
a dense felt-like tomentum

disk flowers 5

involucre with 4
series of 4-5 bracts

head

trunk grayish-brown with small fissures,
bark fibrous and somewhat shreddy

246

Family:	ASTERACEAE
Tribe:	ASTEREAE
Species:	*Chrysothamnus nauseosus* (Pall.) Britt.
Common Name:	Rubber rabbitbrush (chamiso blanco, gray rabbitbrush, golden rabbitbrush)
Life Span:	Perennial
Origin:	Native
Season:	Warm

GROWTH FORM

shrub (to 2 m tall); rounded crown from several erect stems from the base, much branched; flowers June to September, reproduces from seeds and basal sprouts

FLORAL AND FRUIT CHARACTERISTICS

inflorescence: heads in terminal cymes; collective inflorescences rounded

flowers: perfect, heads discoid; involucre (7–8 mm long, 2–3 mm wide) with 4 series of 4–5 bracts; bracts lanceolate to linear (2–8 mm long, 0.5–1 mm wide), imbricate, acute, glabrous, margins ciliate; disk flowers 5, corolla yellowish-green (7–9 mm long), glabrous

fruits: achene (4–5 mm long), linear, 5-angled, pubescent or less often glabrous; pappus of dull white bristles, copious

other: inflorescence and bracts often persisting well into the next year

VEGETATIVE CHARACTERISTICS

leaves: alternate, simple; blades linear to filiform (2–6 cm long, 0.5–3 mm wide), apex acute; margins entire; subglabrous to densely villous, 1- to 3-nerved, rarely 5-nerved; sessile

stems: twigs erect, flexible, yellowish-green with a dense felt-like tomentum; trunk and older twigs grayish-brown with small fissures, bark fibrous and somewhat shreddy

other: highly variable species with many subspecies, entire plant with a nauseous odor

HISTORIC, FOOD, AND MEDICINAL USES: some Native Americans made chewing gum from pulverized wood and bark, also used as tea, cough syrup, yellow dye, and for chest pains; the Hopi stripped bark from the branches and used the branches for basket weaving; small commercial source for rubber

LIVESTOCK LOSSES: occasionally reported to be toxic to livestock

FORAGE VALUE: worthless to poor for livestock, fair for deer on winter range; dense stands may indicate poor range management

HABITAT: dry plains, hillsides, roadsides, and ravine walls in sandy or clayey soils

Douglas rabbitbrush
Chrysothamnus viscidiflorus (Hook.) Nutt.

SYN = *C. lanceolata* Nutt.

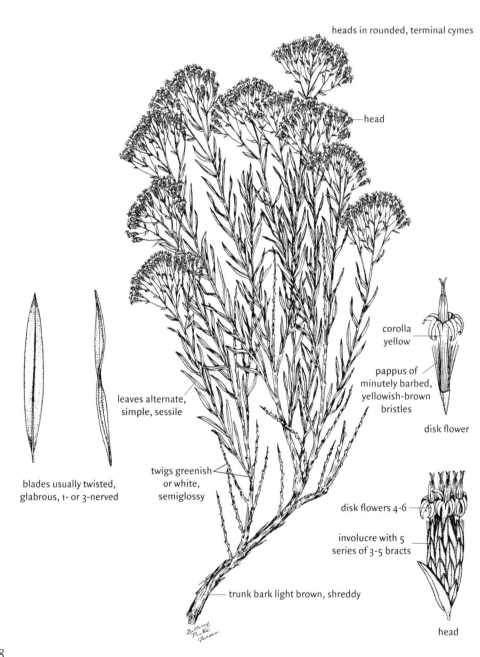

heads in rounded, terminal cymes

head

corolla yellow

pappus of minutely barbed, yellowish-brown bristles

disk flower

leaves alternate, simple, sessile

blades usually twisted, glabrous, 1- or 3-nerved

twigs greenish or white, semiglossy

disk flowers 4-6

involucre with 5 series of 3-5 bracts

trunk bark light brown, shreddy

head

Family:	ASTERACEAE
Tribe:	ASTEREAE
Species:	*Chrysothamnus viscidiflorus* (Hook.) Nutt.
Common Name:	Douglas rabbitbrush (chamisa, green rabbitbrush, yellow rabbitbrush)
Life Span:	Perennial
Origin:	Native
Season:	Warm

GROWTH FORM
shrub (to 1 m tall); erect, branching from near the base, crown rounded; flowers July to September, reproduces from seeds

FLORAL AND FRUIT CHARACTERISTICS
inflorescence: heads in terminal cymes (5–6 cm tall and wide), collective inflorescences moderately rounded to flat-topped

flowers: perfect, heads discoid; involucre (6–8 mm long) with 5 series of 3–5 bracts; bracts linear, obtuse or acute, glabrous or pubescent, yellowish; disk flowers 4–6, corolla yellow (4–7 mm long)

fruits: achene (3–4 mm long), wedge-shaped, flattened, light brown, ribs 5, densely villous; pappus of yellowish-brown bristles (5 mm long), minutely barbed

VEGETATIVE CHARACTERISTICS
leaves: alternate, simple; blades linear to lanceolate (1–5 cm long, 1–4 mm wide), usually twisted, apex acute or obtuse; margins entire, scabrous; glabrous, 1- or 3-nerved; sessile

stems: twigs mostly erect, stiff, brittle, striate, greenish or white, glabrous or with scattered pubescence, semiglossy; trunk bark light brown, shreddy

other: a highly variable species, leaves especially variable, with numerous varieties

HISTORIC, FOOD, AND MEDICINAL USES: roots were chewed as gum in the Southwest; contains rubber, especially when growing in alkali soils

LIVESTOCK LOSSES: none

FORAGE VALUE: poor, occasionally browsed by sheep and cattle when other feed is not available, browsed lightly by deer and pronghorn in the summer and winter, elk utilize it in winter; small mammals eat the heads

HABITAT: dry, open prairies; valleys and hillsides in sagebrush, ponderosa pine, lodgepole pine, or aspen belts

Curlycup gumweed
Grindelia squarrosa (Pursh) Dun.

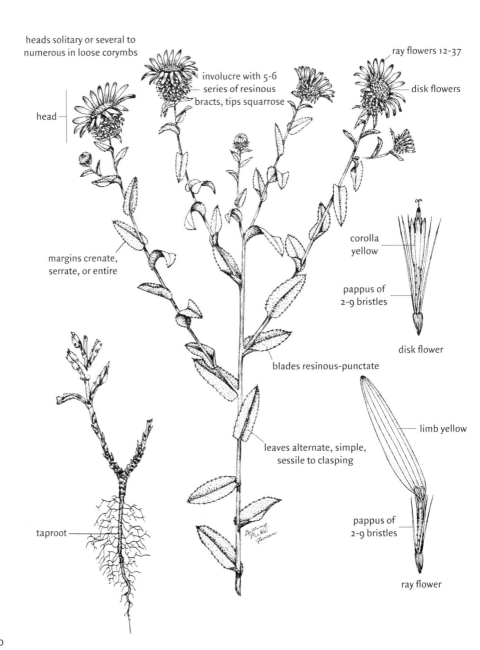

heads solitary or several to
numerous in loose corymbs

involucre with 5-6
series of resinous
bracts, tips squarrose

ray flowers 12-37

disk flowers

head

corolla
yellow

pappus of
2-9 bristles

disk flower

margins crenate,
serrate, or entire

blades resinous-punctate

limb yellow

leaves alternate, simple,
sessile to clasping

pappus of
2-9 bristles

taproot

ray flower

Family:	ASTERACEAE
Tribe:	ASTEREAE
Species:	*Grindelia squarrosa* (Pursh) Dun.
Common Name:	Curlycup gumweed (gumweed, curlytop gumweed)
Life Span:	Perennial (Biennial)
Origin:	Native
Season:	Warm

GROWTH FORM

forb (0.2–1 m tall); erect, 1 to several stems from a taproot, branching above; starts growth in early spring, flowers July to August, reproduces from seeds

FLORAL AND FRUIT CHARACTERISTICS

inflorescence: heads solitary or several to numerous in loose corymbs; heads (0.5–2.5 cm tall, 0.7–3 cm wide) resinous-sticky, darkened with drying

flowers: perfect or some unisexual, heads radiate or discoid; involucre (7–9 mm tall) with 5–6 series of bracts; bracts imbricate, tips squarrose and resinous (especially the lower series); ray flowers 12–37, pistillate, limb yellow (7–15 mm long); disk flowers perfect or staminate, numerous, corolla tubular, yellow

fruits: achene (2.3–3 mm long), oblong, 4-angled, glabrous; pappus of 2–9 bristles, shorter than the disk florets

VEGETATIVE CHARACTERISTICS

leaves: alternate, simple; blades ovate to oblong to oblanceolate (1.5–7 cm long, 4–15 mm wide), thick; apex obtuse to acute; margins crenate, serrate, or entire; resinous-punctate; sessile to clasping

stems: erect, 1 to several

other: aromatic, a species with several varieties due to variable flowers, leaf serration, and leaf size

HISTORIC, FOOD, AND MEDICINAL USES: some Native Americans used the resinous secretions to relieve asthma, bronchitis, and colic; Pawnee Indians boiled leaves and flowering tops to treat saddle sores and raw skin; flower extract is used in modern medicine to treat whooping cough and asthma

LIVESTOCK LOSSES: may accumulate selenium, but the resinous covering usually discourages consumption

FORAGE VALUE: worthless to livestock and most wildlife, although sheep occasionally eat the heads

HABITAT: waste places, overgrazed prairies, and alluvial grounds; increases under drought conditions

Broom snakeweed
Gutierrezia sarothrae (Pursh) Britt. & Rusby

SYN = *Xanthocephalum sarothrae* (Pursh) Shinners

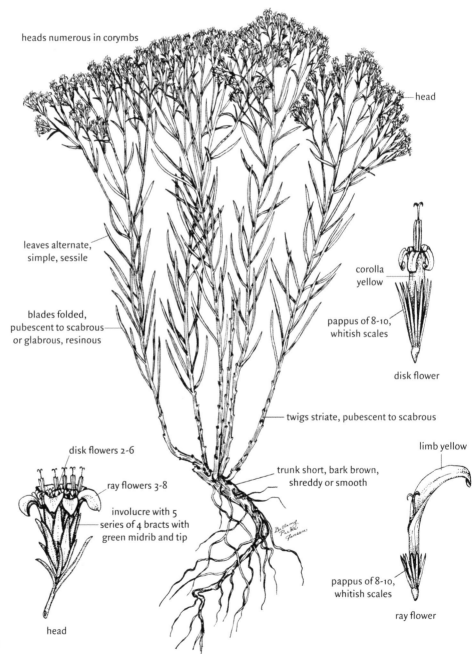

heads numerous in corymbs

head

leaves alternate,
simple, sessile

blades folded,
pubescent to scabrous
or glabrous, resinous

corolla
yellow

pappus of 8-10,
whitish scales

disk flower

twigs striate, pubescent to scabrous

disk flowers 2-6

ray flowers 3-8

involucre with 5
series of 4 bracts with
green midrib and tip

trunk short, bark brown,
shreddy or smooth

limb yellow

pappus of 8-10,
whitish scales

ray flower

head

Family:	ASTERACEAE
Tribe:	ASTEREAE
Species:	*Gutierrezia sarothrae* (Pursh) Britt. & Rusby
Common Name:	Broom snakeweed (escobilla común, yerba de víbora, perennial broomweed, turpentine weed, hierba resinosa)
Life Span:	Perennial
Origin:	Native
Season:	Warm

GROWTH FORM
subshrub (to 60 cm tall); stems bushy, highly branched; flowers July to October, reproduces from seeds

FLORAL AND FRUIT CHARACTERISTICS
inflorescences: heads numerous in corymbs, rounded, loose or dense

flowers: pistillate and perfect, heads inconspicuously radiate; involucre (3–6 mm tall, 2 mm wide) with 5 series of 4 bracts; bracts linear (1.2–3.5 mm long), acute, imbricate, with green midrib and tip; ray flowers 3–8, pistillate, limb yellow (1–3 mm long); disk flowers 2–6, perfect, corolla yellow

fruits: achene (1.7–2 mm long, 0.5 mm wide), cylindric, brown, densely pubescent; pappus of 8–10 scales (0.5–1 mm long), whitish, acute

VEGETATIVE CHARACTERISTICS
leaves: alternate, simple; blades linear to filiform (5–70 mm long, 1–3 mm wide), numerous, folded; margins entire; pubescent to scabrous or glabrous, glandular-resinous; sessile

stems: twigs erect, thin, flexible, green to brown, striate, pubescent to scabrous; trunk short, bark brown, shreddy or smooth

HISTORIC, FOOD, AND MEDICINAL USES: used by southwestern Native Americans and Mexicans as a broom; decoctions were used for indigestion; pieces of the plant were chewed and placed on bee and wasp stings and rattlesnake bites

LIVESTOCK LOSSES: poisonous to sheep and cattle, causing death or abortion; poisonous principle is a saponin that is most toxic during leaf growth; will also accumulate selenium

FORAGE VALUE: fair for sheep and poor for cattle and horses on winter range; otherwise worthless, indicator of poor management

HABITAT: open plains, upland sites, and dry hillsides; adapted to a broad range of soils

Prairie goldenrod
Solidago missouriensis Nutt.

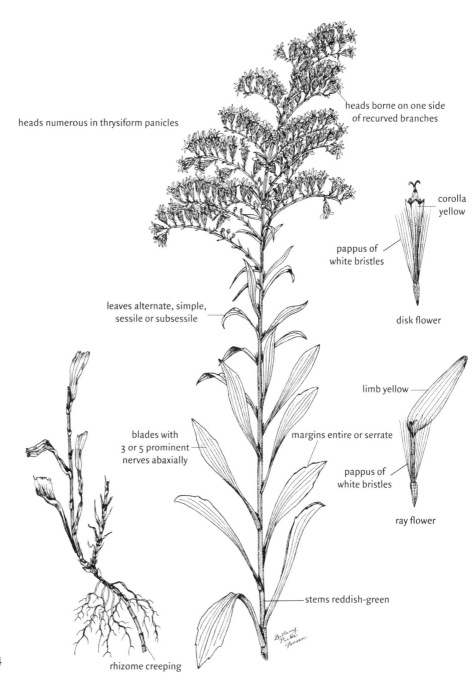

heads numerous in thrysiform panicles

heads borne on one side of recurved branches

corolla yellow

pappus of white bristles

disk flower

leaves alternate, simple, sessile or subsessile

limb yellow

blades with 3 or 5 prominent nerves abaxially

margins entire or serrate

pappus of white bristles

ray flower

stems reddish-green

rhizome creeping

Family:	ASTERACEAE
Tribe:	ASTEREAE
Species:	*Solidago missouriensis* Nutt.
Common Name:	Prairie goldenrod (Missouri goldenrod)
Life Span:	Perennial
Origin:	Native
Season:	Warm

GROWTH FORM
forb (0.2–1 m tall); erect or ascending, arising singularly or as a group from a creeping rhizome or woody caudex; flowers June to October, reproduces from seeds and rhizomes

FLORAL AND FRUIT CHARACTERISTICS
inflorescences: heads numerous in thrysiform panicles; usually broader than tall, heads ascending, borne on one side of the recurved branches

flowers: pistillate and perfect, heads radiate; involucre (3–5 mm tall) with 3 series of 4–6 bracts; bracts lanceolate, obtuse to acute; ray flowers 7–13, pistillate, limb yellow; disk flowers 8–13, perfect, shorter than ray flowers, corolla yellow

fruits: achene (1–2.2 mm long), cylindrical, glabrous to pubescent; pappus of numerous white bristles (2.5–3 mm long)

VEGETATIVE CHARACTERISTICS
leaves: alternate, simple; blades thick, firm; lowermost blades largest, blades reduced upward; blades oblanceolate to elliptic to linear above (4–6 cm long, 0.5–3 cm wide); apex acute, base long-tapered; margins entire or serrate, glabrous, with 3 or 5 prominent nerves abaxially; sessile or subsessile

stems: simple or rarely branched, stiff, glabrous, reddish-green

HISTORIC, FOOD, AND MEDICINAL USES: some Native Americans chewed leaves and flowers to relieve sore throats, roots were chewed to relieve toothache; pollen highly desirable to several bee species

LIVESTOCK LOSSES: may be toxic to sheep

FORAGE VALUE: poor, but will be grazed by cattle and sheep in spring and early summer; deer and pronghorn will graze the lower leaves; generally becomes more abundant in poorly managed areas

HABITAT: upland prairies, plains, meadows, open woods, and roadsides

False dandelion
Agoseris glauca (Pursh) Raf.

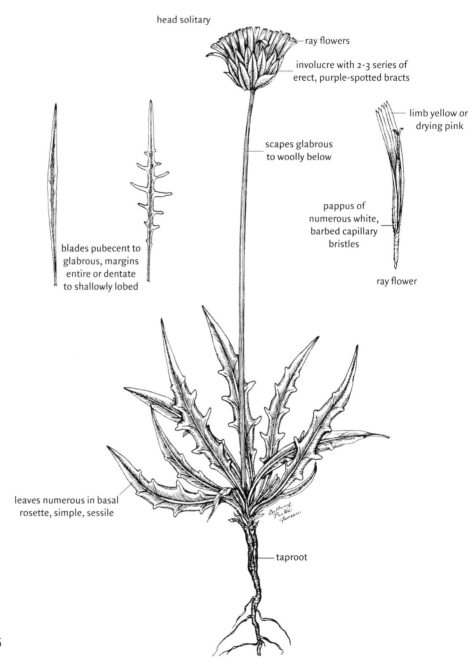

head solitary

ray flowers

involucre with 2-3 series of
erect, purple-spotted bracts

limb yellow or
drying pink

scapes glabrous
to woolly below

pappus of
numerous white,
barbed capillary
bristles

blades pubecent to
glabrous, margins
entire or dentate
to shallowly lobed

ray flower

leaves numerous in basal
rosette, simple, sessile

taproot

Family:	ASTERACEAE
Tribe:	CICHORIEAE
Species:	*Agoseris glauca* (Pursh) Raf.
Common Name:	False dandelion (falso diente de león, mountain dandelion, pale agoseris)
Life Span:	Perennial
Origin:	Native
Season:	Cool

GROWTH FORM

forb (5–60 cm tall); scapose, from a taproot; flowers May to September, reproduces from seeds

FLORAL AND FRUIT CHARACTERISTICS

inflorescences: heads solitary (1–6 cm wide), terminal, erect, stout; scape slender (to 60 cm tall), glabrous

flowers: perfect, heads ligulate; involucre oblong-cylindrical (1–2.5 cm tall), with 2–3 series of bracts; bracts lanceolate, erect, tapering from base, imbricate, acute, frequently purple-spotted, glabrous, outer series of bracts the shortest and broadest; ray flower limbs yellow or drying pink (1.2–1.5 cm long), exceeding bracts; receptacle flat, naked, apex toothed

fruits: achene (5–10 mm long), stout, 10-ribbed, slightly hirsute, not flattened, tapering to a beak; pappus of numerous white, barbed capillary bristles (to 2 cm long)

VEGETATIVE CHARACTERISTICS

leaves: numerous in basal rosettes, simple; blades (5–30 cm long, 2–30 mm wide) highly variable in outline, linear to lanceolate or oblanceolate; margins entire or dentate to shallowly lobed; pubescent to glabrous, glaucous, distinct white midvein; sessile

stems: scape glabrous to woolly below, somewhat glaucous

other: entire plant contains a milky juice; variable species with several varieties

HISTORIC, FOOD, AND MEDICINAL USES: sap was chewed by some Native Americans to clean teeth

LIVESTOCK LOSSES: none

FORAGE VALUE: fair to good for sheep, pronghorn, and deer; lightly grazed by cattle and horses; tends to decrease under grazing by sheep and increase under grazing by cattle; abundance usually indicates rangeland deterioration

HABITAT: moist meadows and prairie swales, common on disturbed or eroded sites, adapted to a wide range of soils

Tapertip hawksbeard
Crepis acuminata Nutt.

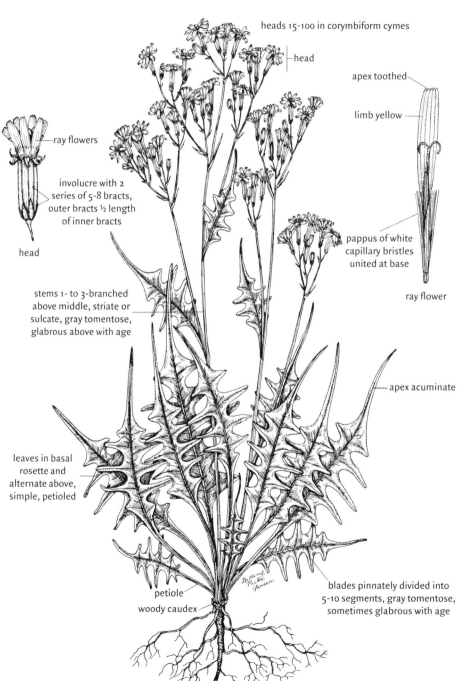

heads 15-100 in corymbiform cymes

head

apex toothed

limb yellow

ray flowers

involucre with 2 series of 5-8 bracts, outer bracts ½ length of inner bracts

head

pappus of white capillary bristles united at base

ray flower

stems 1- to 3-branched above middle, striate or sulcate, gray tomentose, glabrous above with age

apex acuminate

leaves in basal rosette and alternate above, simple, petioled

petiole

woody caudex

blades pinnately divided into 5-10 segments, gray tomentose, sometimes glabrous with age

Family:	ASTERACEAE
Tribe:	CICHORIEAE
Species:	*Crepis acuminata* Nutt.
Common Name:	Tapertip hawksbeard (longleaf hawksbeard)
Life Span:	Perennial
Origin:	Native
Season:	Cool

GROWTH FORM

forb (20–70 cm tall); arising as 1–3 stems from a woody caudex, branched above; flowers May to August, reproduces from seeds

FLORAL AND FRUIT CHARACTERISTICS

inflorescences: heads 15–100 in a corymbiform cyme; heads small or short, slender; peduncles short

flowers: perfect, heads of ray flowers; involucre cylindrical (9–15 mm long, 2.5–4 mm wide), with 2 series of 5–8 bracts; outer bracts one-half the length of inner bracts, lanceolate to deltoid, glabrous or tomentose; inner bracts lanceolate, glabrous or tomentose; ray flower limbs yellow (9–20 mm long), apex toothed

fruits: achene (5.5–9 mm long), linear, yellowish-brown, 12-ribbed; pappus of white capillary bristles, bristles united at base

VEGETATIVE CHARACTERISTICS

leaves: basal rosette and alternate above, simple; basal and lower blades lanceolate (12–40 cm long, 0.5–11 cm wide) with winged-petiole, pinnately divided into 5–10 segments; segments entire to dentate, apex acuminate; middle cauline leaves few and similar to basal leaves; upper cauline leaves reduced to bracts; blade tapers to an indistinct petiole; gray tomentose, sometimes glabrous with age

stems: erect, stout, 1- to 3-branched above the middle, striate or sulcate; gray tomentose especially when young, glabrous above with age but remaining tomentose below

other: entire plant contains a milky juice

HISTORIC, FOOD, AND MEDICINAL USES: none

LIVESTOCK LOSSES: none

FORAGE VALUE: good for deer, pronghorn, cattle, sheep, and horses; preferred forage of sheep; most palatable in late spring and early summer

HABITAT: prairie hillsides and broken slopes; most abundant in open areas with dry, well-drained or shallow soils

Dandelion
Taraxacum officinale Weber

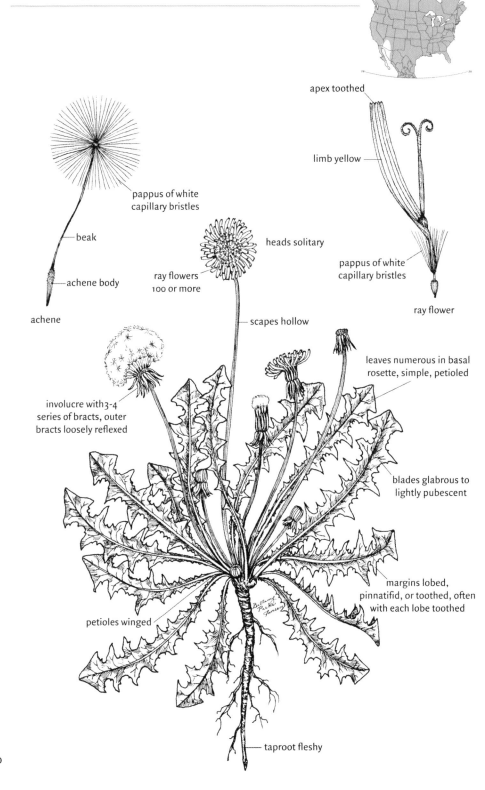

pappus of white capillary bristles

beak

achene body

achene

ray flowers 100 or more

heads solitary

scapes hollow

apex toothed

limb yellow

pappus of white capillary bristles

ray flower

involucre with 3-4 series of bracts, outer bracts loosely reflexed

leaves numerous in basal rosette, simple, petioled

blades glabrous to lightly pubescent

margins lobed, pinnatifid, or toothed, often with each lobe toothed

petioles winged

taproot fleshy

Family:	ASTERACEAE
Tribe:	CICHORIEAE
Species:	*Taraxacum officinale* Weber
Common Name:	Dandelion (diente de león, common dandelion)
Life Span:	Perennial
Origin:	Introduced (from Eurasia)
Season:	Cool

GROWTH FORM

forb (3–60 cm tall); scapose, from a fleshy taproot; flowers April to October, reproduces from seeds and sprouts

FLORAL AND FRUIT CHARACTERISTICS

inflorescences: heads solitary (1–5 cm wide); terminating an erect, scape slender (to 60 cm tall)

flowers: perfect, heads ligulate; involucre obconic (1–2.5 cm long) with 3–4 series of bracts; outer bracts shorter, linear, graduated in length, loosely reflexed; inner bracts in 2 series, linear, erect, spreading with age, imbricate; ray flowers 100 or more, limb yellow, apex toothed

fruits: achene (3–4 mm long), flattened, brown, 5-ribbed, spinose, tipped with a long filiform beak that is longer than achene body; pappus of white capillary bristles (4–5 mm long); bristles abundant

other: heads form at ground level, then scape elongates in about 48 hours

VEGETATIVE CHARACTERISTICS

leaves: numerous in basal rosettes, simple; blades oblanceolate (5–30 cm long, 2–15 cm wide); margins variously lobed, pinnatifid, or toothed, often with each lobe toothed; terminal lobe usually longest; glabrous to lightly pubescent; petiole winged, may be indistinct

stems: scape (to 60 cm tall) glabrous to puberulent, hollow

other: entire plant contains a milky juice

HISTORIC, FOOD, AND MEDICINAL USES: young leaves can be eaten as spring greens; roots can be ground and used as a coffee substitute, mild laxative, or to treat heartburn; good honey plant; tea and wine can be made from the flowers, flowers can be fried in batter and eaten

LIVESTOCK LOSSES: none

FORAGE VALUE: fair to good for livestock and wildlife, readily eaten since it is relatively succulent; generally abundant on poorly managed rangeland, but can also occur on well managed rangeland

HABITAT: rangeland, weedy meadows, open stream banks, disturbed sites, and lawns; common on a wide variety of soils

Dotted gayfeather
Liatris punctata Hook.

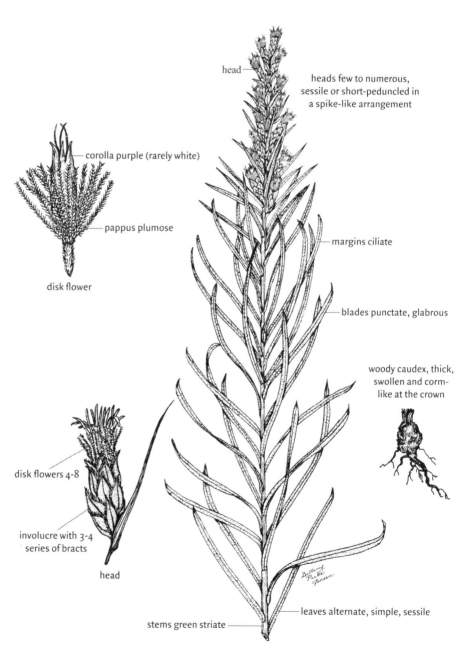

head

heads few to numerous,
sessile or short-peduncled in
a spike-like arrangement

corolla purple (rarely white)

pappus plumose

disk flower

margins ciliate

blades punctate, glabrous

woody caudex, thick,
swollen and corm-
like at the crown

disk flowers 4-8

involucre with 3-4
series of bracts

head

leaves alternate, simple, sessile

stems green striate

Family:	ASTERACEAE
Tribe:	EUPATORIEAE
Species:	*Liatris punctata* Hook.
Common Name:	Dotted gayfeather (blazing star)
Life Span:	Perennial
Origin:	Native
Season:	Warm

GROWTH FORM

forb (10–80 cm tall); arising singularly or clustered from a thick, woody caudex; swollen and corm-like at the crown; flowers July to October, reproduces from seeds and corm-like rootstocks

FLORAL AND FRUIT CHARACTERISTICS

inflorescences: heads few to numerous, sessile or short-peduncled in a spike-like arrangement (6–30 cm long)

flowers: perfect; heads discoid; involucre cylindrical (1.5–2 cm tall, 8–10 mm wide), loose with 3–4 series of bracts; bracts thick, punctate, appressed, apex acute, margins ciliate; disk flowers 4–8; corolla purple (rarely white), pilose inside; stamens and styles short, exserted in late flowering

fruits: achene (6–7.5 mm long), 10-ribbed, pubescent; pappus plumose (9–11 mm long), exceeding corolla, numerous

VEGETATIVE CHARACTERISTICS

leaves: alternate, simple; blades linear (8–15 cm long, 1.5–6 mm wide), reduced above, numerous, imbricate, rigidly ascending or arching; margins thickened, whitish, ciliate; punctate, glabrous; sessile

stems: erect, simple or rarely branched, green striate, glabrous

HISTORIC, FOOD, AND MEDICINAL USES: corm-like rootstock reportedly used by some Native Americans for food; plants of this genus were consumed in New England during the nineteenth century for treatment of gonorrhea

LIVESTOCK LOSSES: none

FORAGE VALUE: fair for cattle; good for sheep, deer, and pronghorn; most valuable when plants are young; disappears with continuous overuse

HABITAT: dry plains, hills, and uplands; most abundant on sandy soils

Triangleleaf bursage
Ambrosia deltoidea (Torr.) Payne

SYN = *Franseria deltoidea* Torr.

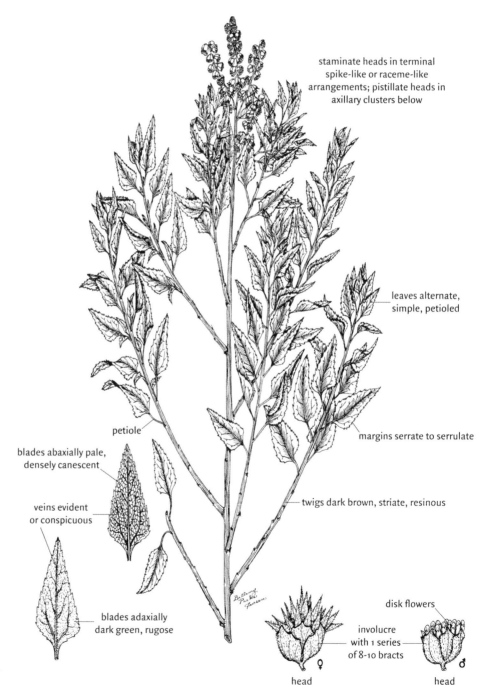

staminate heads in terminal spike-like or raceme-like arrangements; pistillate heads in axillary clusters below

leaves alternate, simple, petioled

margins serrate to serrulate

petiole

blades abaxially pale, densely canescent

veins evident or conspicuous

twigs dark brown, striate, resinous

blades adaxially dark green, rugose

disk flowers

involucre with 1 series of 8–10 bracts

head ♀

head ♂

264

Family:	ASTERACEAE
Tribe:	HELIANTHEAE
Species:	*Ambrosia deltoidea* (Torr.) Payne
Common Name:	Triangleleaf bursage (triangle burroweed, triangle bursage)
Life Span:	Perennial
Origin:	Native
Season:	Cool

GROWTH FORM
monoecious shrub (to 1 m tall); crown rounded to flat-topped, branches erect; flowers March to April, reproduces from seeds

FLORAL AND FRUIT CHARACTERISTICS
inflorescences: staminate heads (6–7 mm in diameter) in spike-like or raceme-like arrangements, terminal; pistillate heads (5–7 mm long and wide) in axillary clusters below

flowers: unisexual, heads discoid; involucre saucer-shaped (3–5 mm wide) with 1 series of 8–10 triangular bracts; bracts tomentose when young, then glabrous; disk flower corolla yellow (2–2.5 mm long), puberulent; staminate heads on peduncles (0.5–3 mm long); pistillate heads sessile

fruits: achene (5–7 mm long), globose or ellipsoidal, glandular, tomentose, with 2–3 flattened beaks; bearing 15–30 flattened spines in 2–3 series; spines puberulent, straight, not hooked

VEGETATIVE CHARACTERISTICS
leaves: alternate, simple; blades narrowly deltoid to ovate (2–3 cm long, 1–1.5 cm wide); apex acute or attenuate; base truncate; margins serrate to serrulate; adaxially dark green and rugose, abaxially pale and densely canescent, veins evident or conspicuous, thick; petioles (up to 1.5 cm long), slightly winged above

stems: twigs erect, dark brown, striate, resinous, rigid, glabrous

HISTORIC, FOOD, AND MEDICINAL USES: none

LIVESTOCK LOSSES: fruit may contaminate fleece and reduce its value, considered to be one of the worst hay fever plants of the Southwest

FORAGE VALUE: worthless to livestock and wildlife

HABITAT: alluvial plains, hillsides, mesas, plains, and gullies; most abundant in rocky or gravelly soils; grows in almost pure stands

White bursage
Ambrosia dumosa (A. Gray) Payne

SYN = *Franseria dumosa* A. Gray

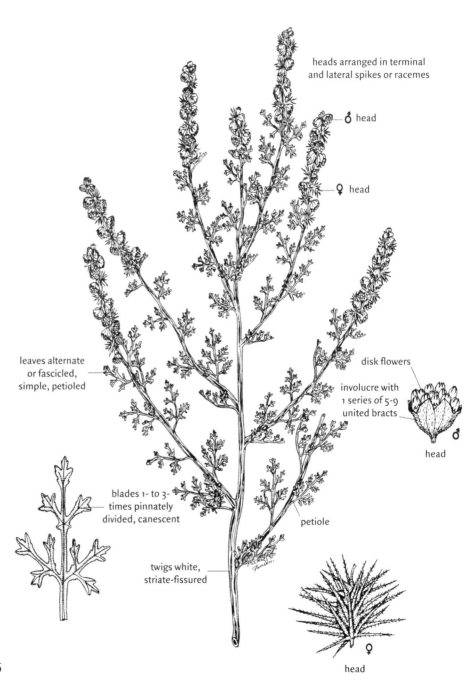

heads arranged in terminal
and lateral spikes or racemes

♂ head

♀ head

leaves alternate
or fascicled,
simple, petioled

disk flowers

involucre with
1 series of 5-9
united bracts

♂
head

blades 1- to 3-
times pinnately
divided, canescent

petiole

twigs white,
striate-fissured

♀
head

Family:	ASTERACEAE
Tribe:	HELIANTHEAE
Species:	*Ambrosia dumosa* (A. Gray) Payne
Common Name:	White bursage (hierba del burro, burroweed, burrobush)
Life Span:	Perennial
Origin:	Native
Season:	Warm

GROWTH FORM

monoecious shrub (to 1 m tall), bushy; crown rounded, compact, much-branched, stiff spinescent; flowers April to November, reproduces from seeds

FLORAL AND FRUIT CHARACTERISTICS

inflorescences: heads arranged in terminal and lateral spikes or racemes; staminate and pistillate heads intermingled within the inflorescence

flowers: unisexual, heads discoid; staminate involucre saucer-shaped (2–2.5 mm long, 4–5 mm wide), with 1 series of 5–9 united bracts; bracts ovate to triangular, canescent, acute; disk flower corolla funnelform, yellow, puberulent; staminate heads numerous (3–5 mm wide), on peduncles (0.2–3 mm long); pistillate heads 1- or 2-flowered, sessile

fruits: achene (4–9 mm long), subglobose, glandular-puberulent, 2-beaked, tips with 15–40 straight spines in 2–3 series (1.5–4 mm long); spines flattened at base, not hooked, striate above

VEGETATIVE CHARACTERISTICS

leaves: alternate or fascicled, simple; blades (1.5–2 cm long, 8–11 mm wide) 1- to 3-times pinnately divided, divisions ovate or obovate (0.5–3 mm long, 0.5–2 mm wide); gray-green canescent; petioled (2–20 mm long)

stems: twigs erect to ascending, rigid, white, somewhat striate-fissured, lightly canescent; inflorescence branches from the previous year forming spines during the current year; trunk tan to light brown, striate

HISTORIC, FOOD, AND MEDICINAL USES: none

LIVESTOCK LOSSES: may accumulate nitrates

FORAGE VALUE: fair for cattle and horses, fair to good for goats, preferred by horses and donkeys

HABITAT: dry plains, mesas, alluvial slopes, and gullies, may become locally abundant

Western ragweed
Ambrosia psilostachya DC.

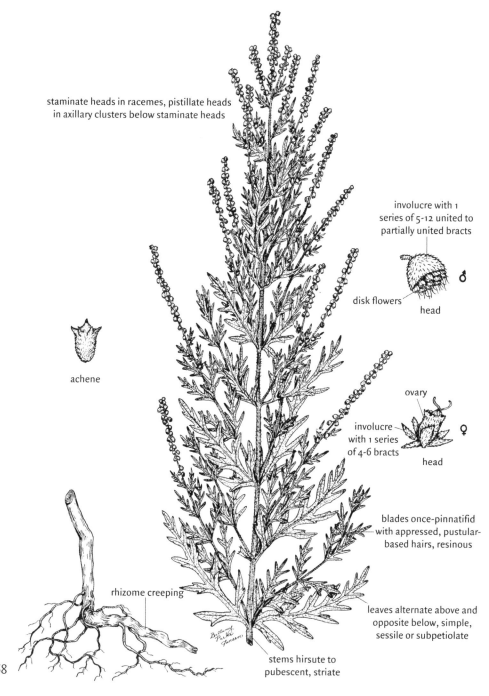

staminate heads in racemes, pistillate heads
in axillary clusters below staminate heads

involucre with 1
series of 5-12 united to
partially united bracts

disk flowers

head ♂

achene

ovary

involucre
with 1 series
of 4-6 bracts

head ♀

blades once-pinnatifid
with appressed, pustular-
based hairs, resinous

rhizome creeping

leaves alternate above and
opposite below, simple,
sessile or subpetiolate

stems hirsute to
pubescent, striate

Family:	ASTERACEAE
Tribe:	HELIANTHEAE
Species:	*Ambrosia psilostachya* DC.
Common Name:	Western ragweed (amargosa, cuman ragweed)
Life Span:	Perennial
Origin:	Native
Season:	Warm

GROWTH FORM

monoecious forb (0.3–1 m tall); erect, branching above; flowers July to October, reproduces from roots and seeds; forms extensive colonies from creeping rhizomes

FLORAL AND FRUIT CHARACTERISTICS

inflorescences: staminate heads in racemes, nodding; pistillate heads in axillary clusters below staminate heads

flowers: unisexual, heads discoid; staminate involucre oblique (2–3 mm tall, 1–5 mm wide), with 1 series of 5–12 united to partially united bracts, hirsute; staminate disk flowers several, corolla absent, anthers yellow; pistillate involucre hirsute (2.5 mm long), with 1 series of 4–6 bracts; bracts short with blunt spines; flowers 1–2, corolla reduced

fruits: achene (3 mm long), enclosed in a hard spiny "bur" formed by the indurate involucre; pappus absent

VEGETATIVE CHARACTERISTICS

leaves: alternate above and opposite below, simple; blades lanceolate to oblong (2–15 cm long, 1–8 cm wide), usually once-pinnatifid, divisions linear to lanceolate; apex acute; margins entire to toothed; covered with appressed, pustular-based hairs, resinous; sessile or subpetiolate with a winged petiole

stems: erect, simple below, branched above, hirsute to pubescent with ascending hairs, striate

HISTORIC, FOOD, AND MEDICINAL USES: leaves were steeped by some Native Americans and used as a treatment for sore eyes; pollen is one of the most important causes of hay fever

LIVESTOCK LOSSES: may accumulate nitrates, milk produced from cows grazing this forb has a bitter taste, resin may induce dermatitis

FORAGE VALUE: worthless for livestock and wildlife, generally unpalatable, but cattle may graze it in early spring and late summer

HABITAT: prairies, barrens, hills, pastures, waste places, and roadsides

Desert marigold
Baileya multiradiata Harv. & A. Gray

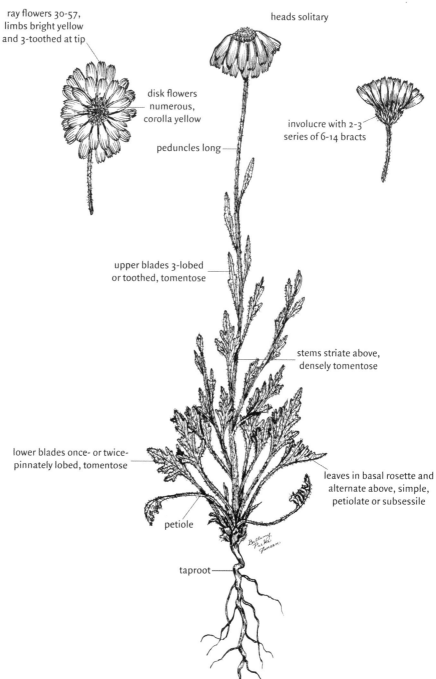

ray flowers 30-57,
limbs bright yellow
and 3-toothed at tip

heads solitary

disk flowers
numerous,
corolla yellow

involucre with 2-3
series of 6-14 bracts

peduncles long

upper blades 3-lobed
or toothed, tomentose

stems striate above,
densely tomentose

lower blades once- or twice-
pinnately lobed, tomentose

leaves in basal rosette and
alternate above, simple,
petiolate or subsessile

petiole

taproot

Family:	ASTERACEAE
Tribe:	HELIANTHEAE
Species:	*Baileya multiradiata* Harv. & A. Gray
Common Name:	Desert marigold (paperdaisy, desert baileya)
Life Span:	Perennial (or biennial)
Origin:	Native
Season:	Warm

GROWTH FORM

forb (10–50 cm tall); erect or ascending, 1 to several stems from a woody branched caudex surmounting a taproot, sparingly branched; short-lived; flowers April to October, reproduces from seeds

FLORAL AND FRUIT CHARACTERISTICS

inflorescences: head solitary (3.5–5 cm wide); terminating long peduncles (10–30 cm long)

flowers: pistillate and perfect, heads radiate; involucre campanulate (6–8 mm long, 1.5–2 cm wide) with 2–3 series of 6–14 bracts; bracts subequal, elliptic to lanceolate, acute; ray flowers 30–57, showy, pistillate, limb linear (1–2 cm long), bright yellow, 3-toothed at tip; disk flowers numerous, perfect, corolla tubular, yellow, 5-toothed, pubescent

fruits: achene (2.5–3.5 mm long), clavate, greenish-white, truncate, with 15–20 ribs, glabrous to resinous; pappus absent

VEGETATIVE CHARACTERISTICS

leaves: basal rosette and alternate above, simple; basal and lower blades ovate (2–15 cm long, 1–5 cm wide), lower blades longest, once- or twice-pinnately lobed; divisions ovate or linear, entire; upper blades linear, 3-lobed or toothed; tomentose; petiolate (2–8 cm long) or subsessile

stems: erect, much-branched after second year, branched only near the base, somewhat striate above, first year a rosette, densely tomentose

HISTORIC, FOOD, AND MEDICINAL USES: cultivated as an ornamental

LIVESTOCK LOSSES: sheep and goats may be killed after consuming large quantities over a relatively long period of time

FORAGE VALUE: poor to worthless for livestock and wildlife, generally will not be grazed if other forage plants are present

HABITAT: dry plains, brushy hills, floodplains, roadsides, and disturbed sites; most abundant in sandy and gravelly soils

Arrowleaf balsamroot
Balsamorhiza sagittata (Pursh) Nutt.

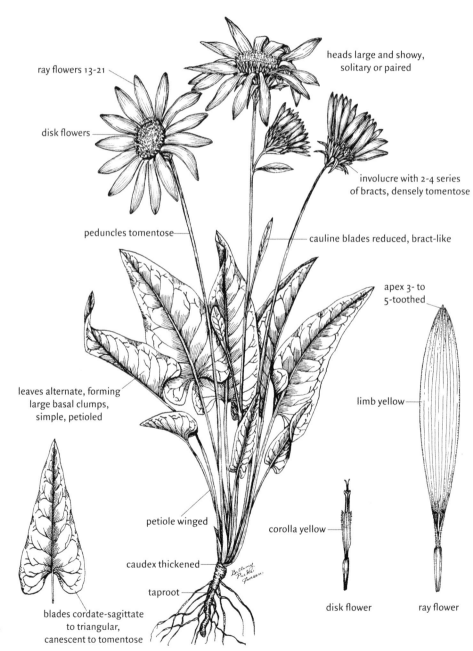

ray flowers 13-21

disk flowers

heads large and showy,
solitary or paired

involucre with 2-4 series
of bracts, densely tomentose

peduncles tomentose

cauline blades reduced, bract-like

apex 3- to
5-toothed

leaves alternate, forming
large basal clumps,
simple, petioled

limb yellow

petiole winged

corolla yellow

caudex thickened

taproot

blades cordate-sagittate
to triangular,
canescent to tomentose

disk flower

ray flower

272

Family:	ASTERACEAE
Tribe:	HELIANTHEAE
Species:	*Balsamorhiza sagittata* (Pursh) Nutt.
Common Name:	Arrowleaf balsamroot (gray dock, breadroot)
Life Span:	Perennial
Origin:	Native
Season:	Cool

GROWTH FORM

forb (20–80 cm tall), scapiform, from a branched thickened caudex and taproot; caudex often with a shreddy appearance due to the many old, persisting leaf petioles; flowers May to August, reproduces from seeds

FLORAL AND FRUIT CHARACTERISTICS

inflorescences: heads large and showy, solitary or paired; long, tomentose peduncle

flowers: heads radiate; involucre (1.5–3 cm wide) with 2–4 series of bracts; bracts subequal, ovate to lanceolate, acuminate, densely tomentose, outer bracts congested (2.5 cm long); ray flowers 13–21, pistillate, limb yellow (2–4 cm long), apex 3- to 5-toothed; disk flowers perfect, corolla tubular, yellow

fruits: achene (7–8 mm long), oblong, glabrous; pappus absent

VEGETATIVE CHARACTERISTICS

leaves: alternate, simple, forming large basal clumps; basal blades cordate-sagittate to triangular (15–50 cm long, 5–15 cm wide), margins entire, silvery canescent to tomentose; cauline blades few, reduced, narrow, bract-like; basal petioles (20–55 cm long), canescent to tomentose

stems: scapiform, but usually with several reduced leaves; tomentose, arising from a large basal cluster of leaves

HISTORIC, FOOD, AND MEDICINAL USES: the Cheyenne boiled roots, stems, and leaves and drank the decoction for stomach pains and headaches; they also steamed the plant and inhaled the vapors for the same purposes; ripe seeds were pounded into flour; roots were commonly eaten raw or boiled

LIVESTOCK LOSSES: none

FORAGE VALUE: good for sheep and big game and fair for cattle when green, flowers are most palatable

HABITAT: plains, valleys, hillsides, and open woods; most abundant in deep, well-drained soils at moderate to high elevations

Tarbush
Flourensia cernua DC.

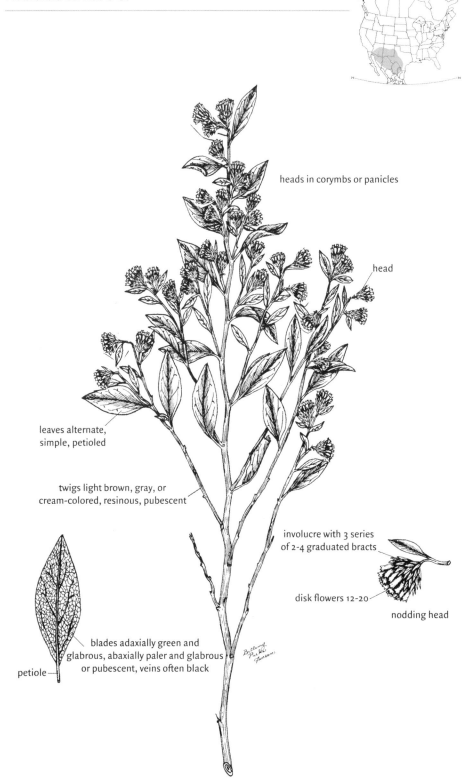

heads in corymbs or panicles

head

leaves alternate,
simple, petioled

twigs light brown, gray, or
cream-colored, resinous, pubescent

involucre with 3 series
of 2-4 graduated bracts

disk flowers 12-20

nodding head

blades adaxially green and
glabrous, abaxially paler and glabrous
or pubescent, veins often black

petiole

Family:	ASTERACEAE
Tribe:	HELIANTHEAE
Species:	*Flourensia cernua* DC.
Common Name:	Tarbush (hojasén, blackbrush, varnishbush, Mexican tarwort, American tarwort)
Life Span:	Perennial
Origin:	Native
Season:	Evergreen

GROWTH FORM

shrub (to 2 m tall), erect to procumbent, densely leafy, highly branched, resinous; flowers July to December, reproduces from seeds

FLORAL AND FRUIT CHARACTERISTICS

inflorescences: heads in corymbs or panicles, eventually nodding; peduncles short and curved, or heads sessile

flowers: perfect, heads discoid; involucre (1 cm long and wide) with 3 series of 2–4 graduated bracts, eventually nodding; bracts linear, with spreading tips, sticky often turning blackish with drying; disk flowers 12–20, dark yellow, resinous

fruits: achene, (5–10 mm long), oblong, compressed, villous; pappus of 2–4 unequal bristles (2.5–3.2 mm long)

VEGETATIVE CHARACTERISTICS

leaves: alternate, simple; blades elliptic or oblong to oval (1.6–2.5 cm long, 6–12 mm wide), acute, thick; margins entire; adaxially green and glabrous, abaxially paler and glabrous or pubescent, veins often black, somewhat sunken; petioles (1–2.5 mm long) puberulent

stems: twigs light brown, gray, or cream-colored, slender, resinous, pubescent; trunk gray to dark gray, glabrous, sparingly striate, resinous

other: entire plant has a tar-like odor

HISTORIC, FOOD, AND MEDICINAL USES: a decoction is made in Mexico from the leaves and flowers for indigestion

LIVESTOCK LOSSES: mature fruits may cause losses of sheep and goats in January to March with the ingestion of 1% of their body weight before the fruits fall from the plant; foliage may be toxic in moderate quantities

FORAGE VALUE: worthless to livestock, generally not utilized, increases with overgrazing; may be utilized by jackrabbits and other wildlife

HABITAT: deserts and dry soils of valleys, mesas, flats, and foothills, especially on limestone soils

Orange sneezeweed
Hymenoxys hoopesii (A. Gray) Bierner

SYN = *Helenium hoopesii* A. Gray

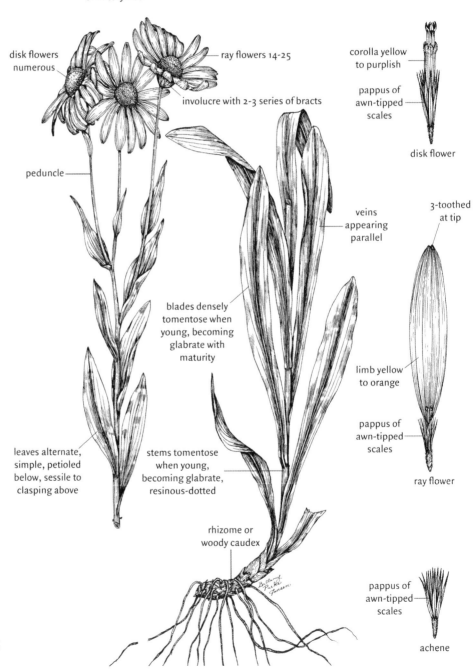

heads solitary
or in corymbs

disk flowers
numerous

ray flowers 14-25

involucre with 2-3 series of bracts

corolla yellow
to purplish

pappus of
awn-tipped
scales

disk flower

peduncle

veins
appearing
parallel

3-toothed
at tip

blades densely
tomentose when
young, becoming
glabrate with
maturity

limb yellow
to orange

pappus of
awn-tipped
scales

ray flower

leaves alternate,
simple, petioled
below, sessile to
clasping above

stems tomentose
when young,
becoming glabrate,
resinous-dotted

rhizome or
woody caudex

pappus of
awn-tipped
scales

achene

Family:	ASTERACEAE
Tribe:	HELIANTHEAE
Species:	*Hymenoxys hoopesii* (A. Gray) Bierner
Common Name:	Orange sneezeweed (owlsclaws, western sneezeweed)
Life Span:	Perennial
Origin:	Native
Season:	Warm

GROWTH FORM

forb (0.3–1 m tall), 1 to several stems from a rhizome or woody caudex; flowers June to September, reproduces from seeds and rhizomes

FLORAL AND FRUIT CHARACTERISTICS

inflorescences: heads solitary or in corymbs with up to 12 heads, large, showy; peduncles erect (4–20 cm long)

flowers: heads radiate; involucre (6–10 mm long) with 2–3 series of bracts; bracts linear to lanceolate (8–11 mm long), acute, pubescent, thin, loosely imbricate, reflexed with age; ray flowers 14–25, pistillate, limb narrow (1.5–3 cm long), yellow to orange, 3-toothed at tip; disk flowers numerous, perfect, corolla tubular (about 5 mm long), yellow to purplish

fruits: achene (2.5–4.5 mm long), obpyramidal, ribs 4–10, pubescent; pappus of awn-tipped scales (2–3.5 mm long), white

VEGETATIVE CHARACTERISTICS

leaves: alternate, simple; basal leaves longest (10–30 cm long, 1–5 cm wide), reduced above, oblanceolate to lanceolate, thick; apex attenuate; principal veins appearing parallel, especially prominent near the base; margins entire; densely tomentose when young, becoming glabrate with maturity; petiolate below with winged petiole, upper leaves sessile to clasping

stems: erect, simple to little-branched, yellowish-green, tomentose when young, becoming glabrate, resinous-dotted

HISTORIC, FOOD, AND MEDICINAL USES: none

LIVESTOCK LOSSES: poisonous to sheep, contains a glycoside, causes "spewing sickness"; considered poisonous to cattle, but it is unpalatable and is seldom grazed

FORAGE VALUE: poor to worthless for livestock and wildlife, most commonly eaten by sheep; will only be consumed when other forage is not readily available

HABITAT: moist slopes, well-drained meadows, stream banks, forest edges, open woods, and valleys

Bitterweed
Hymenoxys odorata DC.

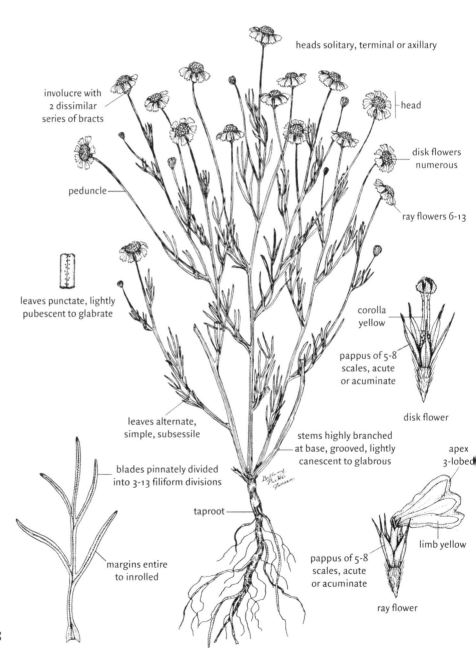

heads solitary, terminal or axillary

involucre with
2 dissimilar
series of bracts

head

peduncle

disk flowers
numerous

ray flowers 6-13

leaves punctate, lightly
pubescent to glabrate

corolla
yellow

pappus of 5-8
scales, acute
or acuminate

disk flower

leaves alternate,
simple, subsessile

stems highly branched
at base, grooved, lightly
canescent to glabrous

apex
3-lobed

blades pinnately divided
into 3-13 filiform divisions

taproot

limb yellow

margins entire
to inrolled

pappus of 5-8
scales, acute
or acuminate

ray flower

Family: ASTERACEAE
Tribe: HELIANTHEAE
Species: *Hymenoxys odorata* DC.
Common Name: Bitterweed (limoncillo, bitter rubberweed)
Life Span: Annual
Origin: Native
Season: Cool

GROWTH FORM
forb (7–50 cm tall); highly-branched producing a bushy appearance, arising from a taproot; flowers May to August, reproduces from seeds

FLORAL AND FRUIT CHARACTERISTICS
inflorescences: heads solitary; terminal or axillary, numerous, small (3–12 mm wide), elevated on peduncles (3–15 cm long)

flowers: heads radiate; involucre campanulate (4–6.5 mm tall), with 2 dissimilar series of bracts; outer 8–13 bracts united and thickened at base, lanceolate, acute; inner bracts long convergent; exceeding the outer bracts; ray flowers 6–13, pistillate, limbs yellow (5–11 mm long), apex 3-lobed, reflexed, persistent; disk flowers perfect, numerous, corolla yellow

fruits: achene (1.5–2 mm long), turbinate, 4-angled, pubescent; pappus of 5–8 scales (1.5–2.5 mm long), acute or acuminate

VEGETATIVE CHARACTERISTICS
leaves: alternate, simple; initially in a basal rosette but the basal leaves caducous; cauline leaves numerous (2–10 cm long), pinnately divided into 3–13 filiform divisions; margins entire to inrolled; punctate, lightly pubescent to glabrate; subsessile

stems: erect, highly branched at base, granulate, grooved, lightly canescent to glabrous

other: entire plant aromatic and somewhat resinous

HISTORIC, FOOD, AND MEDICINAL USES: none

LIVESTOCK LOSSES: poisonous to sheep especially in winter, toxicity increases with water stress, poison accumulates in animals; toxin is water soluble, identity is unknown; plants retain toxicity when dry

FORAGE VALUE: poor to worthless for all classes of livestock and wildlife

HABITAT: disturbed sites, deteriorated pastures and rangeland, and roadsides; most abundant in limestone soils

Prairie coneflower
Ratibida columnifera (Nutt.) Woot. & Standl.

SYN = *R. columnaris* (Sims) D. Don

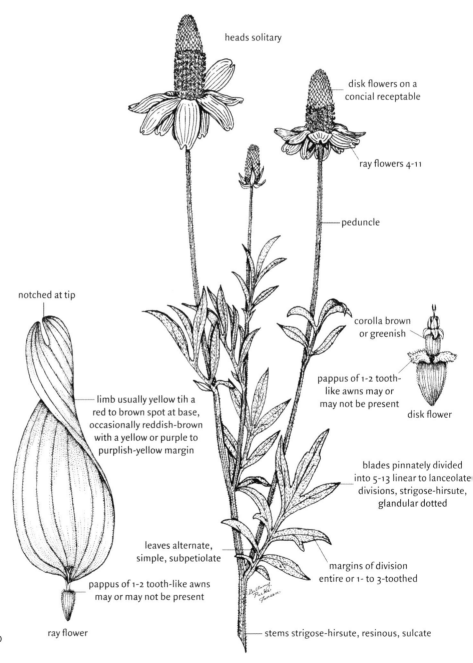

heads solitary

disk flowers on a
concial receptable

ray flowers 4-11

peduncle

notched at tip

corolla brown
or greenish

pappus of 1-2 tooth-
like awns may or
may not be present

disk flower

limb usually yellow tih a
red to brown spot at base,
occasionally reddish-brown
with a yellow or purple to
purplish-yellow margin

blades pinnately divided
into 5-13 linear to lanceolate
divisions, strigose-hirsute,
glandular dotted

leaves alternate,
simple, subpetiolate

margins of division
entire or 1- to 3-toothed

pappus of 1-2 tooth-like awns
may or may not be present

ray flower

stems strigose-hirsute, resinous, sulcate

Family:	ASTERACEAE
Tribe:	HELIANTHEAE
Species:	*Ratibida columnifera* (Nutt.) Woot. & Standl.
Common Name:	Prairie coneflower (upright prairieconeflower, Mexican hat)
Life Span:	Perennial
Origin:	Native
Season:	Warm

GROWTH FORM
forb (0.2–1.2 m tall); solitary to clustered stems from a woody caudex surmounting a taproot; flowers June to September, reproduces from seeds

FLORAL AND FRUIT CHARACTERISTICS
inflorescences: heads solitary; several to many per stem, cylindrical (1–5.5 cm long, 2–3 cm wide), terminal, on naked peduncles (6–25 cm long)

flowers: heads radiate; involucre short, with 2 series of 5–14 bracts; bracts linear, acuminate, hirsute, outer bract longer (4–12 mm long) than the inner (about 3 mm long); ray flowers 4–11, sterile, limb oblong (1–3.5 cm long), usually yellow with a red to brown spot at base, occasionally reddish-brown with a yellow or purple to purplish-yellow margin, drooping, notched at tip; disk flowers on a conical receptacle (1.5–5 cm long, 8–12 mm wide), perfect, corolla brown or greenish (1.5–2.5 mm long)

fruits: achene (1.5–3 mm long), oblong, compressed, gray, ciliate on inner edge; pappus of 1–2 tooth-like awns may or may not be present

VEGETATIVE CHARACTERISTICS
leaves: alternate, simple; blades oblong (2.5–15 cm long, 2–6 cm wide), pinnately divided into 5–13 linear to lanceolate divisions (most 1.5 cm or more long), often very unequal, divided nearly to the midvein; margins of divisions entire or 1- to 3-toothed; strigose-hirsute, glandular dotted; subpetiolate

stems: erect, branched, leafy, strigose-hirsute, resinous, sulcate

HISTORIC, FOOD, AND MEDICINAL USES: the Cheyenne boiled leaves and stems to make a yellow solution applied externally to draw poison out of rattlesnake bites and for relief from poison ivy; other Native American tribes made tea from the flowers and leaves; planted as an ornamental

LIVESTOCK LOSSES: none

FORAGE VALUE: fair to good for sheep and wildlife, fair for cattle

HABITAT: prairies, plains, roadsides, and disturbed sites; adapted to a broad range of soils

Mulesears
Wyethia amplexicaulis (Nutt.) Nutt.

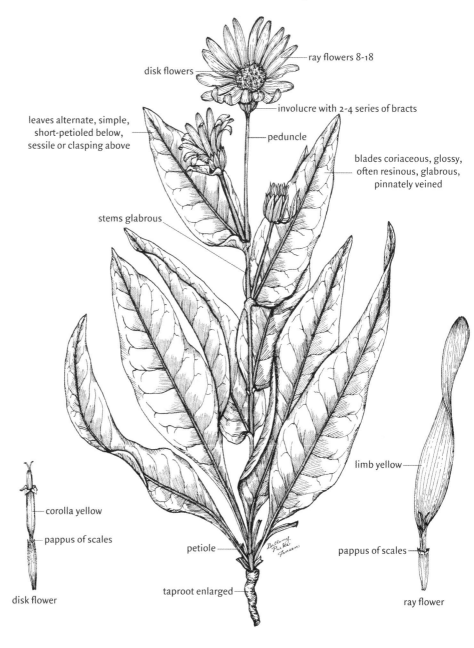

heads solitary or several

ray flowers 8-18

disk flowers

involucre with 2-4 series of bracts

leaves alternate, simple,
short-petioled below,
sessile or clasping above

peduncle

blades coriaceous, glossy,
often resinous, glabrous,
pinnately veined

stems glabrous

corolla yellow

pappus of scales

petiole

taproot enlarged

limb yellow

pappus of scales

disk flower

ray flower

Family:	ASTERACEAE
Tribe:	HELIANTHEAE
Species:	*Wyethia amplexicaulis* (Nutt.) Nutt.
Common Name:	Mulesears (pe-ik, dock)
Life Span:	Perennial
Origin:	Native
Season:	Cool

GROWTH FORM
forb (30–80 cm tall); erect or ascending, solitary to few stems from a simple or branching caudex surmounting an enlarged taproot; flowers May to July, reproduces from seeds and branching caudex

FLORAL AND FRUIT CHARACTERISTICS
inflorescences: heads solitary or several; terminal and axillary, large and showy (2–4 cm tall, 4–10 cm in diameter), terminal heads larger than axillary heads, peduncled

flowers: heads radiate; involucre (1.8–4 cm long) with 2–4 series of bracts; bracts few to numerous, lanceolate to obovate, acute or obtuse, broad, glabrous; ray flowers 8–18, pistillate, limb (2–6 cm long), yellow; disk flowers perfect, corolla yellow

fruits: achene (6–15 mm long), oblong, 4-angled, glabrous or puberulent; pappus of scales (about 2 mm long), 1–2 awns (1 mm long)

VEGETATIVE CHARACTERISTICS
leaves: alternate, simple, mostly basal; blades oblong-lanceolate (15–50 cm long, 5–15 cm wide), coriaceous, reduced above; apex acute; margins entire or denticulate; glossy, often resinous, glabrous, pinnately veined; short-petioled below, sessile or clasping above

stems: simple or rarely branched, leafy, glabrous

other: entire plant aromatic

HISTORIC, FOOD, AND MEDICINAL USES: some Native Americans fermented roots for 2 days in a pit heated with hot stones to develop a sweet-flavored food

LIVESTOCK LOSSES: none

FORAGE VALUE: poor for cattle, horses, deer, and elk; poor to fair for sheep; immature foliage is most frequently consumed; heads are eaten by all classes of livestock and big game

HABITAT: rangeland, hillsides, open woods, dry meadows, and moist draws

Woolly mulesears
Wyethia mollis A. Gray

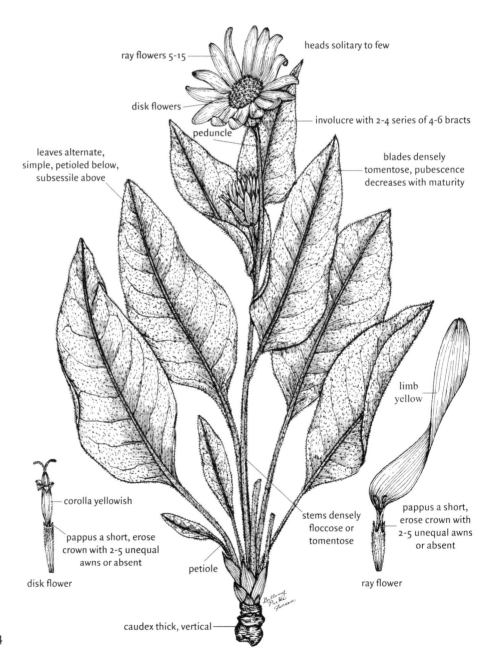

ray flowers 5-15

heads solitary to few

disk flowers

involucre with 2-4 series of 4-6 bracts

peduncle

leaves alternate, simple, petioled below, subsessile above

blades densely tomentose, pubescence decreases with maturity

limb yellow

corolla yellowish

pappus a short, erose crown with 2-5 unequal awns or absent

disk flower

stems densely floccose or tomentose

pappus a short, erose crown with 2-5 unequal awns or absent

ray flower

petiole

caudex thick, vertical

Family:	ASTERACEAE
Tribe:	HELIANTHEAE
Species:	*Wyethia mollis* A. Gray
Common Name:	Woolly mulesears (woolly wyethia)
Life Span:	Perennial
Origin:	Native
Season:	Cool

GROWTH FORM
forb (0.4–1 m tall); cespitose; solitary to a few stems arising from a thick, vertical caudex surmounting a taproot; forms dense stands; flowers May to August, reproduces from seeds

FLORAL AND FRUIT CHARACTERISTICS
inflorescences: heads solitary to few (1–4), broad (4–9 cm in diameter), showy; terminal and axillary, peduncled

flowers: heads radiate; involucre (1.3–3 cm tall) with 2–4 series of 4–6 bracts; bracts erect, triangular to oblong, acute, tomentose; ray flowers 5–15, pistillate, limbs yellow (1.5–4.5 cm long, 1–1.5 cm wide), acute; disk flowers, perfect, corolla yellowish

fruits: achene (8–11 mm long), oblong, 4-angled, glaucous, glandular-dotted, pubescent above; pappus a short, erose crown with 2–5 unequal awns (about 7 mm long) or absent

VEGETATIVE CHARACTERISTICS
leaves: alternate, simple, principal ones basal; blades lanceolate to oblong-ovate (9–50 cm long, 6–17 cm wide), narrow at the base, reduced above; apex acute; margins entire; densely tomentose, especially when young, becoming more green with maturity as pubescence decreases; resinous, pinnately veined; petiolate below, petiole about as long as the blade, subsessile above

stems: erect or ascending, simple, leafy, densely floccose or tomentose

HISTORIC, FOOD, AND MEDICINAL USES: roots and fruits were eaten by some Native Americans

LIVESTOCK LOSSES: none

FORAGE VALUE: poor for cattle and fair for sheep and deer; most palatable in early spring immediately following snow melt; heads readily eaten by livestock and big game

HABITAT: open wooded slopes, rocky openings, and ridges; most abundant on dry, open sites

Threadleaf groundsel
Senecio douglasii DC.

SYN = S. *flaccidus* Less., S. *longilobus* Benth.

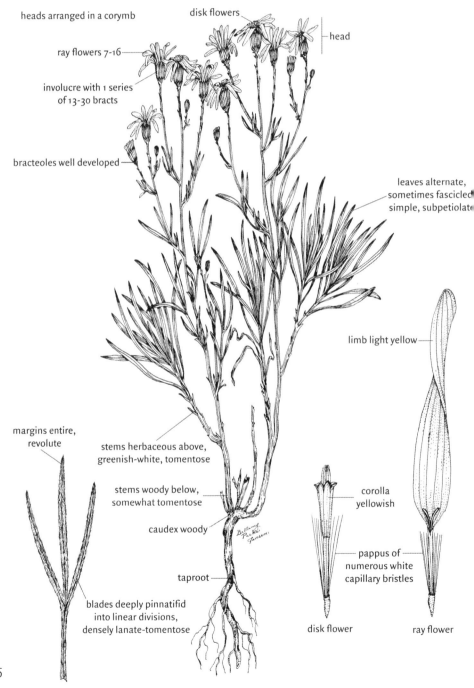

heads arranged in a corymb

disk flowers

head

ray flowers 7-16

involucre with 1 series of 13-30 bracts

bracteoles well developed

leaves alternate, sometimes fascicled, simple, subpetiolate

limb light yellow

margins entire, revolute

stems herbaceous above, greenish-white, tomentose

stems woody below, somewhat tomentose

caudex woody

corolla yellowish

pappus of numerous white capillary bristles

taproot

blades deeply pinnatifid into linear divisions, densely lanate-tomentose

disk flower

ray flower

Family:	ASTERACEAE
Tribe:	SENECIONEAE
Species:	*Senecio douglasii* DC.
Common Name:	Threadleaf groundsel (cenicillo, shrubby butterweed, threadleaf ragwort)
Life Span:	Perennial
Origin:	Native
Season:	Warm

GROWTH FORM
subshrub (to 2 m tall); woody stems arising from a woody caudex surmounting a taproot, erect to ascending, few-branched, herbaceous above; flowers April to November, reproduces from seeds

FLORAL AND FRUIT CHARACTERISTICS
inflorescences: heads (3–20) in corymbs, campanulate in outline (1–1.5 cm wide); clusters of lowering branches often aggregated to form large and showy inflorescences

flowers: heads radiate; bracteoles well developed, unequal; involucre (7–11 mm tall) with 1 series of 13–30 bracts; bracts linear (5–8 mm long), narrow, herbaceous; ray flowers 7–16, pistillate, limb light yellow (1–1.5 cm long); disk flowers perfect, numerous, corolla yellowish

fruits: achene (1.5–2.5 mm long), narrowly cylindrical, ribs 5–10, canescent; pappus of numerous white capillary bristles

VEGETATIVE CHARACTERISTICS
leaves: alternate, sometimes with fasiscles of small leaves in the axils, simple; many deeply pinnatifid; blades and divisions linear or linear-filiform; divisions unequal (4–12 cm long, 0.5–5 mm wide); apex acute or obtuse; margins entire and revolute; densely lanate-tomentose; subpetiolate

stems: woody below, gray, somewhat tomentose; herbaceous above, greenish-white, tomentose

HISTORIC, FOOD, AND MEDICINAL USES: the Navajo Indians drank tea made from the whole plant to ensure a good voice for ceremonial songs, used for numerous medicinal purposes by other Native Americans

LIVESTOCK LOSSES: poisonous to cattle, horses, and sheep (to a lesser extent); contains several alkaloids, poison accumulates in animals

FORAGE VALUE: worthless to poor for cattle; sheep and goats may lightly browse it

HABITAT: plains, mesas, sandy washes, gravelly streambeds, open slopes, and grasslands; most abundant in well-drained sandy or rocky soils

Sawtooth butterweed
Senecio serra Hook.

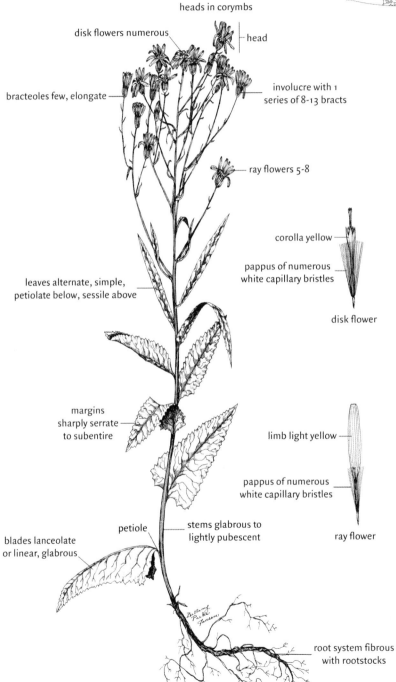

heads in corymbs

disk flowers numerous

head

bracteoles few, elongate

involucre with 1
series of 8-13 bracts

ray flowers 5-8

corolla yellow

pappus of numerous
white capillary bristles

disk flower

leaves alternate, simple,
petiolate below, sessile above

margins
sharply serrate
to subentire

limb light yellow

pappus of numerous
white capillary bristles

petiole

stems glabrous to
lightly pubescent

ray flower

blades lanceolate
or linear, glabrous

root system fibrous
with rootstocks

Family:	ASTERACEAE
Tribe:	SENECIONEAE
Species:	*Senecio serra* Hook.
Common Name:	Sawtooth butterweed (butterweed groundsel, tall groundsel)
Life Span:	Perennial
Origin:	Native
Season:	Warm

GROWTH FORM
forb (0.6–2 m tall); erect to ascending, stout, several stems from a fibrous root system with rootstocks; flowers June to September, reproduces from seeds and rootstocks

FLORAL AND FRUIT CHARACTERISTICS
inflorescences: heads in corymbs of a few rather large heads (11–13 mm tall, 5–8 mm wide) and many smaller heads (6–9 mm tall, 3–5 mm wide); numerous, erect, on slender peduncles

flowers: heads radiate; bracteoles few, elongate; involucre (5–9 mm long) with 1 series of 8–13 bracts; bracts narrow (4–7 mm long), linear, apex acute, black-tipped; ray flowers mostly 5–8, pistillate, limbs light yellow (5–10 mm long); disk flowers perfect, numerous, corolla yellow

fruits: achene flattened, ribs 5–10, glabrous; pappus of numerous white capillary bristles

VEGETATIVE CHARACTERISTICS
leaves: alternate, simple; blades lanceolate or linear (7–15 cm long, 1–4 cm wide), numerous, lower leaves caducous; apex long-acute or acuminate; margins sharply serrate to subentire; glabrous; petiolate below, sessile above

stems: erect, simple below, branching near inflorescence, glabrous to lightly pubescent

HISTORIC, FOOD, AND MEDICINAL USES: none

LIVESTOCK LOSSES: none

FORAGE VALUE: good to excellent for sheep, good for elk and deer, fair to poor for cattle; most palatable during spring and summer

HABITAT: meadows, damp ground, woodlands, and moist stream banks; most abundant in rich, well-drained, sandy or gravelly loams

Gray horsebrush
Tetradymia canescens DC.

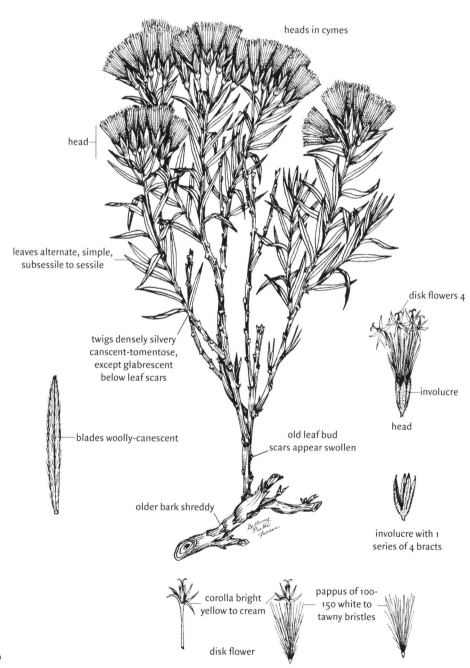

heads in cymes

head

leaves alternate, simple, subsessile to sessile

twigs densely silvery canscent-tomentose, except glabrescent below leaf scars

blades woolly-canescent

older bark shreddy

old leaf bud scars appear swollen

disk flowers 4

involucre

head

involucre with 1 series of 4 bracts

corolla bright yellow to cream

disk flower

pappus of 100-150 white to tawny bristles

Family:	ASTERACEAE
Tribe:	SENECIONEAE
Species:	*Tetradymia canescens* DC.
Common Name:	Gray horsebrush (spineless horsebrush, common horsebrush)
Life Span:	Perennial
Origin:	Native
Season:	Warm

GROWTH FORM
shrub (to 1 m tall); highly branched, spreading to ascending; flowers May to September, reproduces from seeds

FLORAL AND FRUIT CHARACTERISTICS
inflorescences: heads (5–20) in cymes; compact, terminal

flowers: heads discoid; involucre with 1 series of 4 bracts (6–15 mm tall); bracts linear to oblong, acute or obtuse, imbricate, thickened, tomentose; disk flowers 4 per head; corolla funnelform (4–7 mm long), bright yellow to cream, lobes often longer than the tube

fruits: achene (2.5–5 mm long), linear to slightly obovoid, ribs 5, densely silky; pappus of 100–150 white to tawny bristles (6–11 mm long), in 2–3 series

VEGETATIVE CHARACTERISTICS
leaves: alternate, simple; blades narrowly lanceolate to oblanceolate or spatulate (1–4 cm long, 1–6 mm wide); apex acute; margins entire; midvein prominent, woolly-canescent; subsessile to sessile

stems: erect, short, stout, highly branched; new growth densely silvery canescent-tomentose, except glabrescent below leaf scars; older bark gray, glabrescent, shreddy; old leaf bud scars appear swollen

HISTORIC, FOOD, AND MEDICINAL USES: the Hopi Indians made a tonic from leaves and roots for uterine disorders

LIVESTOCK LOSSES: causes photosensitization in sheep, symptoms are called "big head" or "swell head" from swelling of the head and facial features; alkaloids may also cause liver damage followed by death in sheep

FORAGE VALUE: poor to worthless for cattle, sheep, goats, and big game; consumed only when other forage is unavailable

HABITAT: barren plains, foothills, and deserts; most abundant on sandy or rocky soils

Mountain alder
Alnus incana (L.) Moench

SYN = *A. rugosa* (Du Roi) Spreng., *A. tenuifolia* Nutt.

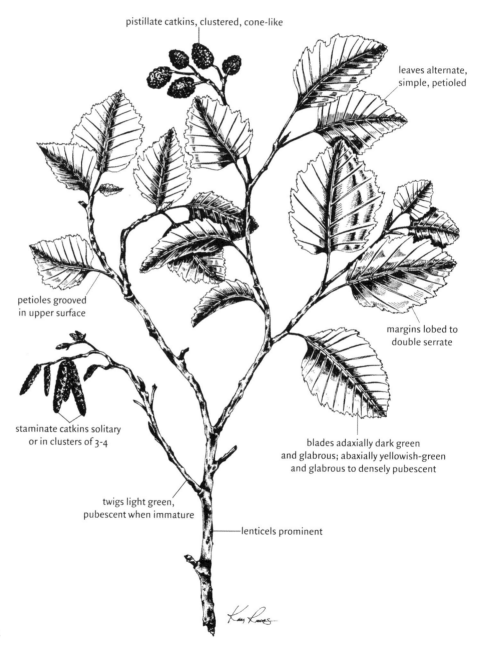

pistillate catkins, clustered, cone-like

leaves alternate, simple, petioled

petioles grooved in upper surface

margins lobed to double serrate

staminate catkins solitary or in clusters of 3-4

blades adaxially dark green and glabrous; abaxially yellowish-green and glabrous to densely pubescent

twigs light green, pubescent when immature

lenticels prominent

Family: BETULACEAE
Species: *Alnus incana* (L.) Moench
Common Name: Mountain alder (thinleaf alder, speckled alder, gray alder)
Life Span: Perennial
Origin: Native
Season: Cool

GROWTH FORM

monoecious shrub to small tree (to 10 m tall), often with clustered trunks; flowers April to June; fruits mature in September; reproduces from seed and basal sprouts

FLORAL AND FRUIT CHARACTERISTICS

inflorescences: catkins; staminate catkins solitary or in clusters of 3–4 (4–9 cm long, 2–9 mm wide), pendulous; pistillate catkins cone-like (1.0–1.7 cm long, 8–12 mm wide), ovoid, clustered, brown, comprised of woody scales

flowers: unisexual, apetalous; staminate flowers in clusters of 3–4, each cluster subtended by a purplish, peltate bracteole on 2 mm long stalk; perianth 4-lobed, lobes obovate (1 mm long), cupped; stamens 4, attached to each calyx lobe; pistillate flowers paired, subtended by a fleshy obovate bract (1 mm long), persistent, becoming woody

fruits: samara, elliptic to obovate (2–3 mm long and wide), brown, flat

VEGETATIVE CHARACTERISTICS

leaves: alternate, simple; blades highly variable, ovate to oblong-ovate (5–10 cm long, 2.5–5 cm wide), apex acute to obtuse; base cuneate to rounded; margins lobed to doubly serrate, teeth unequal; adaxially dark green, glabrous; abaxially yellowish-green glabrous to densely pubescent; petioles (1–3 cm long), grooved on upper surface; stipules oblong (6–10 mm long)

stems: twigs light green, pubescent when immature, becoming reddish-brown; lenticels prominent; trunk bark grayish-brown with whitish lenticels

HISTORIC, FOOD, AND MEDICINAL USES: some Native Americans made a red dye from the bark

LIVESTOCK LOSSES: none

FORAGE VALUE: limited value for cattle; fair to good for deer, elk, and moose

HABITAT: wet to moist soils, usually sandy or gravelly soils along streams, rivers, ponds, or in swamps and moist woodlands

Tansymustard
Descurainia pinnata (Walt.) Britt.

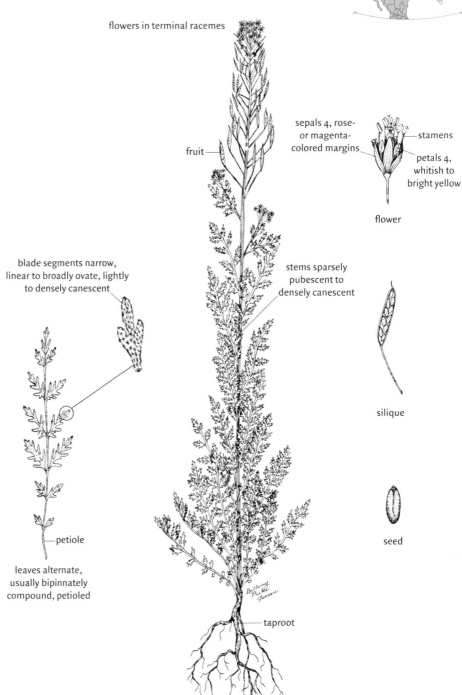

flowers in terminal racemes

sepals 4, rose- or magenta- colored margins

stamens

petals 4, whitish to bright yellow

fruit

flower

blade segments narrow, linear to broadly ovate, lightly to densely canescent

stems sparsely pubescent to densely canescent

silique

petiole

leaves alternate, usually bipinnately compound, petioled

seed

taproot

Family:	BRASSICACEAE
Species:	*Descurainia pinnata* (Walt.) Britt.
Common Name:	Tansymustard (mostácilla, western tansymustard, green tansymustard)
Life Span:	Annual
Origin:	Native
Season:	Cool

GROWTH FORM

forb (5 to 80 cm tall); erect to ascending, simple or branched, often branched above; 1 to several stems from a taproot; flowers March to August, reproduces from seeds

FLORAL AND FRUIT CHARACTERISTICS

inflorescences: racemes; terminal, elongating with maturity

flowers: perfect, regular; sepals 4 (1–1.5 mm long), oblong to ovate, petal-like, margins membranous and rose- or magenta-colored; petals 4, obovate to spatulate (1–1.5 mm long), clawed, whitish to bright yellow

fruits: siliques; clavate (4–20 mm long, 1–1.5 mm wide at maturity), 2-celled with several seeds per cell; pedicels (to 2.5 cm long) in fruit erect or ascending

VEGETATIVE CHARACTERISTICS

leaves: alternate, usually bipinnately compound (1.5–10 cm long, including petiole); blades above reduced in size and usually simply pinnate; segments narrow, linear to broadly ovate; lightly to densely canescent; petioled

stems: sparsely pubescent to densely canescent with usually branched trichomes, sometimes glandular

other: a variable species with several subspecies

HISTORIC, FOOD, AND MEDICINAL USES: seeds used by Native Americans to make pinole flour, young plants were used as potherbs

LIVESTOCK LOSSES: known to be poisonous to cattle in the Southwest, symptoms in cattle include "paralyzed tongue" or an inability to swallow food and water; generally, it is from consuming large quantities of plant material; however, this may not be easily avoided since it is green earlier in the spring than most forage species

FORAGE VALUE: poor for cattle, fair to good for sheep and goats, unpalatable to horses; palatability declines with maturity

HABITAT: waste places, prairies, open woods, disturbed sites, and roadsides; adapted to a broad range of soils; most abundant in dry or sandy soils

Desert princesplume
Stanleya pinnata (Pursh) Britt.

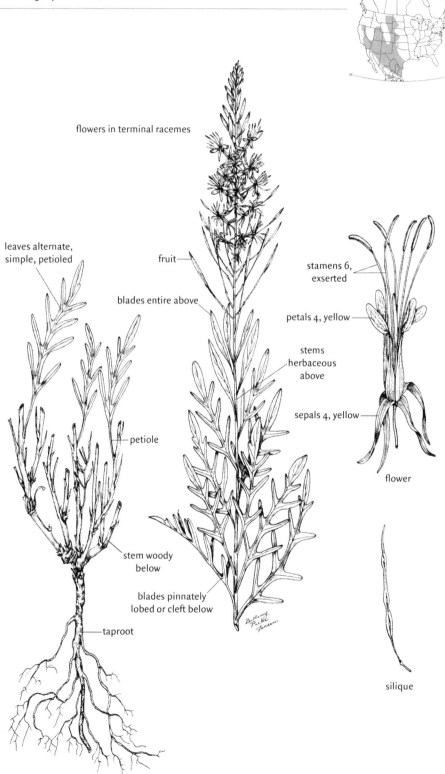

flowers in terminal racemes

leaves alternate,
simple, petioled

fruit

blades entire above

stamens 6,
exserted

petals 4, yellow

stems
herbaceous
above

sepals 4, yellow

petiole

flower

stem woody
below

blades pinnately
lobed or cleft below

taproot

silique

Family:	BRASSICACEAE
Species:	*Stanleya pinnata* (Pursh) Britt.
Common Name:	Desert princesplume
Life Span:	Perennial
Origin:	Native
Season:	Cool

GROWTH FORM
forb (0.6–1.5 m tall), often suffrutescent; 1 to several stems from a woody base surmounting a taproot, simple to branched above; flowers April to August, reproduces from seeds

FLORAL AND FRUIT CHARACTERISTICS
inflorescences: racemes (10–35 cm long); terminal, many-flowered, showy, elongated in fruit (to 50 cm)

flowers: perfect, regular, pedicellate (about 1 cm long); sepals 4 (7–17 mm long), linear-oblong, spreading, yellow; petals 4 (9–17 mm long), pilose on inner surface, yellow, claw brownish; stamens 6, exserted, nearly equal

fruits: siliques; linear (2–8 cm long), nearly terete, seeds in 1 row, many-seeded; pedicel elongate (1–3 cm long)

VEGETATIVE CHARACTERISTICS
leaves: alternate, simple; blades lanceolate to elliptic or obovate (5–12 cm long, 3.5–5 cm wide), thick; margins pinnately lobed or cleft below, entire above; glabrous or glaucous or sparsely pubescent; cauline leaves petioled; basal leaves frequently absent by flowering

stems: erect, glabrous, green, glaucous; woody below, herbaceous above

other: a variable species with several forms

HISTORIC, FOOD, AND MEDICINAL USES: pioneers and Native Americans cooked and ate the stems and leaves

LIVESTOCK LOSSES: poisonous throughout the growing season, accumulates selenium and is indicative of selenium in the soil; rarely consumed when other forage is available in early spring; poisoning on rangeland is rare, but force feeding animals has caused poisoning

FORAGE VALUE: worthless to livestock and wildlife

HABITAT: dry hills, plains, valleys, and desert washes; reliable indicator of seleniferous soils; adapted to a broad range of soil textures; most abundant in dry soils on abused rangeland

Snowberry
Symphoricarpos albus (L.) Blake

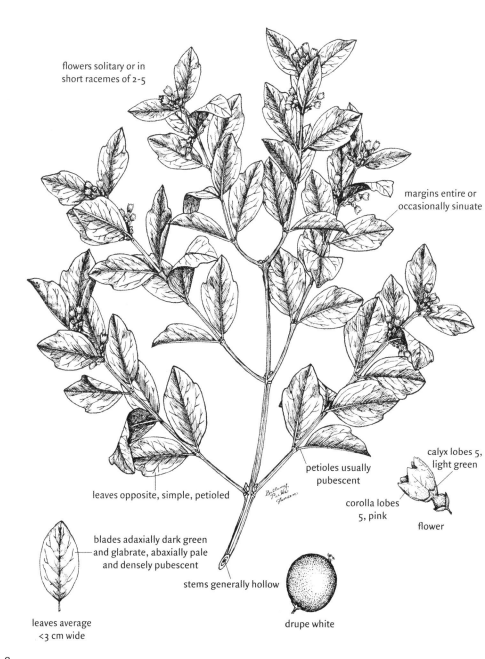

flowers solitary or in
short racemes of 2-5

margins entire or
occasionally sinuate

petioles usually
pubescent

leaves opposite, simple, petioled

calyx lobes 5,
light green

corolla lobes
5, pink

flower

blades adaxially dark green
and glabrate, abaxially pale
and densely pubescent

stems generally hollow

leaves average
<3 cm wide

drupe white

Family:	CAPRIFOLIACEAE
Species:	*Symphoricarpos albus* (L.) Blake
Common Name:	Snowberry (common snowberry, white coralberry)
Life Span:	Perennial
Origin:	Native
Season:	Cool

GROWTH FORM

shrub (to 1 m tall); stems few to several from a woody base, occasionally from creeping rootstocks, thicket-forming; flowers May to July, reproduces from seeds and rhizomes

FLORAL AND FRUIT CHARACTERISTICS

inflorescences: solitary or in short racemes; terminal and in upper axils, 2- to 5-flowered

flowers: perfect, regular; calyx lobes 5 (0.4–0.9 mm long), triangular to lanceolate, light green, glabrous to ciliate; corolla campanulate; lobes 5 (5–7 mm long, 3–5 mm wide), pink, rounded, slightly spreading, outer surface of lobes glabrous, inner surface villous; stamens 5; stamens and style not exceeding corolla; pedicellate

fruits: drupes ovoid (7–9 mm long, 6–8 mm wide), fleshy, white; calyx persistent; solitary or few in upper leaf axils; 2 nutlets per drupe, 1 seed per nutlet

VEGETATIVE CHARACTERISTICS

leaves: opposite, simple; blades oval or ovate (1–4 cm long, 8–25 mm wide, averaging less than 3 cm wide); apex acute to obtuse; base cuneate to rounded; margins entire or occasionally sinuate, ciliate when immature; adaxially dark green and glabrate, abaxially pale and densely pubescent; petiole (1–3 mm long) usually pubescent

stems: twigs erect, slender, generally hollow, yellowish-brown, pubescent when young, especially at nodes, glabrous with age; bark of trunk thin, grayish or reddish-brown, shreddy; winter buds with pubescent scales (2 mm long)

HISTORIC, FOOD, AND MEDICINAL USES: Native Americans made a tonic from the roots, an eyewash from the bark, and all parts were crushed and applied to wounds; consumption of relatively large quantities of fruits may cause vomiting and diarrhea

LIVESTOCK LOSSES: leaves contain saponins which are cathartic

FORAGE VALUE: fair for sheep and goats in winter, otherwise worthless for livestock; important food and cover for song and game birds; occasionally browsed by mule deer; ornamental

HABITAT: wooded hillsides, prairies, and open slopes; present in both moist and dry soils

Western snowberry
Symphoricarpos occidentalis Hook.

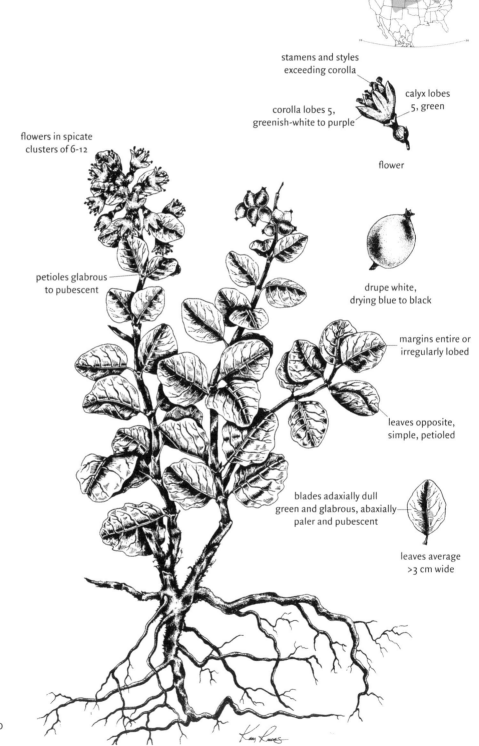

stamens and styles
exceeding corolla

corolla lobes 5,
greenish-white to purple

calyx lobes
5, green

flower

flowers in spicate
clusters of 6-12

drupe white,
drying blue to black

petioles glabrous
to pubescent

margins entire or
irregularly lobed

leaves opposite,
simple, petioled

blades adaxially dull
green and glabrous, abaxially
paler and pubescent

leaves average
>3 cm wide

Family:	CAPRIFOLIACEAE
Species:	*Symphoricarpos occidentalis* Hook.
Common Name:	Western snowberry (wolfberry)
Life Span:	Perennial
Origin:	Native
Season:	Cool

GROWTH FORM
shrub (to 1.5 m tall); few to several stems from creeping rootstocks, branches numerous; flowers June to August, reproduces from seeds and rhizomes; forms large colonies

FLORAL AND FRUIT CHARACTERISTICS
inflorescences: spicate clusters of 6–12 flowers in upper leaf axils, dense

flowers: perfect, regular, sessile; calyx lobes 5, triangular (less than 1 mm long), acute, green; corolla campanulate (5–8 mm long), wider than long; lobes 5, rounded, greenish-white to purple, outer surface of lobes glabrous, inner surface hirsute; stamens 5; stamens and style exceeding the corolla

fruits: drupes globose (6–9 mm wide), numerous, fleshy, white but drying blue to black, smooth, calyx persistent; 2 nutlets per drupe, each nutlet with 1 seed

VEGETATIVE CHARACTERISTICS
leaves: opposite, simple; blades ovate to elliptic or suborbicular (2–6 cm long, 1–4 cm wide, averaging more than 3 cm wide); apex acute to obtuse (sometimes rounded); base cuneate, rounded or truncate; margins entire or irregularly lobed, typically ciliate; adaxially dull green and glabrous, abaxially paler and pubescent; veins indented on adaxial surface; petiole (2–7 mm long) glabrous to pubescent

stems: twigs slender, brown, pubescent to glabrate; trunk bark grayish-brown, shreddy; buds small (0.5 mm long)

other: juvenile shoots sometimes produce large leaves (to 10 cm long, 8 cm wide)

HISTORIC, FOOD, AND MEDICINAL USES: leaves were steeped by the Blackfoot Indians to make a wash for sore eyes, fruits were used as famine food, and boiled fruits were given to horses as a diuretic; Lakota children made lightweight arrows from the stems to use in play

LIVESTOCK LOSSES: leaves contain saponins which are cathartic

FORAGE VALUE: poor for cattle, fair for sheep and goats; good for deer, pronghorn, and occasionally important for other big game; food and cover for small mammals and several song and game birds; ornamental; often considered to be a weed

HABITAT: prairie, ravines, rocky and gravely hillsides, open woods

Fourwing saltbush
Atriplex canescens (Pursh) Nutt.

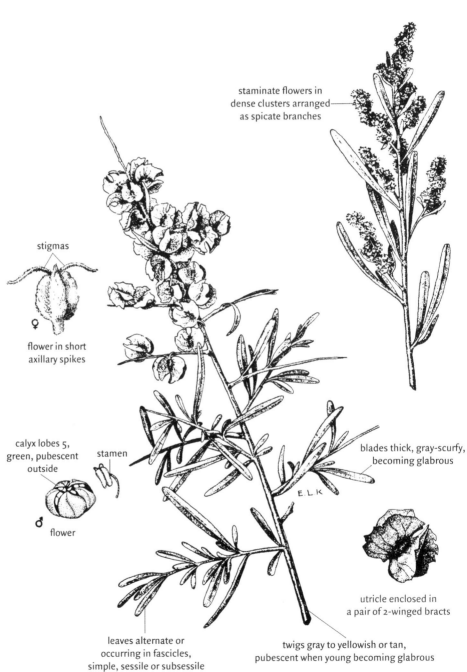

staminate flowers in
dense clusters arranged
as spicate branches

stigmas

♀

flower in short
axillary spikes

calyx lobes 5,
green, pubescent
outside

stamen

♂

flower

blades thick, gray-scurfy,
becoming glabrous

E.L.K

utricle enclosed in
a pair of 2-winged bracts

leaves alternate or
occurring in fascicles,
simple, sessile or subsessile

twigs gray to yellowish or tan,
pubescent when young becoming glabrous

Family:	CHENOPODIACEAE
Species:	*Atriplex canescens* (Pursh) Nutt.
Common Name:	Fourwing saltbush (costilla de vaca, chamizo, wingscale, chamiza)
Life Span:	Perennial
Origin:	Native
Season:	Evergreen

GROWTH FORM

dioecious or occasionally monoecious shrub (to 3 m tall); erect, stout, much branched; flowers May to September, reproduces from seeds

FLORAL AND FRUIT CHARACTERISTICS

inflorescences: panicles; staminate flowers in dense clusters arranged as spicate branches in a terminal panicle; pistillate flowers in short axillary spikes, appearing as a terminal panicle (5–10 cm long)

flowers: unisexual, apetalous; calyx lobes 5; lobes ovate (less than 1 mm long and wide), obtuse, green, pubescent outside; stamens 5; pistillate flowers subtended by 2 bracts (each about 1.6 mm long)

fruits: utricles enclosed in a pair of 2-winged bracts (8–10 mm long); wings lacerate, entire or undulate, apex entire or bifid, sessile or short-pedicled; pairs of bracts enclosing utricle giving a 4-winged appearance

VEGETATIVE CHARACTERISTICS

leaves: alternate or occurring in fascicles, simple; blades oblong to obovate or lanceolate (2.5–4.5 cm long, 3–7 mm wide), thick; apex obtuse; base cuneate; margins entire, may be inrolled; gray-scurfy, becoming glabrous; sessile or subsessile

stems: twigs rigid, erect, gray to yellowish or tan, pubescent when young becoming glabrous; bark thin, tight, gray, lightly sulcate

other: plant highly variable in size, leaf shape, and fruiting bract size and shape

HISTORIC, FOOD, AND MEDICINAL USES: some Native Americans ground seeds to make flour for bread or to mix with sugar and water for a drink; the Navajos used an infusion of leaves and stems to make a yellow dye; pollen commonly causes hay fever

LIVESTOCK LOSSES: concentrated feeding causes scours in cattle, unsubstantiated reports of toxicity to goats

FORAGE VALUE: good for cattle, sheep, goats, pronghorn, and deer; furnishes valuable browse in winter; fruits provide good food for wildlife

HABITAT: bluffs, hillsides, deserts, and saline or alkali flats; common in many different soil types

Shadscale saltbush
Atriplex confertifolia (Torr. & Frém.) S. Wats.

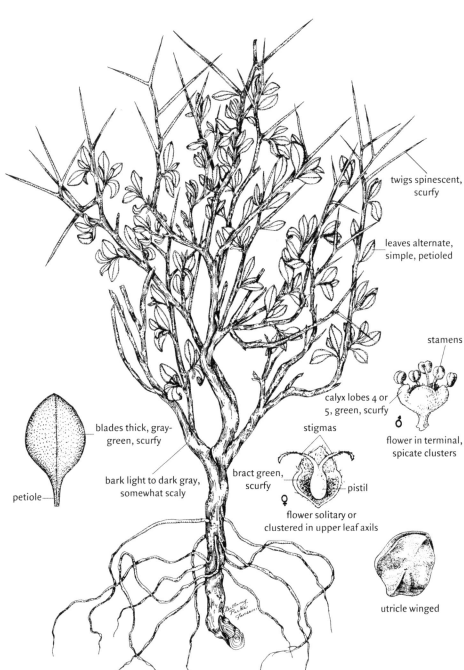

twigs spinescent,
scurfy

leaves alternate,
simple, petioled

stamens

calyx lobes 4 or
5, green, scurfy

♂

flower in terminal,
spicate clusters

blades thick, gray-
green, scurfy

stigmas

bark light to dark gray,
somewhat scaly

bract green,
scurfy

pistil

petiole

♀

flower solitary or
clustered in upper leaf axils

utricle winged

Family:	CHENOPODIACEAE
Species:	*Atriplex confertifolia* (Torr. & Frém.) S. Wats.
Common Name:	Shadscale saltbush (saladillo, spiny saltbrush, roundleaf saltbrush)
Life Span:	Perennial
Origin:	Native
Season:	Cool

GROWTH FORM

dioecious shrub (to 90 cm tall); late deciduous, crown rounded, densely branched, rigid, spinescent; flowers July to August, reproduces from seeds

FLORAL AND FRUIT CHARACTERISTICS

inflorescences: spicate clusters; staminate flowers terminal, dense, subtended by leafy bracts; pistillate flowers solitary or clustered in upper leaf axils

flowers: unisexual, apetalous; staminate calyx lobes 4 or 5; lobes oblong (1–1.5 mm long and wide), green, scurfy; stamens 5; sessile; pistillate flowers subtended by 2 fleshy bracts, bracts ovate (4–12 mm long, 4–12 mm wide), green, scruffy, united one-third the length

fruits: utricles enclosed in wings comprised of persisting pistillate bracts; bracts similar to leaves, margins herbaceous, entire to undulate or serrulate to denticulate, yellowish-brown

VEGETATIVE CHARACTERISTICS

leaves: alternate, simple; blades ovate, triangular, or elliptic (9–20 mm long, 5–15 mm wide), thick; apex obtuse or rounded; base attenuate; margins entire; gray-green, scurfy; petioled (3–5 mm long)

stems: twigs rigid, erect, stout, spinescent, yellow-brown, smooth, scurfy; trunk irregular, appearing as a cluster of older branches, bark light to dark gray, somewhat scaly

HISTORIC, FOOD, AND MEDICINAL USES: some Native Americans ground the fruits into flour

LIVESTOCK LOSSES: none

FORAGE VALUE: good to fair for all classes of livestock, pronghorn, and mule deer; provides winter and spring browse; fruits provide food for game and song birds; resistant to heavy grazing

HABITAT: desert valleys, hills, and bluffs in stony or clayey alkaline soils; forming almost pure stands in some locations; frequently grows in association with *Stipa hymenoides*

Saltbush
Atriplex gardneri (Moq.) D. Dietr.

SYN = in part *A. nuttallii* S. Wats.

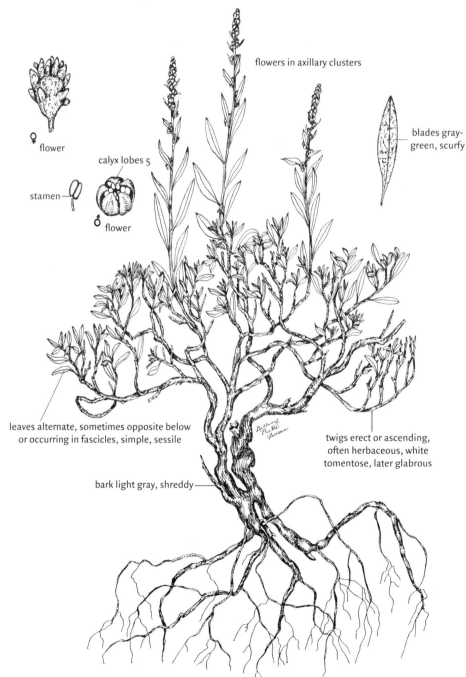

♀ flower

stamen

calyx lobes 5

♂ flower

flowers in axillary clusters

blades gray-green, scurfy

leaves alternate, sometimes opposite below or occurring in fascicles, simple, sessile

twigs erect or ascending, often herbaceous, white tomentose, later glabrous

bark light gray, shreddy

Family: CHENOPODIACEAE
Species: *Atriplex gardneri* (Moq.) D. Dietr.
Common Name: Saltbush (saladillo, Gardner saltbush)
Life Span: Perennial
Origin: Native
Season: Evergreen

GROWTH FORM
dioecious or rarely monoecious subshrub (to 50 cm tall); ascending from a decumbent spreading base, much-branched; flowers June to August, reproduces from seeds and rootstocks

FLORAL AND FRUIT CHARACTERISTICS
inflorescences: clustered, axillary; staminate and pistillate flowers in dense clusters toward the branch tips giving the appearance of a terminal inflorescence

flowers: unisexual, apetalous; staminate calyx lobes 5 (sometimes 3); lobes ovate (less than 1 mm long and wide), obtuse; stamens 5; pistillate flowers subtended by 2 fleshy bracts; bracts ovate to orbicular (3–5 mm long), united to middle, indurate, enclosing ovary

fruits: utricles enclosed in pistillate bracts (5–8 mm long), margins dentate, surface smooth or with various linear appendages, sessile or short-pedicellate (4–6 mm long)

VEGETATIVE CHARACTERISTICS
leaves: alternate, sometimes opposite below or occurring in fascicles, simple; blades narrowly oblong-linear or oblanceolate (2.5–5 cm long, 3–10 mm wide), thick; apex rounded; base narrowed or cuneate; margins entire, may be inrolled; gray-green, scurfy; sessile

stems: twigs slender, erect or ascending, often herbaceous; white tomentose, later glabrous and dark; trunk bark light gray, smooth, somewhat shreddy

other: highly variable species

HISTORIC, FOOD, AND MEDICINAL USES: some Native Americans ground parched fruits to make pinole flour

LIVESTOCK LOSSES: none

FORAGE VALUE: good for cattle, sheep, deer, and pronghorn; important winter browse on western rangeland, withstands heavy grazing

HABITAT: plains, valleys, and badlands; usually in saline or alkaline soils

Winterfat
Ceratoides lanata (Pursh) J. T. Howell

SYN = *Eurotia lanata* (Pursh) Moq., *Krascheninnikovia lanata* (Pursh) A. D. J. Meeuse & Smit

flowers clustered, spiciform

stigmas

♀ flower

calyx lobes 4, white with green stripe, woolly outside, glabrous inside

stamens

♂ flower

margins entire, revolute

leaves alternate or in fascicles, simple, sessile or petioled

blades covered with dense red or white stellate hairs

twigs gray to reddish-brown, covered with dense stellate hairs

trunk bark gray-brown, exfoliating

Family:	CHENOPODIACEAE
Species:	*Ceratoides lanata* (Pursh) J. T. Howell
Common Name:	Winterfat (hierba lanosa, lambstail, winter sage, white sage)
Life Span:	Perennial
Origin:	Native
Season:	Cool

GROWTH FORM
monoecious or rarely dioecious subshrub (to 1 m tall); highly branched, herbaceous branches ascending; flowers April to September, reproduces from seeds

FLORAL AND FRUIT CHARACTERISTICS
inflorescences: clustered, spiciform; staminate flowers terminal and axillary in clusters of 6–8 flowers; pistillate flowers axillary, solitary or in clusters of 2–4 flowers

flowers: unisexual, apetalous; staminate calyx lobes 4; lobes obovate (1.5–2 mm long, 0.7–1.2 mm wide), white with green stripe, woolly outside, glabrous inside; stamens 4; pistillate flowers subtended by 2 bracts; bracteoles ovate, green, pubescent

fruits: utricles enclosed in 2 bracts; bracts lanceolate (5–6 mm long), 2-horned above, covered with 4 dense tufts of long, white hair

VEGETATIVE CHARACTERISTICS
leaves: alternate or in fascicles, simple; blades linear to narrowly lanceolate (1–4 cm long, 1.7–2.5 mm wide); apex obtuse (rarely acute); margins entire, revolute; covered with dense red or white stellate and simple hairs, midrib prominent; sessile or petioled (1–3 mm long)

stems: twigs gray to reddish-brown, stout, ascending, covered with dense stellate and simple hairs; trunk bark gray-brown, exfoliating

HISTORIC, FOOD, AND MEDICINAL USES: the Blackfoot soaked the leaves in warm water to make a hair wash, others treated fever with a decoction from the leaves

LIVESTOCK LOSSES: none

FORAGE VALUE: good for sheep, pronghorn, elk, and mule deer; fair for cattle; most valuable in winter

HABITAT: hillsides, mesas, foothills, and plains; most abundant in subalkaline or chalky medium to fine soils

Spiny hopsage
Grayia spinosa (Hook.) Moq.

SYN = *Atriplex grayi* Collotzi *ex* Weber, *A. spinosa* (Hook.) Collotzi

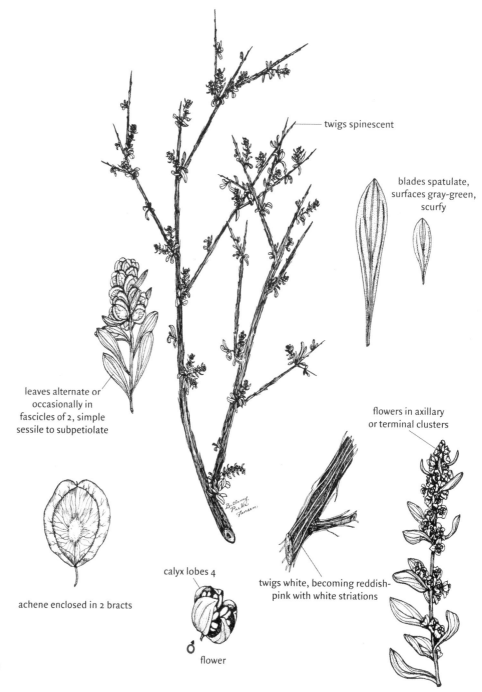

twigs spinescent

blades spatulate,
surfaces gray-green,
scurfy

leaves alternate or
occasionally in
fascicles of 2, simple
sessile to subpetiolate

flowers in axillary
or terminal clusters

achene enclosed in 2 bracts

calyx lobes 4

♂ flower

twigs white, becoming reddish-
pink with white striations

310

Family:	CHENOPODIACEAE
Species:	*Grayia spinosa* (Hook.) Moq.
Common Name:	Spiny hopsage (Grays saltbush, spiny sage)
Life Span:	Perennial
Origin:	Native
Season:	Warm

GROWTH FORM
dioecious or rarely monoecious shrub (to 1.5 m tall); erect, highly branched; branches divergent, spinescent; flowers April to July, reproduces from seeds

FLORAL AND FRUIT CHARACTERISTICS
inflorescences: clustered; axillary or terminal clusters

flowers: unisexual, apetalous; staminate calyx lobes 4, stamens 4 or 5; pistillate flowers subtended by 2 persistent bracts; bracts ovate or orbicular (5–12 mm long, 6 mm wide), enclosing ovary

fruits: achenes enclosed in 2 bracts, bracts thin, glabrous, greenish-white or reddish, entire, sessile, each with a wing on the back or midrib, appearing as a sac-like structure covering the ovary

VEGETATIVE CHARACTERISTICS
leaves: alternate or occasionally in fascicles of 2, simple; blades oblanceolate, spatulate, or obovate (1–3 cm long), fleshy; apex obtuse to subacute; margins entire; gray-green, scurfy, becoming glabrous with maturity; sessile to subpetiolate

stems: twigs rigid, erect to ascending, white, becoming reddish-pink with white striations formed by stringy exfoliating bark; scurfy when young, turning glabrous, spinescent; trunk bark reddish-gray, exfoliating into thin strips, white striations may persist, glabrous

other: bark can be quite variable, from smooth and yellowish-white to red and striate

HISTORIC, FOOD, AND MEDICINAL USES: some Native Americans ground parched seeds to make pinole flour

LIVESTOCK LOSSES: spines may cause minor injury

FORAGE VALUE: good to fair for sheep, goats, deer, and pronghorn; fair to poor for cattle; poor for horses; browsed in fall, winter, and spring; fruits valuable for fattening sheep

HABITAT: mesas, flats, and valleys; adapted to alkaline, limestone, gravelly, and dry heavy clay soils

Halogeton
Halogeton glomeratus (Bieb.) C. A. Mey.

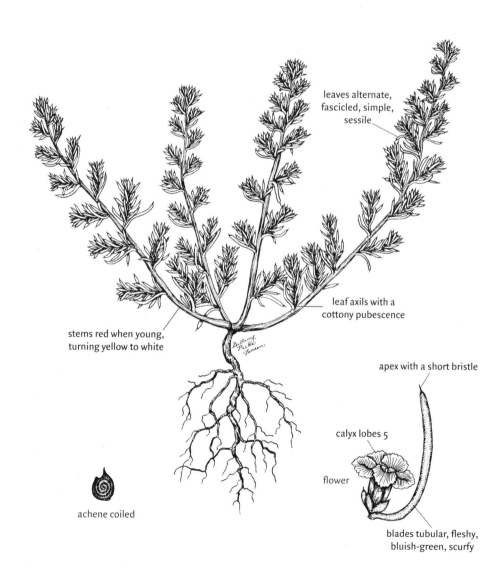

leaves alternate, fascicled, simple, sessile

leaf axils with a cottony pubescence

stems red when young, turning yellow to white

apex with a short bristle

calyx lobes 5

flower

achene coiled

blades tubular, fleshy, bluish-green, scurfy

Family:	CHENOPODIACEAE
Species:	*Halogeton glomeratus* (Bieb.) C. A. Mey.
Common Name:	Halogeton
Life Span:	Annual
Origin:	Introduced (from Asia)
Season:	Warm

GROWTH FORM

forb (5–30 cm tall); few or highly branched from base, stems spreading first then erect or ascending; flowers July to September, reproduces from seeds

FLORAL AND FRUIT CHARACTERISTICS

inflorescences: axillary clusters; distributed throughout the aerial portions of the plant

flowers: perfect or pistillate only, apetalous, small and usually inconspicuous; calyx lobes 5; lobes thin (1.5 mm long), rounded, persistent; stamens 3–5

fruits: achenes coiled; calyx lobes enlarging with maturity to form yellowish or reddish fan-like wings; sessile

other: persistent calyx is often misidentified as the fruiting body

VEGETATIVE CHARACTERISTICS

leaves: alternate, fascicled, simple; blades linear (5–10 mm long), tubular, fleshy, smooth; rounded at apex with a short bristle; margins entire; bluish-green, scurfy; sessile

stems: ascending, red when young, turning yellow to white; few to many primary branches from the base; numerous secondary branches short, glaucous

other: plant is quite striking when growing with its bright red stem and blue-green leaves; leaf axils with a cottony pubescence

HISTORIC, FOOD, AND MEDICINAL USES: introduced into North America as late as 1930 and has rapidly spread becoming a serious weed on rangeland

LIVESTOCK LOSSES: poisonous, contains toxic amounts of sodium, potassium, and calcium oxalates; sheep are the most susceptible; first signs of poisoning occur 2–6 hours after an animal ingests a fatal amount, and death occurs in 9 to 11 hours

FORAGE VALUE: poor to fair for both cattle and sheep; provides usable forage only when mixed in small quantities with other forage plants

HABITAT: dry deserts, barren areas, heavily grazed prairie, roadsides, and other disturbed sites; especially abundant in alkaline or saline soils

Greenmolly summercypress
Kochia americana S. Wats.

SYN = K. *vestita* (S. Wats.) Rydb.

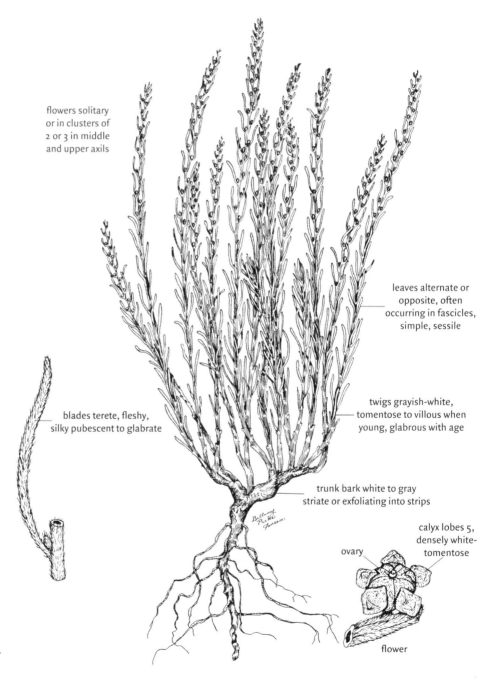

flowers solitary
or in clusters of
2 or 3 in middle
and upper axils

leaves alternate or
opposite, often
occurring in fascicles,
simple, sessile

blades terete, fleshy,
silky pubescent to glabrate

twigs grayish-white,
tomentose to villous when
young, glabrous with age

trunk bark white to gray
striate or exfoliating into strips

calyx lobes 5,
densely white-
tomentose

ovary

flower

Family:	CHENOPODIACEAE
Species:	*Kochia americana* S. Wats.
Common Name:	Greenmolly summercypress (perennial summercypress, red sage, green molly)
Life Span:	Perennial
Origin:	Native
Season:	Warm

GROWTH FORM

subshrub (to 50 cm tall), late deciduous, woody and much branched below, simple and herbaceous above from the branching crown of a woody root; flowers July to September, reproduces from seeds

FLORAL AND FRUIT CHARACTERISTICS

inflorescences: spikes; flowers solitary or in clusters of 2 or 3 in middle and upper axils

flowers: perfect or pistillate only, regular (1.2–2.6 mm in diameter); calyx lobes 5; lobes incurved, herbaceous, persistent, densely white-tomentose; stamens 3–5 exserted or aborted in the pistillate flowers

fruits: utricles depressed globose (2 mm long and wide); persistent calyx forming 5 wedge-shaped, membranous, horizontal wings (1.5–2 mm long), margins scarious and toothed or erose

other: flowers often dry a dark color and readily fall from the plant upon drying

VEGETATIVE CHARACTERISTICS

leaves: alternate or opposite, often occurring in fascicles, simple; blades narrow, linear (6–30 mm long), erect or ascending, terete, fleshy, flat when dried; apex acute; margins entire; dark green, silky pubescent to glabrate; sessile

stems: twigs simple, erect or ascending, grayish-white, numerous, tomentose to villous when young, glabrous with age; trunk bark white to gray striate or exfoliating into strips

other: new growth, especially leaves often drying black; two varieties are glabrate, a third densely and permanently villous

HISTORIC, FOOD, AND MEDICINAL USES: none

LIVESTOCK LOSSES: may accumulate nitrates

FORAGE VALUE: excellent for sheep and goats, poor for cattle and deer; often used as winter forage for sheep; high in protein during the fall

HABITAT: desert valleys, flats, marshes, roadsides, and foothills of the cold desert region; usually found growing in alkaline soils

Kochia
Kochia scoparia (L.) Schrad.

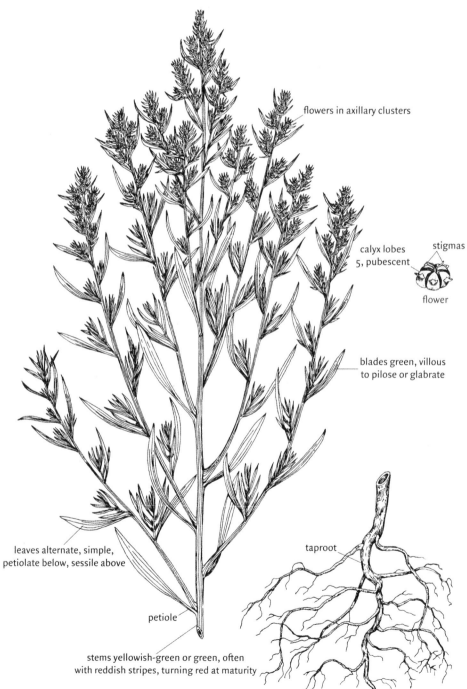

flowers in axillary clusters

calyx lobes 5, pubescent

stigmas

flower

blades green, villous to pilose or glabrate

leaves alternate, simple, petiolate below, sessile above

petiole

taproot

stems yellowish-green or green, often with reddish stripes, turning red at maturity

Family:	CHENOPODIACEAE
Species:	*Kochia scoparia* (L.) Schrad.
Common Name:	Kochia (cochia, coquia, fireweed, Mexican fireweed, summercypress, belvedere)
Life Span:	Annual
Origin:	Introduced (from Eurasia)
Season:	Warm

GROWTH FORM

forb (0.1–1 m tall); erect and spreading, much-branched from the base; flowers June to August, reproduces from seeds

FLORAL AND FRUIT CHARACTERISTICS

inflorescences: clusters; axillary, remotely long-spiciform or densely long-spiciform to compact cylindric or oblong-claviform; some plants floriferous for most of their height

flowers: perfect or unisexual, apetalous; paired, subtended by pubescent leafy bracts (3–18 mm long); calyx campanulate (0.3–0.6 mm long, 1.5–2 mm wide), lobes 5; lobes oblong (2.3–3 mm wide), pubescent, bifid, persistent; stamens 5; sessile

fruits: utricles depressed-globose, enclosed in a 5-winged persistent pericarp

other: some plants floriferous for most of their height

VEGETATIVE CHARACTERISTICS

leaves: alternate, simple; blades linear or lanceolate to narrow obovate (2–10 cm long, 1–12 mm wide), flat; apex acute or obtuse; margins entire; green, villous to pilose or glabrate; petiolate below, sessile above

stems: erect or spreading, much-branched, yellowish-green or green, often with reddish stripes, turning red with maturity, glabrous or glaberescent

other: plant size is moisture dependent; plant is often called a tumble weed, breaking off at ground level and rolling with the wind when it grows in a cylindrical to globose shape

HISTORIC, FOOD, AND MEDICINAL USES: escaped ornamental, common cause of hay fever and other respiratory problems

LIVESTOCK LOSSES: may cause nitrate poisoning, bloat, and photosensitization

FORAGE VALUE: good for livestock and wildlife when immature, quality and palatability rapidly decline with maturity; generally considered a weed, however, is palatable to livestock and can be highly productive and nutritious

HABITAT: roadsides, pastures, wastelands, disturbed sites, and fields; adapted to a broad range of soils

Russian thistle
Salsola iberica Sennen & Pau

SYN = *S. kali* L., *S. pestifera* A. Nels., *S. tragus* L.

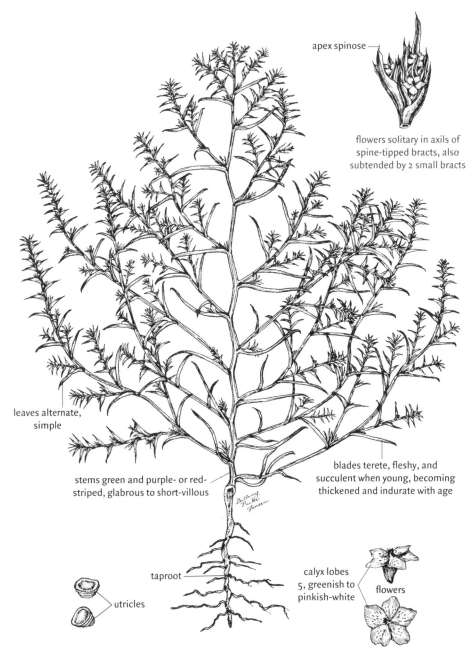

apex spinose

flowers solitary in axils of
spine-tipped bracts, also
subtended by 2 small bracts

leaves alternate,
simple

stems green and purple- or red-
striped, glabrous to short-villous

blades terete, fleshy, and
succulent when young, becoming
thickened and indurate with age

taproot

utricles

calyx lobes
5, greenish to
pinkish-white

flowers

Family:	CHENOPODIACEAE
Species:	*Salsola iberica* Sennen & Pau
Common Name:	Russian thistle (rodadora, tumbleweed, cardo ruso, maromera, prickly russianthistle)
Life Span:	Annual
Origin:	Introduced (from Eurasia)
Season:	Warm

GROWTH FORM

forb (30–80 cm tall); much-branched, ascending or spreading, rounded in shape; flowers July to October, reproduces from seeds

FLORAL AND FRUIT CHARACTERISTICS

inflorescences: flowers solitary; in axils of spine-tipped bracts, also sub-tended by 2 small bracts

flowers: perfect, apetalous; calyx lobes 5 (2.5–3.5 mm long, 3–6 mm wide); lobes ovate or oblong, greenish to pinkish-white, acute, entire, persistent; stamens 3–5, exserted with age

fruits: utricles ovoid or orbicular (1.5–2.5 mm wide), flattened, apex concave or convex; persistent calyx forming 5 wings (3–6 mm wide), white to pink

VEGETATIVE CHARACTERISTICS

leaves: alternate, simple; blades linear to filiform (2–8 cm long, 1 mm wide) reduced above, terete, fleshy, and succulent when young, becoming thickened and indurate with age, becoming spinose at the apex; base broad; margins entire to denticulate; upper leaves eventually thickening and enclosing fruit

stems: ascending or spreading, freely branching, green and purple- or red-striped, glabrous to short-villous

other: forming a rounded plant that breaks off at ground level to tumble along with the wind, thus a tumbleweed

HISTORIC, FOOD, AND MEDICINAL USES: young shoots can be used as a potherb; seeds can be ground into meal; first introduced into Dakota Territory in flax seed in about 1873

LIVESTOCK LOSSES: mechanical injury from sharp-pointed leaves, may accumulate nitrates, may contain oxalates

FORAGE VALUE: fair for cattle and sheep in early spring, becoming worthless with maturity because of the sharp-pointed leaves; young plants can be cut for hay

HABITAT: cultivated fields, heavily grazed pastures, roadsides, waste places, and disturbed sites in nearly all soil types

Black greasewood
Sarcobatus vermiculatus (Hook.) Torr.

SYN = *S. baileyi* Coville

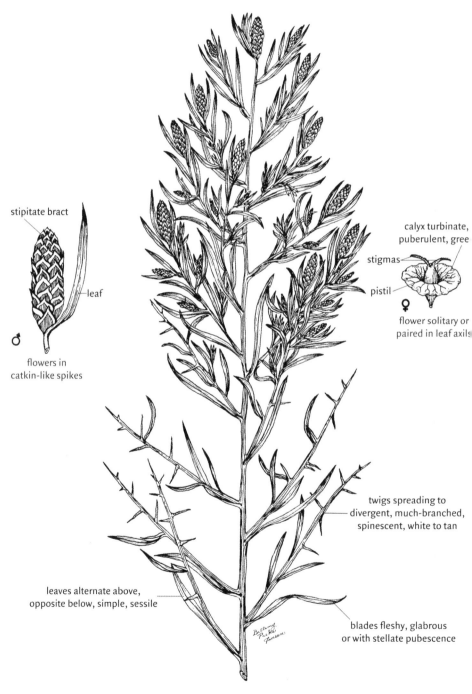

stipitate bract

leaf

♂

flowers in
catkin-like spikes

calyx turbinate,
puberulent, gree

stigmas

pistil

♀

flower solitary or
paired in leaf axils

twigs spreading to
divergent, much-branched,
spinescent, white to tan

leaves alternate above,
opposite below, simple, sessile

blades fleshy, glabrous
or with stellate pubescence

Family:	CHENOPODIACEAE
Species:	*Sarcobatus vermiculatus* (Hook.) Torr.
Common Name:	Black greasewood (chico, greasewood, chicobush)
Life Span:	Perennial
Origin:	Native
Season:	Warm

GROWTH FORM

monoecious shrub (to 2.5 m tall); late deciduous, erect or spreading, rounded, much-branched, rigid; flowers May to August, reproduces from seeds and sprouts

FLORAL AND FRUIT CHARACTERISTICS

inflorescences: staminate flowers in terminal and axillary catkin-like spikes (1.5–3 cm long, 4–5 mm wide); pistillate flowers solitary or paired in leaf axils below staminate flowers; sessile or short-peduncled

flowers: unisexual, apetalous; staminate flowers subtended by a scale or stipitate bract; scale rhombic, pointed, green; stamens 3; calyx of pistillate flowers turbinate (1.5 mm long, 1 mm wide), puberulent, green

fruits: achenes (4–5 mm long, 2.5–3.5 mm wide) turbinate, enclosed in the persistent perianth which forms a wing (to 1 cm wide); wing around the middle scarious, veined, green to tan or reddish

VEGETATIVE CHARACTERISTICS

leaves: alternate above, opposite below, simple; blades linear to filiform (1–4 cm long, 1.5–2.5 mm wide) elliptical or 4-angled in cross-section, fleshy; apex obtuse or acute; base cuneate; glabrous or with stellate pubescence; sessile

stems: twigs spreading to divergent, much-branched, rigid, spinescent, white to tan; trunk bark yellowish-gray to light brown, fissured and exfoli-ating, shiny, with elliptic pits

HISTORIC, FOOD, AND MEDICINAL USES: wood is sometimes used for fuel; American Indians used sharpened branches as planting tools

LIVESTOCK LOSSES: soluble oxalates have caused mass mortality in flocks of sheep; cattle and horses are rarely poisoned; young twigs are especially toxic

FORAGE VALUE: good to fair for sheep, goats, and big game in winter; important food and cover for small mammals, jackrabbits, and birds

HABITAT: dry plains, slopes, eroded hills, and flats; especially in alkaline or saline soils

St. Johnswort
Hypericum perforatum L.

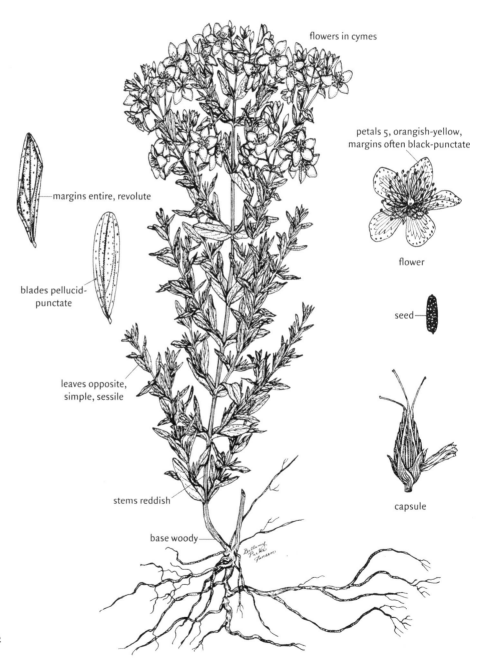

flowers in cymes

petals 5, orangish-yellow,
margins often black-punctate

flower

margins entire, revolute

blades pellucid-
punctate

seed

leaves opposite,
simple, sessile

stems reddish

capsule

base woody

Family:	CLUSIACEAE
Species:	*Hypericum perforatum* L.
Common Name:	St. Johnswort (klamathweed, goatweed)
Life Span:	Perennial
Origin:	Introduced (from Europe and Africa)
Season:	Warm

GROWTH FORM

forb (0.3–1.5 m tall); erect, somewhat clustered, much-branched above and often from the base; often with small, sterile shoots basally; flowers June to August, reproduces from seeds and rhizomes

FLORAL AND FRUIT CHARACTERISTICS

inflorescences: cymes; collectively flat-topped or rounded, leafy-bracteate at branch tips, densely flowered

flowers: perfect, regular; sepals 5, linear-lanceolate (4–6 mm long), unequal; petals 5, orangish-yellow, obovate to cuneate (8–12 mm long), twisting when dry, margins often black-punctate; stamens numerous, about as long as the petals

fruits: capsules globose (1–2 cm long), exceeding the sepals, 3-celled, tipped by 3 spreading styles, glandular; several-seeded

VEGETATIVE CHARACTERISTICS

leaves: opposite, simple; blades elliptic to linear or oblong (1–3 cm long, 2–8 mm wide), reduced above, diverging from the stem at about 90°angles; apex acute to rounded; base acuminate; margins entire, revolute; pellucid-punctate; sessile

stems: reddish, woody at base, may appear jointed due to opposite leaf scars

HISTORIC, FOOD, AND MEDICINAL USES: the Menominee Indians mixed St. Johnswort with black raspberry root in hot water and drank the tea for tuberculosis; acts as a diuretic, may kill internal worms

LIVESTOCK LOSSES: poisonous to livestock; causes a photosensitizing reaction resulting in dermatitis to nonpigmented skin of horses, cattle, and sheep; sunlight acts as a catalyst, therefore, symptoms only occur when livestock are exposed to strong sunlight after grazing; frequently fatal

FORAGE VALUE: poor to worthless for cattle, sheep, horses, and wildlife; fair for goats; mourning doves and quail eat the seeds

HABITAT: prairies, pastures, waste areas, disturbed fields, and roadsides; most abundant in sandy soils

Redosier dogwood
Cornus sericea L.

SYN = C. alba L., C. stolonifera Michx.

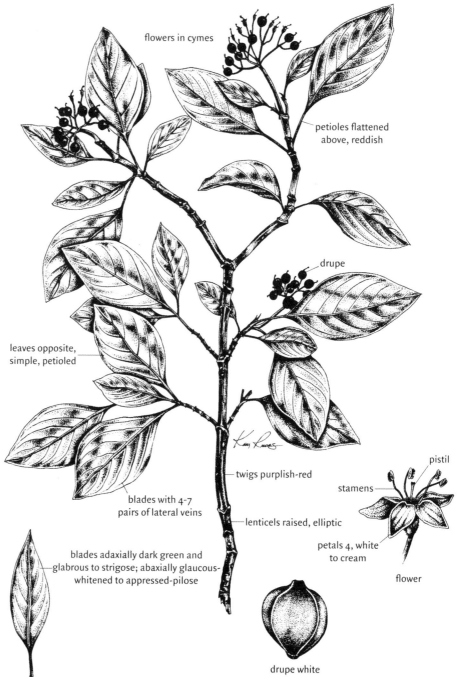

flowers in cymes

petioles flattened
above, reddish

drupe

leaves opposite,
simple, petioled

twigs purplish-red

blades with 4-7
pairs of lateral veins

lenticels raised, elliptic

pistil

stamens

petals 4, white
to cream

flower

blades adaxially dark green and
glabrous to strigose; abaxially glaucous-
whitened to appressed-pilose

drupe white

Family:	CORNACEAE
Species:	*Cornus sericea* L.
Common Name:	Redosier dogwood (American dogwood, red osier, red dogwood)
Life Span:	Perennial
Origin:	Native
Season:	Cool

GROWTH FORM

shrub (1–4 m tall); highly branched; erect to procumbent, stoloniferous; flowers May to July; fruits mature August to September, reproduces from seeds and stolons

FLORAL AND FRUIT CHARACTERISTICS

inflorescence: cymes, flat-topped to convex (2–6 cm wide), 4–6 main branches, densely flowered; pedicels reddish (4–6 cm long), tomentose, becoming glabrous

flowers: perfect, regular; sepal lobes 4, triangular (to 0.5 mm long), erect to spreading, green, pubescent; corolla ovate (1.5–2.5 mm long); petals 4, triangular (2–4 mm long), white to cream in color, erect to spreading at anthesis, margins inrolled, outer surfaces pubescent, inner surfaces papillose; stamens 4, opposite the sepals

fruits: drupes globose (6–9 mm wide), white, few appressed hairs, depressed at the pedicle, style persistent; 1 seed per drupe; seeds pale to dark brown (5 mm long, 4–6 mm wide) with 8 white to yellow stripes

VEGETATIVE CHARACTERISTICS

leaves: opposite, simple; blades oblong-lanceolate to ovate (5–11 cm long, 2.5–5.5 cm wide); apex usually acute or acuminate; base cuneate; margins entire; adaxially dark green, glabrous to strigose; abaxially glaucous-whitened to appressed-pilose, occasionally with reddish hairs; lateral veins 4–7 pairs, evenly spaced; petioles (0.5–2.5 cm long) flattened above, reddish

stems: twigs purplish-red, glabrous to appressed or spreading pubescent; lenticels raised, elliptic, light in color; leaf scars crescent-shaped, dark; terminal buds ovoid, compressed; trunk bark red to occasionally greenish, smooth

HISTORIC, FOOD, AND MEDICINAL USES: some Native Americans smoked the inner bark layers; a bark extract was used as an emetic in treating coughs and fevers; a red dye was made by boiling the small roots; a tea from the leaves has been used as a quinine substitute

LIVESTOCK LOSSES: none

FORAGE VALUE: not utilized by livestock, nearly worthless to big game, fruits are eaten by birds

HABITAT: moist soil along streams, meadows, and lake shores in wooded or open areas

Oneseed juniper
Juniperus monosperma (Engelm.) Sarg.

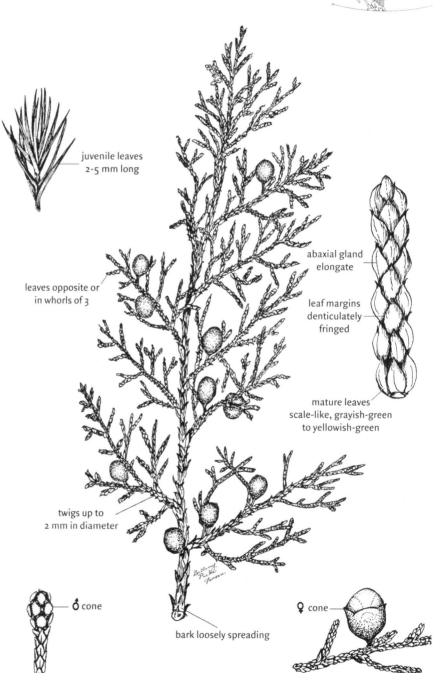

juvenile leaves
2-5 mm long

abaxial gland
elongate

leaves opposite or
in whorls of 3

leaf margins
denticulately
fringed

mature leaves
scale-like, grayish-green
to yellowish-green

twigs up to
2 mm in diameter

♂ cone

♀ cone

bark loosely spreading

326

Family:	CUPRESSACEAE
Species:	*Juniperus monosperma* (Engelm.) Sarg.
Common Name:	Oneseed juniper (táscate, cherrystone juniper, enebro, sabina)
Life Span:	Perennial
Origin:	Native
Season:	Evergreen

GROWTH FORM

dioecious shrub or small tree (to 12 m tall); crown rounded, branching near the ground, often with several trunks; pollinates March to April, cones mature September, reproduces from seeds

CONE CHARACTERISTICS

cones: axillary and terminal, borne on branches of the previous year; staminate cones solitary, oblong-ovoid (3–4 mm long, 2 mm in diameter), brownish or yellowish; ovulate cones clustered, globose

mature ovulate cones: berry-like; subglobose (3–7 mm in diameter), fleshy, blue to brownish or copper-colored, infrequently blue, 1-seeded or occasionally 2-seeded

VEGETATIVE CHARACTERISTICS

leaves: opposite or in whorls of 3, scale-like; imbricate and appressed against the branches, ovate (1–3 mm long, 0.5–1.2 mm wide), fleshy, thickened, rounded; apex acute or acuminate; margins denticulately fringed; grayish-green to yellowish-green, old leaves reddish-brown; abaxial gland elongate

stems: twigs slender (up to 2 mm in diameter), reddish-brown; bark loosely spreading

other: leaves on juvenile branchlets often longer (2–5 mm long) than on mature branchlets

HISTORIC, FOOD, AND MEDICINAL USES: berry-like cones were gathered and eaten by Native Americans; wood used for prayer sticks and bows; green dye was obtained from bark; cones were ground to make flour; fibrous bark was used for mats, saddles, and breechcloths; cones of *Juniperus* species are used to flavor gin

LIVESTOCK LOSSES: may cause abortion in livestock

FORAGE VALUE: worthless to cattle and sheep, seldom consumed; occasionally browsed by goats; mature cones are eaten by deer, quail, foxes, chipmunks, squirrels, songbirds, and coyotes

HABITAT: hillsides, canyons, flats, and exposed ledges; adapted to a broad range of soil types

Rocky Mountain juniper
Juniperus scopulorum Sarg.

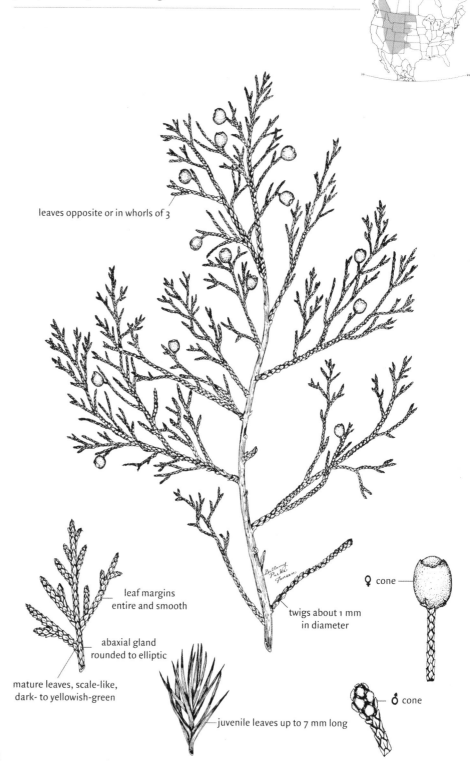

leaves opposite or in whorls of 3

leaf margins
entire and smooth

abaxial gland
rounded to elliptic

mature leaves, scale-like,
dark- to yellowish-green

juvenile leaves up to 7 mm long

twigs about 1 mm
in diameter

♀ cone

♂ cone

Family:	CUPRESSACEAE
Species:	*Juniperus scopulorum* Sarg.
Common Name:	Rocky Mountain juniper (Rocky Mountain cedar)
Life Span:	Perennial
Origin:	Native
Season:	Evergreen

GROWTH FORM

dioecious shrub or small tree (to 12 m tall); crown rounded to pyramidal, irregular, branching above and near the ground; pollinates April to May, cones mature October to December of the second year following pollination, reproduces from seeds

CONE CHARACTERISTICS

cones: inconspicuous, solitary at tips of branchlets; staminate cone oblong (2–4 mm long, 1–2 mm in diameter), sessile, yellowish-brown; ovulate cone globose

mature ovulate cones: berry-like; subglobose (5–8 mm in diameter), fleshy, resinous-juicy, bright blue to purplish, mostly 2-seeded

VEGETATIVE CHARACTERISTICS

leaves: opposite or in whorls of 3, scale-like; closely appressed, not or slightly imbricate, rhombic-ovate to ovate-elliptic (1–4 mm long, 0.7–1.5 mm wide), fleshy, thickened, rounded; apex obtuse to weakly acute; margins entire and smooth; dark- to yellowish-green, often pale and glaucous; abaxial gland rounded to elliptic

stems: twigs slender (about 1 mm in diameter), scaly; flattened at first, then becoming round; trunk bark grayish- to reddish-brown, shreddy, furrowed

other: leaves on juvenile branchlets acerose (to 7 mm long)

HISTORIC, FOOD, AND MEDICINAL USES: some Native Americans ate the cones raw or cooked, used as flavoring for meat, cones and young shoots were boiled for tea, cones were ground for mush and cakes, wax from berries used in candles; currently used as an ornamental, in shelterbelts, for fence posts, and fuel; reported to have insect repellent properties; cones of *Juniperus* species are used to flavor gin

LIVESTOCK LOSSES: may cause abortions in livestock

FORAGE VALUE: worthless to livestock, generally not consumed; important browse plant for pronghorn, mule deer, and bighorn sheep; birds utilize the cones

HABITAT: ridges, bluffs, canyons, hillsides, and wash areas; often on undeveloped, erodible soils; most abundant on calcareous and somewhat alkaline soils

Bracken fern
Pteridium aquilinum (L.) Kuhn

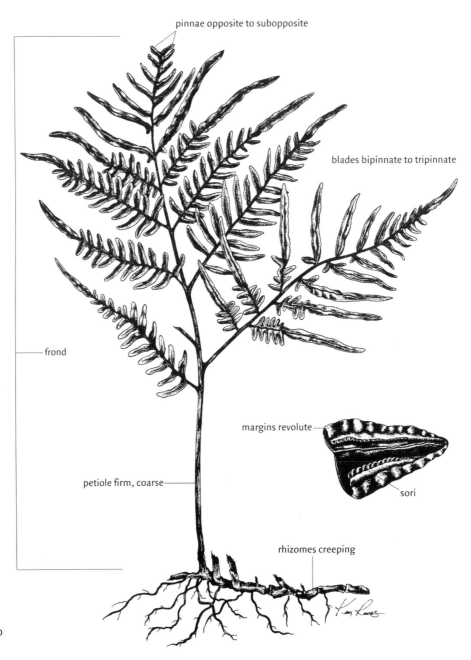

pinnae opposite to subopposite

blades bipinnate to tripinnate

frond

margins revolute

sori

petiole firm, coarse

rhizomes creeping

Family:	DENNSTAEDTIACEAE
Species:	*Pteridium aquilinum* (L.) Kuhn
Common Name:	Bracken fern (helecho, western bracken, western brackenfern, brake)
Life Span:	Perennial
Origin:	Native
Season:	Warm

GROWTH FORM
fern (to 2 m tall); forming dense colonies or thickets from creeping rhizomes; spores produced July to September; reproduces from spores and rhizomes

SORI AND SPORE CHARACTERISTICS:
sori: marginal, more or less continuous borne on vascular strand connecting the vein ends of the leaf blades, often obscured by recurved false outer indusium; indusium double, outer false and reflexed; inner true and developed or obsolescent

spores: tetrahedral-globose, brown, minute, without perispores

VEGETATIVE CHARACTERISTICS
fronds: widely spaced along rhizome; blades bipinnate to tripinnate (20–80 cm long, 20–90 cm wide), broadly deltoid, glabrous to pubescent, coriaceous; pinnae opposite to subopposite, lower pair deltoid to ovate, nearly as long as upper portion of blade, upper pinnae smaller, deltoid to lanceolate; margins revolute, veins free or joined at margin; petiole firm, coarse, ascending to erect, dark at the base

stems: rhizomes (to 1 cm thick), deeply subterranean, creeping and branched, blackish, pilose to glabrate; true vessels present

other: variable plant with four varieties recognized in North America

HISTORIC, FOOD, AND MEDICINAL USES: some Native Americans ate young fronds raw or boiled and ground rhizomes into a meal; sometimes eaten to control internal parasites

LIVESTOCK LOSSES: toxic to horses and to some extent to cattle; toxicity due to a breakdown of thiamine in the blood; cumulative poison

FORAGE VALUE: cattle, elk, and deer may graze it in early spring, unpalatable throughout the rest of the year

HABITAT: pastures, open forests, cut-over areas, and abandoned fields; poor prairie and woodland soils; less frequent in rich, moist woodlands

Russet buffaloberry
Shepherdia canadensis (L.) Nutt.

SYN = *Elaeagnus canadensis* (L.) A. Nels.

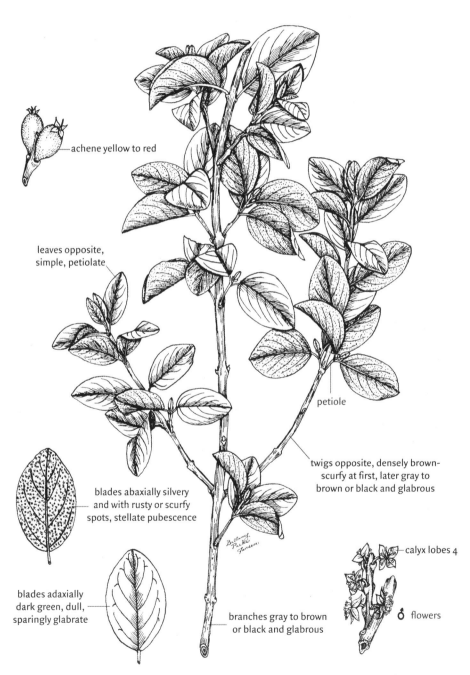

achene yellow to red

leaves opposite,
simple, petiolate

petiole

twigs opposite, densely brown-
scurfy at first, later gray to
brown or black and glabrous

blades abaxially silvery
and with rusty or scurfy
spots, stellate pubescence

calyx lobes 4

♂ flowers

blades adaxially
dark green, dull,
sparingly glabrate

branches gray to brown
or black and glabrous

Family:	ELAEAGNACEAE
Species:	*Shepherdia canadensis* (L.) Nutt.
Common Name:	Russet buffaloberry (Canadian buffaloberry, rabbitberry)
Life Span:	Perennial
Origin:	Native
Season:	Cool

GROWTH FORM

dioecious shrub (to 3 m tall); much-branched, spreading, crown rounded; flowers April to June, fruits mature July to September, reproduces from seeds

FLORAL AND FRUIT CHARACTERISTICS

inflorescences: flowers solitary or in clusters of 2–4; appearing before the leaves at nodes on 1-year-old twigs

flowers: unisexual, apetalous; calyx lobes of staminate flowers 4 (1–2 mm long), adaxial surface yellow, abaxial surface brown; stamens usually 8; pistillate flowers with united urn-shaped calyx enclosing the ovary; lobes 4, spreading (0.8–1.5 mm long) with 3 prominent veins on the surface

fruits: achenes; drupe-like, oval to ellipsoidal (3–8 mm long), enveloped by fleshy base of hypanthium, yellow to red, juicy, insipid, bitter

VEGETATIVE CHARACTERISTICS

leaves: opposite, simple; blades oval to ovate or elliptic (2–6 cm long); apex obtuse; base rounded to subcordate; margins entire; adaxially dark green, dull, sparingly glabrate; abaxially silvery and with rusty or scurfy spots, stellate pubescence; petiolate (3–5 mm long)

stems: twigs opposite, densely brown-scurfy when young, older branches gray to brown or black and glabrous, unarmed

HISTORIC, FOOD, AND MEDICINAL USES: fruits are edible, cooked or raw, although insipid; fruits may be whipped with sugar for a dessert; commonly causes diarrhea

LIVESTOCK LOSSES: none

FORAGE VALUE: fair for sheep before frost, otherwise considered worthless; seldom browsed by cattle, horses, or big game animals; fruits are consumed by many kinds of birds and small mammals

HABITAT: limestone slopes, ledges, riverbanks, and open wooded slopes; adapted to a broad range of soils, most abundant in moist soils

Longleaf ephedra
Ephedra trifurca Torr. ex S. Wats.

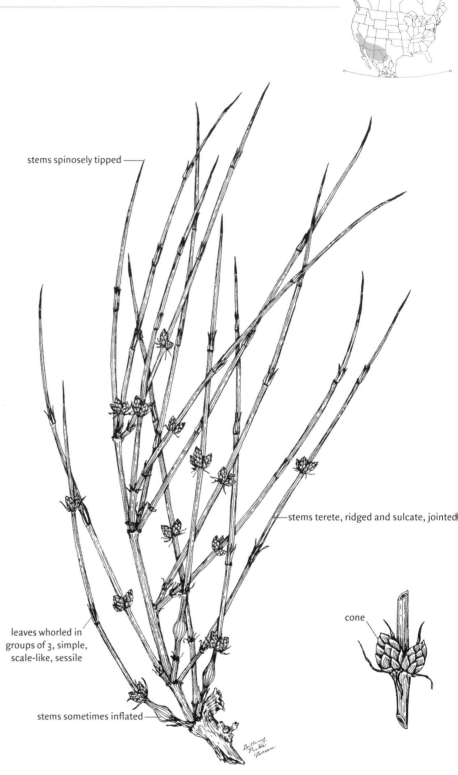

stems spinosely tipped

stems terete, ridged and sulcate, jointed

cone

leaves whorled in
groups of 3, simple,
scale-like, sessile

stems sometimes inflated

Family:	EPHEDRACEAE
Species:	*Ephedra trifurca* Torr. *ex* S. Wats.
Common Name:	Longleaf ephedra (cañatillo, Mormon tea, itamo real, jointfir, longleaf jointfir)
Life Span:	Perennial
Origin:	Native
Season:	Evergreen

GROWTH FORM
dioecious shrub (to 2 m tall); erect, branches solitary or whorled at the nodes; pollinates March to April, seeds mature April to May, reproduces from seeds

CONE CHARACTERISTICS
cones: staminate cones solitary or numerous at stem nodes, obovate (5–6 mm long), subtended by 2 bracts; ovulate cones of 2 erect ovules enclosed in an urceolate involucre of scales (10–11 mm long); staminate cones with 2–8 stamens protruding beyond their united bracts; pistillate cones with scales in many whorls of 3

mature ovulate cones: elliptic to obovate (7–12 mm long, 2–4 mm wide), solitary or 2–3 at upper nodes, reddish-brown; margins of the cone bracts conspicuously membranous

VEGETATIVE CHARACTERISTICS
leaves: whorled in groups of 3, simple, scale-like (5–13 mm long); united for one-half to three-fourths of their total length, persistent and spinose on older branches; apex subspinosely tipped; sheath membranous, becoming fibrous with age; sessile

stems: rather rigid, terete, ridged and sulcate, jointed (internodes 3–9 cm long), spinosely tipped, pale yellowish-green; portions of stems may be inflated

HISTORIC, FOOD, AND MEDICINAL USES: some Native Americans and Mexicans have long used stem decoctions as a cooling beverage and have eaten the cones; common name—Mormon tea—is derived from use as a beverage by Latter-day Saints pioneers in the American West; contains the drug ephedrine

LIVESTOCK LOSSES: none

FORAGE VALUE: poor for livestock, heavily browsed in emergency situations, may be important winter forage; browsed by bighorn sheep and jackrabbits; cones are eaten by scaled quail

HABITAT: mesas, foothills, open sites in valleys, and dry creek beds; most abundant in coarse soils

Pointleaf manzanita
Arctostaphylos pungens Kunth

SYN = *A. chaloneorum* J. B. Roof, *A. pseudopungens* J. B. Roof

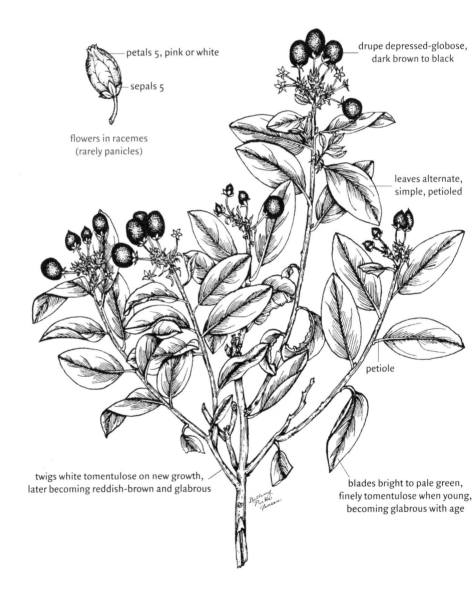

petals 5, pink or white

sepals 5

flowers in racemes
(rarely panicles)

drupe depressed-globose,
dark brown to black

leaves alternate,
simple, petioled

petiole

twigs white tomentulose on new growth,
later becoming reddish-brown and glabrous

blades bright to pale green,
finely tomentulose when young,
becoming glabrous with age

Family:	ERICACEAE
Species:	*Arctostaphylos pungens* Kunth
Common Name:	Pointleaf manzanita (manzanilla, Mexican manzanita, pingüica, madroño)
Life Span:	Perennial
Origin:	Native
Season:	Evergreen

GROWTH FORM

shrub (to 3 m tall); erect or spreading, branched from the base; forming extensive, dense thickets; flowers January to March, fruits mature April to July, reproduces from seeds and rootsprouts

FLORAL AND FRUIT CHARACTERISTICS

inflorescences: racemes (rarely panicles); cylindric, congested, nodding; rachis often distinctly thickened and club-shaped at the apex

flowers: perfect, regular; calyx persistent; sepals 5, short-deltoid (2–3 mm long, 1–2 mm wide), firm, thick, pubescent or tomentose; corolla urceolate (5–7 mm long); petals 5, pink or white, recurved at summit; stamens 10, included; ovary superior

fruits: drupes depressed-globose (5–8 mm in diameter), fleshy, dark brown to black, glabrous; nutlets 5, ridged dorsally, each 1-seeded

VEGETATIVE CHARACTERISTICS

leaves: alternate, simple; blades oblong to oblanceolate (1.5–3 cm long, 9–15 mm wide); apex acute and mucronate or mucronulate; base rounded or subcuneate; margins entire; surfaces bright to pale green, finely tomentulose when young, becoming glabrous with age; petioled (to 1 cm long)

stems: twigs rigid, white tomentulose on new growth, later becoming reddish-brown and glabrous; sometimes rooting where branches touch the ground

HISTORIC, FOOD, AND MEDICINAL USES: fruit is sold in Mexican markets and used for jellies; leaves and fruits are used in Mexico as a diuretic and household remedy for dropsy, bronchitis, and venereal diseases

LIVESTOCK LOSSES: none

FORAGE VALUE: worthless for cattle and sheep, poor to fair for goats; goats consume leaves and browse young twigs; goats will peel back bark in the spring, presumably for sap; fruits eaten by grouse, skunks, deer, quail, bears, and coyotes

HABITAT: rocky mesas and mountain slopes; most abundant on dry, well-drained soils

Bearberry
Arctostaphylos uva-ursi (L.) Spreng.

SYN = *A. adenotricha* (Fern. & J. F. Macbr.) A. & D. Löve & Kapoor

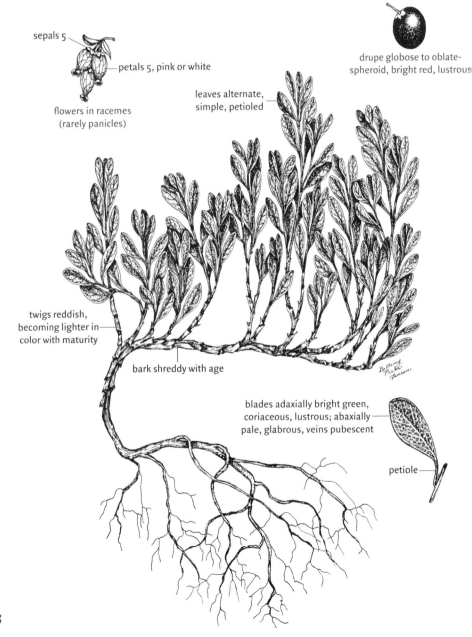

sepals 5

petals 5, pink or white

flowers in racemes
(rarely panicles)

drupe globose to oblate-
spheroid, bright red, lustrous

leaves alternate,
simple, petioled

twigs reddish,
becoming lighter in
color with maturity

bark shreddy with age

blades adaxially bright green,
coriaceous, lustrous; abaxially
pale, glabrous, veins pubescent

petiole

338

Family:	ERICACEAE
Species:	*Arctostaphylos uva-ursi* (L.) Spreng.
Common Name:	Bearberry (manzanita, kinnikinnick)
Life Span:	Perennial
Origin:	Native
Season:	Evergreen

GROWTH FORM

shrub with depressed or trailing stems forming prostrate mats (to 2 m in diameter); much-branched, terminal portions erect (to 20 cm tall); flowers April to July, fruits mature July to October, reproduces from seeds

FLORAL AND FRUIT CHARACTERISTICS

inflorescences: racemes (rarely panicles); dense, terminal, few- to several-flowered, flowers pendulous

flowers: perfect, regular; calyx persistent; sepals 5 (1–1.5 mm long), pink or white, distinct; corolla ovoid to urceolate (4–8 mm long); petals 5, pink or white, reflexed; stamens 10, shorter than the corolla; ovary superior

fruits: drupes; globose to oblate-spheroid (4–10 mm in diameter), fleshy, bright red, lustrous, somewhat persistent; nutlets 5, ridged dorsally, each 1-seeded

VEGETATIVE CHARACTERISTICS

leaves: alternate, simple; blades spatulate to obovate (1–3 cm long, 5–10 mm wide); apex obtuse, retuse, or rounded; base cuneate or narrowed; margins entire, revolute; adaxially bright green, coriaceous, lustrous; abaxially pale, glabrous, veins pubescent; petioled (to 1 cm long)

stems: young twigs reddish, becoming lighter in color with maturity; bark becomes shreddy with age

HISTORIC, FOOD, AND MEDICINAL USES: fruit is insipid but edible if cooked; fruits used by early settlers for treating urinary disorders, leaves were used as a tobacco substitute

LIVESTOCK LOSSES: none

FORAGE VALUE: worthless to poor for livestock, browsed by big horn sheep and deer; fruits provides food for grouse, wild turkeys, and other birds

HABITAT: woodlands, hillsides, and mountain slopes to slightly above timberline; grows in well-drained soils in either open or shaded areas; most abundant in sandy and rocky soils

Guajillo
Acacia berlandieri Benth.

SYN = *A. emoryana* Benth.

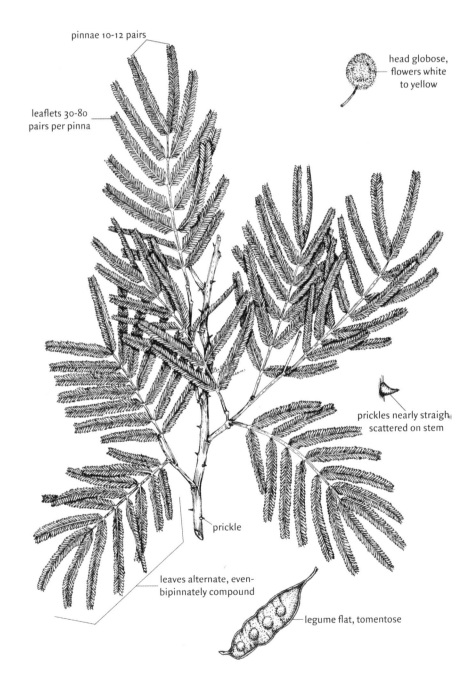

pinnae 10-12 pairs

leaflets 30-80
pairs per pinna

head globose,
flowers white
to yellow

prickles nearly straight,
scattered on stem

prickle

leaves alternate, even-
bipinnately compound

legume flat, tomentose

Family:	FABACEAE
Species:	*Acacia berlandieri* Benth.
Common Name:	Guajillo (Berlandier acacia)
Life Span:	Perennial
Origin:	Native
Season:	Warm

GROWTH FORM

shrub (to 4 m tall); several main stems from the base, ascending, sparingly branched; flowers November to March, fruits mature June to July, reproduces from seeds

FLORAL AND FRUIT CHARACTERISTICS

inflorescences: heads globose (8–15 mm in diameter), rarely elongated (2 cm long, 1 cm wide), axillary, solitary or in clusters, many-flowered; peduncles pubescent (2–5 cm long)

flowers: perfect, regular; calyx lobes 5, pubescent; petals 5, white to yellow, pubescent; stamens numerous, exserted, distinct

fruits: legumes (8–15 cm long, 1.5–2.5 cm wide), oblong, flat, thin, straight or somewhat curved, apex obtuse or apiculate, margins thickened, velvety-tomentose when mature, dark brown, several-seeded; one margin of the legume is generally straighter than the other

VEGETATIVE CHARACTERISTICS

leaves: alternate, even-bipinnately compound (9–18 cm long), delicate, almost fern-like in appearance; pinnae 10–12 pairs; leaflets 30–80 pairs per pinna, crowded; leaflets linear to oblong (2–6 mm long), oblique, apex acute, margins entire, tomentose when immature and glabrate when mature, prominently nerved; stipules small, caducous

stems: twigs gray to white; generally armed with nearly straight, scattered prickles (1–3 mm long)

HISTORIC, FOOD, AND MEDICINAL USES: important honey plant, gums and dyes have been extracted from this shrub; used as an ornamental either as a hedge or specimen planting

LIVESTOCK LOSSES: may cause hydrocyanic acid poisoning in livestock when extremely large amounts are consumed

FORAGE VALUE: fair for livestock and wildlife, seeds furnish important food for birds and small mammals

HABITAT: limestone ridges and caliche hills; most abundant in sandy soils, uncommon in deep soils

Huisache
Acacia farnesiana (L.) Willd.

SYN = *A. smallii* Isely

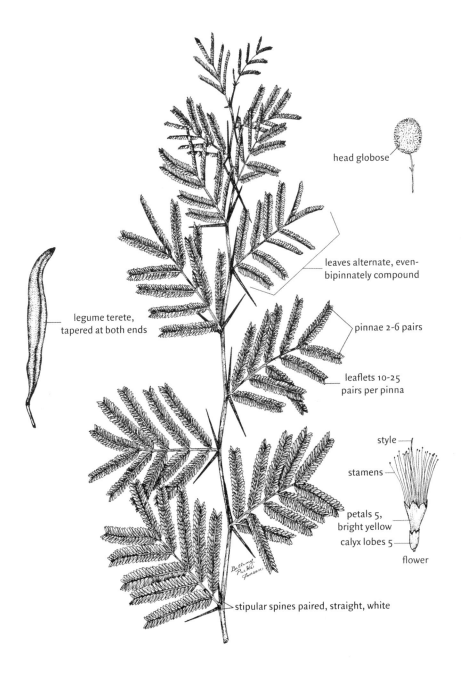

head globose

leaves alternate, even-bipinnately compound

pinnae 2-6 pairs

leaflets 10-25 pairs per pinna

legume terete, tapered at both ends

style

stamens

petals 5, bright yellow

calyx lobes 5

flower

stipular spines paired, straight, white

Family:	FABACEAE
Species:	*Acacia farnesiana* (L.) Willd.
Common Name:	Huisache (huizache, sweet acacia)
Life Span:	Perennial
Origin:	Native
Season:	Warm

GROWTH FORM
shrub or small tree (to 9 m tall); often with several trunks, ascending; densely branched crown; flowers February to March, may flower whenever rain occurs during dry years, reproduces from seeds

FLORAL AND FRUIT CHARACTERISTICS
inflorescences: heads globose (about 1 cm in diameter), axillary, solitary to paired, many-flowered; peduncle slender (1–4 cm long), pubescent

flowers: perfect, regular; calyx lobes 5, minute (1–2 mm long), pubescent; corolla funnelform; petals 5, bright yellow; stamens numerous, exserted

fruits: legumes (2–8 cm long), terete, tapered at both ends, straight or curved, woody and stout, coriaceous, reddish-brown to purple or black, pulpy within, very tardily dehiscent, several-seeded; seeds in 2 rows

other: flowers intensely fragrant

VEGETATIVE CHARACTERISTICS
leaves: alternate, even-bipinnately compound (3–9 cm long); pinnae 2–6 pairs; leaflets 10–25 pairs per pinna, linear to oblong (3–5 mm long), apex acute or obtuse with a minute mucro, base unequal, margins entire, grayish-green; petiolate gland borne in the middle of the petiole or absent

stems: twigs rigid, slender, numerous, striate, glabrous or puberulent; stipular spines paired (6–14 mm long), straight, rigid, needle-like, white

HISTORIC, FOOD, AND MEDICINAL USES: formerly source of oils for perfumes; wood has been used for fence posts, farm tools, and smaller wooden items; gummy sap can be used for manufacturing mucilage; ornamental; honey plant

LIVESTOCK LOSSES: spines may cause injury to mouths and legs of animals

FORAGE VALUE: poor for livestock and wildlife, young leaves are occasionally eaten; legumes and seeds furnish valuable food for wildlife

HABITAT: brushy areas, open woods, hummocks, and disturbed areas; found in a broad range of soils; most abundant in dry, sandy soils

Catclaw acacia
Acacia greggii A. Gray

SYN = *A. wrightii* Benth.

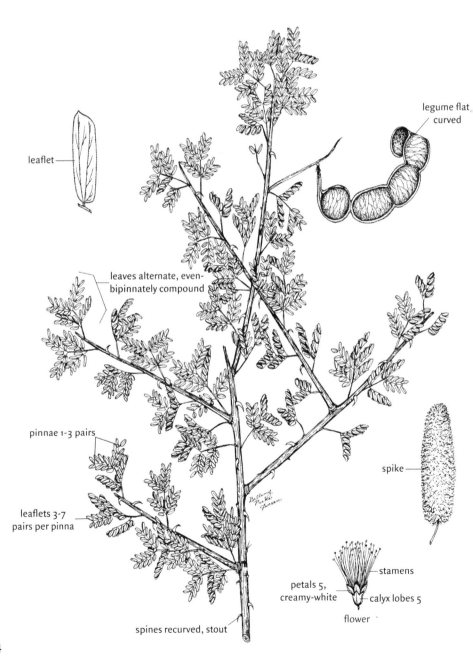

leaflet

legume flat, curved

leaves alternate, even-bipinnately compound

pinnae 1-3 pairs

spike

leaflets 3-7 pairs per pinna

stamens

petals 5, creamy-white

calyx lobes 5

flower

spines recurved, stout

Family:	FABACEAE
Species:	*Acacia greggii* A. Gray
Common Name:	Catclaw acacia (uña de gato, Gregg acacia, gatuño)
Life Span:	Perennial
Origin:	Native
Season:	Warm

GROWTH FORM

shrub or small tree (usually 1–2 m tall, rarely to 8 m); crown rounded, forming thickets; flowers April to October, may flower later following rain in dry years, reproduces from seeds

FLORAL AND FRUIT CHARACTERISTICS

inflorescences: spikes; oblong (2–6 cm long, 1 cm wide), axillary, solitary or paired, many-flowered

flowers: perfect, regular; calyx lobes 5, green, obscure (2–3 mm long), puberulent; petals 5, creamy-white; stamens numerous, long-exserted

fruits: legumes (5–8 cm long, 1.5–2 cm broad), flat, thin, curved or often curled and contorted, usually flexible, tardily becoming rigid, margins thickened, light brown to reddish-brown, few- to several-seeded, constricted between the flat seeds

VEGETATIVE CHARACTERISTICS

leaves: alternate, even-bipinnately compound; pinnae 1–3 pairs; leaflets 3–7 pairs per pinna, obovate to narrowly oblong (2–7 mm long), apex obtuse, base unequal, with a short petiole, margins entire, pubescent, slightly reticulate

stems: twigs much-branched, pale or reddish-brown to gray, with recurved spines (5–9 mm long) on internodes; spines similar in appearance to a cat's claws, stout, dark brown or gray

HISTORIC, FOOD, AND MEDICINAL USES: wood used for fuel; the Pimas and Papagos made pinole flour from the pods and prepared it as a mush; important honey plant; host plant for several insects which produce lac, a material used in varnish and shellac

LIVESTOCK LOSSES: spines may cause injury

FORAGE VALUE: poor for livestock and wildlife, foliage is seldom browsed; plants provide cover for wildlife, seeds are an important food source for many species of wildlife, main food source for some species of quail

HABITAT: valleys, ravines, mesas, washes, and canyon slopes; most abundant in dry, coarse-textured soils

Blackbrush acacia
Acacia rigidula Benth.

SYN = A. *amentacea* DC.

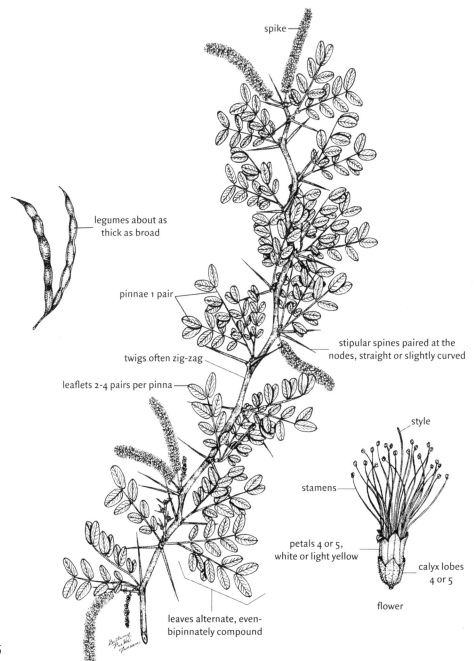

spike

legumes about as
thick as broad

pinnae 1 pair

twigs often zig-zag

leaflets 2-4 pairs per pinna

stipular spines paired at the
nodes, straight or slightly curved

style

stamens

petals 4 or 5,
white or light yellow

calyx lobes
4 or 5

flower

leaves alternate, even-
bipinnately compound

Family:	FABACEAE
Species:	*Acacia rigidula* Benth.
Common Name:	Blackbrush acacia (chaparro prieto, blackbrush)
Life Span:	Perennial
Origin:	Native
Season:	Warm

GROWTH FORM
shrub (to 5 m tall); several branches from the base, branches divaricate; much-branched above, often forming impenetrable thickets; flowers April to May, reproduces from seeds

FLORAL AND FRUIT CHARACTERISTICS
inflorescences: spikes oblong (2–6 cm long, 1 cm wide), numerous, axillary, many-flowered

flowers: perfect, regular; calyx lobes 4 or 5; petals 4 or 5, white or light yellow; stamens several, exserted

fruits: legumes (6–8 cm long, 4–7 mm wide), linear, about as thick as broad, often falcate, apex acuminate, puberulent, reddish-brown to black, bivalved, dehiscent, may or may not be constricted between seeds, several-seeded

other: flowers fragrant

VEGETATIVE CHARACTERISTICS
leaves: alternate, even-pinnately compound; usually only 1 pair of pinnae; leaflets 2–4 pairs per pinna, oblong (6–15 mm long, less than 7 mm wide), oblique, apex rounded and mucronate (sometimes notched), base asymmetrical, margins entire, dark green, glabrous, glossy, nerves conspicuous

stems: twigs rigid often zig-zag; bark glabrous, tight, reddish-gray to light or dark gray, some whitish bark on most branches; stipular spines paired at the nodes, straight or slightly curved (8–25 mm long)

HISTORIC, FOOD, AND MEDICINAL USES: sometimes used as an ornamental, flowers a source for honey

LIVESTOCK LOSSES: spines may cause injury to the mouths and legs of animals

FORAGE VALUE: poor for livestock, fair for wildlife; plants provide cover for wildlife; seeds are an important food source for some species of birds and small mammals

HABITAT: plains, mesas, and ridge tops; most abundant in sandy and limestone soils

Leadplant
Amorpha canescens Pursh

SYN = *A. brachycarpa* Palmer

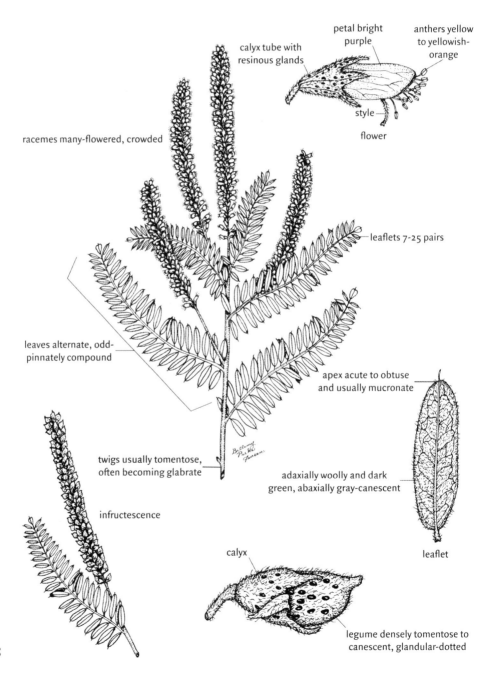

calyx tube with resinous glands

petal bright purple

anthers yellow to yellowish-orange

style

flower

racemes many-flowered, crowded

leaflets 7-25 pairs

leaves alternate, odd-pinnately compound

apex acute to obtuse and usually mucronate

twigs usually tomentose, often becoming glabrate

adaxially woolly and dark green, abaxially gray-canescent

infructescence

leaflet

calyx

legume densely tomentose to canescent, glandular-dotted

Family:	FABACEAE
Species:	*Amorpha canescens* Pursh
Common Name:	Leadplant (prairie shoestring)
Life Span:	Perennial
Origin:	Native
Season:	Warm

GROWTH FORM

shrub (to 1.2 m tall, many reach 2 m in ungrazed areas); erect or ascending; stems 1 to several, often branched; flowers June through early August, reproduces from seeds and rhizomes

FLORAL AND FRUIT CHARACTERISTICS

inflorescences: racemes (5–15 cm long); 1 to several, spicate, terminal and from upper axils, many-flowered, crowded; rachis densely villous

flowers: perfect, irregular; calyx tube turbinate (3–5 mm long), glands resinous; petals 1, broadly obovate (4–5 mm long, 3 mm wide), bright purple (occasionally light blue to violet blue); stamens 10, exserted; anthers yellow to yellowish-orange, conspicuous

fruits: legumes (3–5 mm long, 1.6–2 mm wide), curved, densely tomentose to canescent, glandular-dotted, 1-seeded; seeds elliptic (2–3 mm long), with a slight beak, orangish-brown, smooth

VEGETATIVE CHARACTERISTICS

leaves: alternate, odd-pinnately compound (3.5–10 cm long); leaflets 7–25 pairs, crowded to imbricate, elliptic to oblong (7–18 mm long, 3–6 mm wide), apex acute to obtuse and usually mucronate, rounded at base, margins entire, adaxially woolly and dark green, abaxially gray-canescent; petioles (0.5–3.5 mm long) pubescent; stipules subulate (1–3 mm long), caducous

stems: twigs usually tomentose, often becoming glabrate

HISTORIC, FOOD, AND MEDICINAL USES: some Native Americans smoked dried leaves and made tea from leaves; they treated neuralgia and rheumatism by cutting stems into small pieces, attaching them to the skin by wetting them, and then lighting them and allowing them to burn down into the skin as a counterirritant; cultivated ornamental

LIVESTOCK LOSSES: none

FORAGE VALUE: excellent, highly nutritive, and palatable for livestock and wildlife; commonly selected over most other species

HABITAT: prairies, open woods, dry plains, and hills; adapted to a broad range of soil types; rarely abundant on improperly managed rangeland

Woolly loco
Astragalus mollissimus Torr.

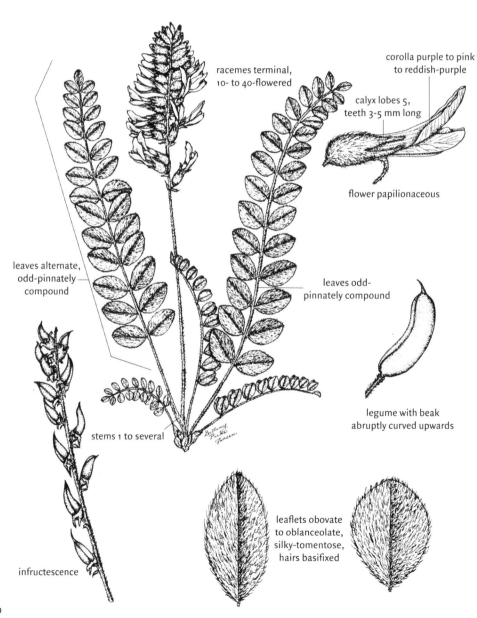

racemes terminal,
10- to 40-flowered

corolla purple to pink
to reddish-purple

calyx lobes 5,
teeth 3-5 mm long

flower papilionaceous

leaves alternate,
odd-pinnately
compound

leaves odd-
pinnately compound

legume with beak
abruptly curved upwards

stems 1 to several

leaflets obovate
to oblanceolate,
silky-tomentose,
hairs basifixed

infructescence

Family:	FABACEAE
Species:	*Astragalus mollissimus* Torr.
Common Name:	Woolly loco (hierba loca, purple locoweed, poisonvetch, hierba plata, chinchin)
Life Span:	Perennial
Origin:	Native
Season:	Cool

GROWTH FORM
forb (10–30 cm tall), 1 to several stems from a woody taproot; densely to loosely tufted, prostrate at maturity, often robust and leafy; starts growth in March, flowers April to June, fruits mature July to August, reproduces from seeds

FLORAL AND FRUIT CHARACTERISTICS
inflorescences: racemes (4–10 cm long), oblong, terminal, 10- to 40-flowered; peduncles naked (5–20 cm long)

flowers: perfect, papilionaceous; calyx tube cylindric (5–10 mm long), somewhat oblique; lobes 5 (teeth 3–5 mm long), long-acuminate, silky; corolla purple to pink to reddish-purple (rarely yellow or white), often drying blue (1.7–2.2 cm long), banner moderately reflexed, keel rounded

fruits: legumes (1–2.5 cm long, 4–9 mm wide), plump-ellipsoid to oblong-ellipsoid, spreading or ascending, beaked; beak abruptly curved upwards; bilocular for most of the length, glabrous to puberulent, many-seeded

VEGETATIVE CHARACTERISTICS
leaves: alternate, odd-pinnately compound (5–22 cm long); leaflets 5–7 pairs, obovate to oblanceolate (5–25 mm long, 2–15 mm wide), apex obtuse to acute and rarely mucronate, margins entire, silky-tomentose, hairs basifixed; stipules distinct (up to 1.5 cm long), triangular, silky

stems: outer stems prostrate, inner stems ascending

HISTORIC, FOOD, AND MEDICINAL USES: famous in western history as one of the causes of "locoed" animals

LIVESTOCK LOSSES: poisonous; can cause loco disease in livestock; toxic principles are the alkaloid locoine and selenium which is accumulated by the plants; large amounts must be consumed for a lethal dose; both poisons are cumulative; green and dry plants are poisonous

FORAGE VALUE: poor to worthless for livestock and wildlife, unpalatable and consumed only when other forage is not available; some animals (especially horses) may become addicted and refuse to eat better forage

HABITAT: dry prairies, hillsides, road rights-of-way, stream valleys, and uplands; most abundant in sandy or rocky soils

Purple prairieclover
Dalea purpurea Vent.

SYN = *Petalostemon purpureus* (Vent.) Rydb.

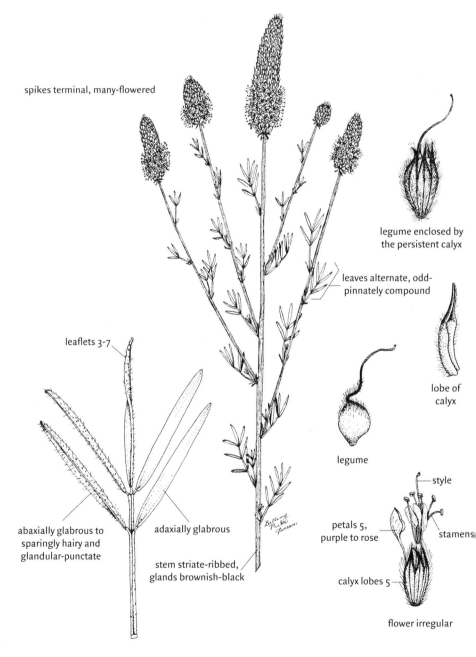

spikes terminal, many-flowered

legume enclosed by the persistent calyx

leaves alternate, odd-pinnately compound

leaflets 3-7

lobe of calyx

legume

abaxially glabrous to sparingly hairy and glandular-punctate

adaxially glabrous

stem striate-ribbed, glands brownish-black

style

petals 5, purple to rose

stamens

calyx lobes 5

flower irregular

Family:	FABACEAE
Species:	*Dalea purpurea* Vent.
Common Name:	Purple prairieclover (violet prairieclover)
Life Span:	Perennial
Origin:	Native
Season:	Warm

GROWTH FORM

forb (20–90 cm tall); erect or ascending; stems few to many, simple or branched from a thick caudex; flowers June to August, seeds mature July to September, reproduces from seeds and rootstocks

FLORAL AND FRUIT CHARACTERISTICS

inflorescences: spikes (1–7 cm long, 7–14 mm wide), oblong to ovate, dense, many-flowered, terminal, numerous

flowers: perfect, irregular, calyx tube silky-villous (2.5–4 mm long), lobes 5; petals 5, purple to rose (5–7 mm long); 4 of the 5 petals and the 5 stamens joined to near the tip of the calyx, banner separate

fruits: legumes (2–2.5 mm long) obliquely ovate, enclosed by the persistent calyx, 1-seeded

VEGETATIVE CHARACTERISTICS

leaves: alternate, odd-pinnately compound (1–4 cm long); leaflets 3–7 (mostly 5), linear (5–25 mm long, 0.5–1.5 mm wide), margins entire, involute, adaxially glabrous, abaxially glabrous to sparingly hairy and glandular-punctate, midrib not visible on adaxial surface of leaflets; petiole similar in appearance to leaflets

stems: thinly pilosulous to glabrous, striate-ribbed, glands brownish-black

HISTORIC, FOOD, AND MEDICINAL USES: some Native Americans ate fresh leaves and boiled leaves, bruised leaves were steeped in water and applied to wounds, roots were chewed for their pleasant flavor, stems were used to make brooms

LIVESTOCK LOSSES: may cause bloat, but it is seldom abundant enough to be a problem

FORAGE VALUE: excellent for livestock and wildlife, important component of prairie hay; high in protein, highly palatable and nutritious

HABITAT: prairies, plains, and hills; adapted to a broad range of soils, most abundant in dry soils

Tailcup lupine
Lupinus caudatus Kellogg

SYN = L. argentinus Rydb., L. argophyllus (A. Gray) Cockerell, L. cutleri Eastw.,
L. montigenus Heller

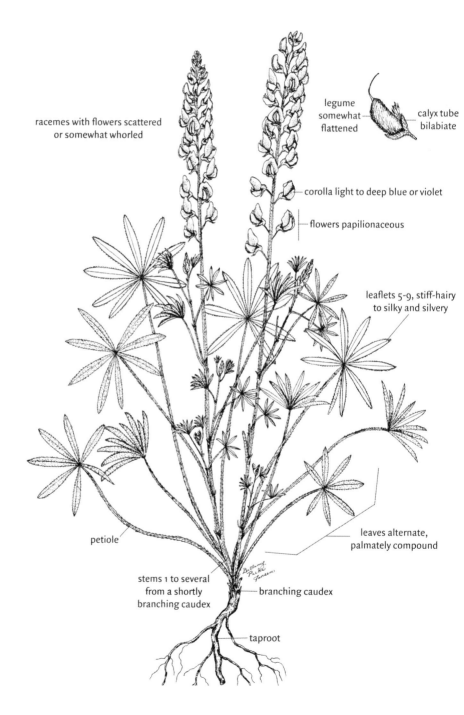

racemes with flowers scattered
or somewhat whorled

legume
somewhat
flattened

calyx tube
bilabiate

corolla light to deep blue or violet

flowers papilionaceous

leaflets 5-9, stiff-hairy
to silky and silvery

petiole

leaves alternate,
palmately compound

stems 1 to several
from a shortly
branching caudex

branching caudex

taproot

354

Family:	FABACEAE
Species:	*Lupinus caudatus* Kellogg
Common Name:	Tailcup lupine (spurred lupine)
Life Span:	Perennial
Origin:	Native
Season:	Cool

GROWTH FORM

forb (20–60 cm tall); erect to ascending, stems 1 to several from a shortly branching caudex surmounting a taproot; branched or simple; flowers June to August, fruits mature July to September, reproduces from seeds

FLORAL AND FRUIT CHARACTERISTICS

inflorescences: racemes (6–20 cm long), oblong, terminal, many-flowered; flowers scattered or somewhat whorled

flowers: perfect, papilionaceous; calyx tube asymmetrical (3–4 mm long), bilabiate, spurred at the base; corolla light to deep blue or violet (1–1.4 cm long); banner petal reflexed, often lighter blue or whitish in the middle, silky on back

fruits: legumes (2–3 cm long), somewhat flattened, silky, may be slightly constricted between the seeds, several-seeded

VEGETATIVE CHARACTERISTICS

leaves: alternate, palmately compound; leaflets 5–9, oblanceolate (2–15 cm long, 5–15 mm wide), apex acute to rarely rounded, margins entire, stiff-hairy to silky and silvery; long-petiolate

stems: somewhat cespitose, strigose

HISTORIC, FOOD, AND MEDICINAL USES: a drug has been extracted for management of cardiac arrhythmias

LIVESTOCK LOSSES: poisonous, especially to sheep and horses, causing weakness and muscular trembling; alkaloids are concentrated in the seeds and occasionally in the young plants, plants are poisonous either green or dry, poisoning seldom occurs when other forage is adequate

FORAGE VALUE: poor for cattle and fair for sheep before the legumes develop; cattle may be attracted to the legumes and graze them selectively; fair to good for elk and deer

HABITAT: dry prairies, foothills, mesas, and stream valleys; adapted to a broad range of soil textures, but most abundant in coarse-textured and well-drained soils

Burclover
Medicago polymorpha L.

SYN = M. *hispida* Gaertn.

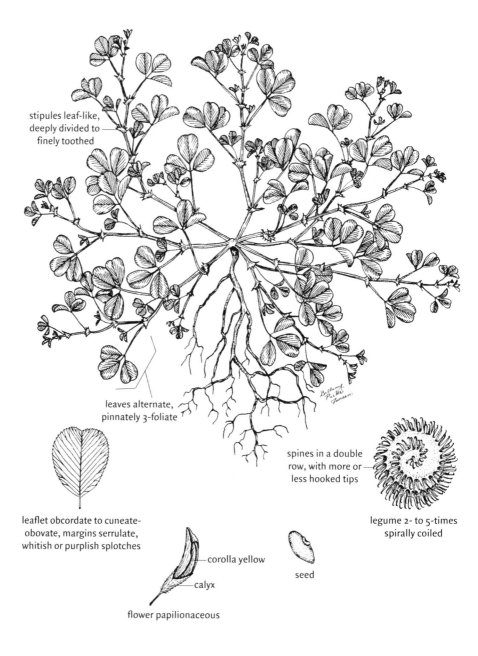

stipules leaf-like, deeply divided to finely toothed

leaves alternate, pinnately 3-foliate

leaflet obcordate to cuneate-obovate, margins serrulate, whitish or purplish splotches

spines in a double row, with more or less hooked tips

legume 2- to 5-times spirally coiled

corolla yellow

calyx

seed

flower papilionaceous

Family:	FABACEAE
Species:	*Medicago polymorpha* L.
Common Name:	Burclover (carretilla, medic, toothed burclover)
Life Span:	Annual
Origin:	Introduced (from Mediterranean region)
Season:	Cool

GROWTH FORM
forb; procumbent to ascending (stems 5–50 cm long), several stems from the base surmounting a taproot; flowers March to June, seeds mature in about 8 weeks, reproduces from seeds

FLORAL AND FRUIT CHARACTERISTICS
inflorescences: heads (5–15 mm long), rounded, axillary, 2- to 8-flowered, borne on a peduncle (1–3 cm long)

flowers: perfect, papilionaceous; calyx tube campanulate (about 1 mm long), pubescent; corolla yellow (3–5 mm long); pedicellate (pedicel 0.5–1 mm long)

fruits: legumes (4–6 mm in diameter, excluding the spines), 2- to 5-times spirally coiled; spines (2–3 mm long) in a double row, with more or less hooked tips; several-seeded

VEGETATIVE CHARACTERISTICS
leaves: alternate, pinnately 3-foliate; leaflets obcordate to cuneate-obovate (6–15 mm long, mostly longer than broad), margins serrulate, sometimes with whitish or purplish splotches; petiolate; stipules leaf-like (6–10 mm long) deeply divided to finely toothed

stems: nearly glabrous to puberulent

HISTORIC, FOOD, AND MEDICINAL USES: young leaves may be used to garnish salads

LIVESTOCK LOSSES: excessive grazing of fresh herbage may cause bloat; fleece may become contaminated with burs (legumes)

FORAGE VALUE: excellent to good for livestock and wildlife, most valuable in spring; burs are highly nutritious and are consumed during the dry season, somewhat resistant to heavy grazing; burs are an important food for birds and small mammals

HABITAT: valleys, plains, lower slopes of foothills, waste ground, and a common weed in lawns and rights-of-way; thrives in rich soils and may occur in poor soils where few other species grow

Lambert crazyweed
Oxytropis lambertii Pursh

SYN = *Aragallus articulatus* Greene, *O. involuta* (A. Nels.) K. Schum., *O. patens* (Rydb.)
    A. Nels.

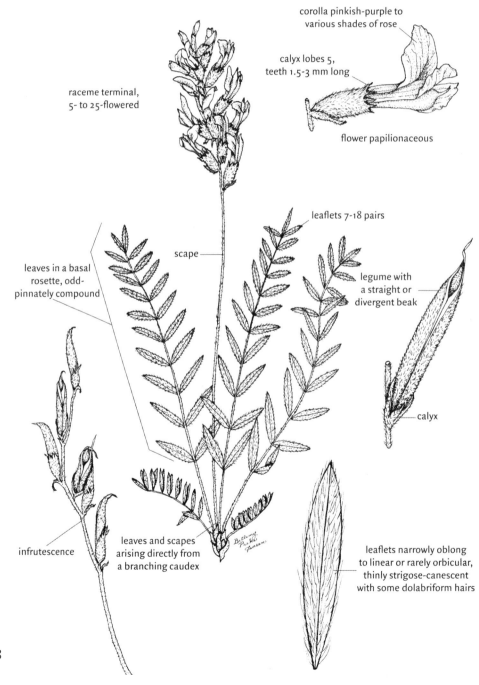

corolla pinkish-purple to
various shades of rose

calyx lobes 5,
teeth 1.5-3 mm long

raceme terminal,
5- to 25-flowered

flower papilionaceous

leaflets 7-18 pairs

scape

leaves in a basal
rosette, odd-
pinnately compound

legume with
a straight or
divergent beak

calyx

infrutescence

leaves and scapes
arising directly from
a branching caudex

leaflets narrowly oblong
to linear or rarely orbicular,
thinly strigose-canescent
with some dolabriform hairs

Family:	FABACEAE
Species:	*Oxytropis lambertii* Pursh
Common Name:	Lambert crazyweed (hierba loca, purple locoweed, white loco, whitepoint loco)
Life Span:	Perennial
Origin:	Native
Season:	Cool

GROWTH FORM
acaulescent forb (scapes 10–35 cm tall); flowers April to August, fruits mature June to October, reproduces from seeds

FLORAL AND FRUIT CHARACTERISTICS
inflorescences: racemes (5–40 mm long, 2–7 mm wide), oblong or capitate, terminal, 5- to 25-flowered, elevated above the leaves on scapes

flowers: perfect, papilionaceous; calyx tube campanulate (5–9 mm long), lobes 5 (teeth 1.5–3 mm long), silky pilose; corolla (1.5–2.5 cm long) pinkish-purple to various shades of rose (white not uncommon), keel petal with an abrupt point (0.5–2.5 mm long)

fruits: legumes (7–30 mm long, 5–6 mm wide), oblong to oblong-ovoid, sessile or nearly so, with a straight or divergent beak (3–7 mm long), silky-strigose, soon glabrous, many-seeded

VEGETATIVE CHARACTERISTICS
leaves: alternate in a basal rosette, odd-pinnately compound, dimorphic; principal leaves (4–20 cm long) with 7–19 leaflet pairs; leaflets narrowly oblong to linear or rarely orbicular (5–40 mm long, 2–7 mm wide), apex acute, margins entire, thinly strigose-canescent with some dolabriform hairs; petioles pubescent; stipules adnate to the petioles

stems: leaves and scape arising directly from a branching caudex

HISTORIC, FOOD, AND MEDICINAL USES: famous in western history as one of the causes of "locoed" animals

LIVESTOCK LOSSES: poisonous; can cause loco disease in horses, cattle, sheep, and goats (see *Astragalus mollissimus* for a discussion of loco disease)

FORAGE VALUE: poor to worthless for livestock and wildlife; generally unpalatable and consumed only when other forage is not available, some animals may develop a craving for the plants

HABITAT: dry upland prairies and plains, adapted to a broad range of soil types

Honey mesquite
Prosopis glandulosa **Torr.**

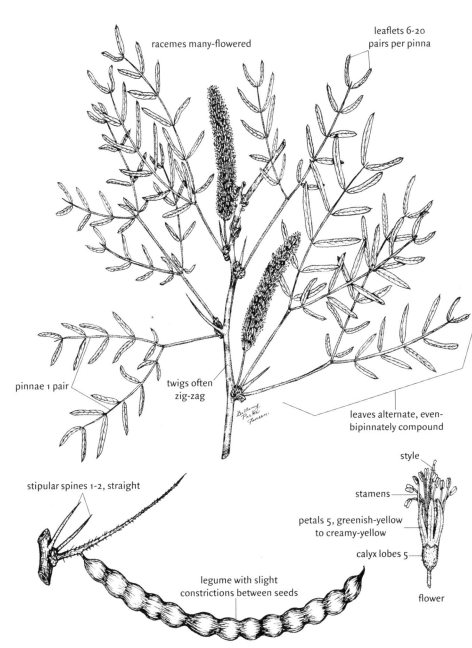

racemes many-flowered

leaflets 6-20 pairs per pinna

pinnae 1 pair

twigs often zig-zag

leaves alternate, even-bipinnately compound

style

stamens

petals 5, greenish-yellow to creamy-yellow

calyx lobes 5

flower

stipular spines 1-2, straight

legume with slight constrictions between seeds

Family:	FABACEAE
Species:	*Prosopis glandulosa* Torr.
Common Name:	Honey mesquite (mezquite, mesquite, glandular mesquite)
Life Span:	Perennial
Origin:	Native
Season:	Warm

GROWTH FORM
shrub or small tree (to 6 m tall); single or multiple trunks, much-branched, crown rounded; growth begins in late spring, flowers May, fruits mature June to August, reproduces from seeds and basal shoots

FLORAL AND FRUIT CHARACTERISTICS
inflorescences: racemes (7–9 cm long), axillary, many-flowered

flowers: perfect, regular; calyx tube campanulate (1 mm long), pubescent, lobes 5; lobes triangular (0.3–0.4 mm long); petals 5 (3 mm long), greenish-yellow to creamy-yellow, distinct, pubescent within; pedicel glandular (0.5 mm long)

fruits: legumes (10–20 cm long, 1 cm wide), in clusters of 2 or 3, linear, straight or nearly so, glabrous, slight constrictions between seeds, several-seeded, dehiscent

other: flowers fragrant

VEGETATIVE CHARACTERISTICS
leaves: alternate, even-bipinnately compound (6–15 cm long); pinnae usually 1 pair, leaflets 6–20 pairs per pinna, linear to linear-oblong (2–6 cm long, 2–3 mm wide), terminal pair usually curved, apex acute and mucronate, margins entire, glabrous or nearly so; sessile or nearly so

stems: twigs rigid, often zig-zag, reddish-brown or grayish-brown; armed with 1–2 stipular spines (to 5 cm long), stout, straight, rarely spineless

HISTORIC, FOOD, AND MEDICINAL USES: wood is used for fuel and lumber; beans were ground into flour which was important in the diets of some Native Americans, flour could be fermented into alcohol

LIVESTOCK LOSSES: ingestion of large amounts may result in rumen stasis and impaction

FORAGE VALUE: poor to good for livestock, deer, and javelina; seeds are an important food for numerous wildlife species and as emergency feed for livestock

HABITAT: plains and prairies, on dry sandy or gravelly soils; especially abundant on abused rangeland and where fire is controlled

Slimflower scurfpea
Psoralidium tenuiflorum (Pursh) Rydb.

SYN = *Psoralea floribunda* Torr. & A. Gray, *Psoralea obtusiloba* Torr. & A. Gray, *Psoralea tenuiflora* Pursh

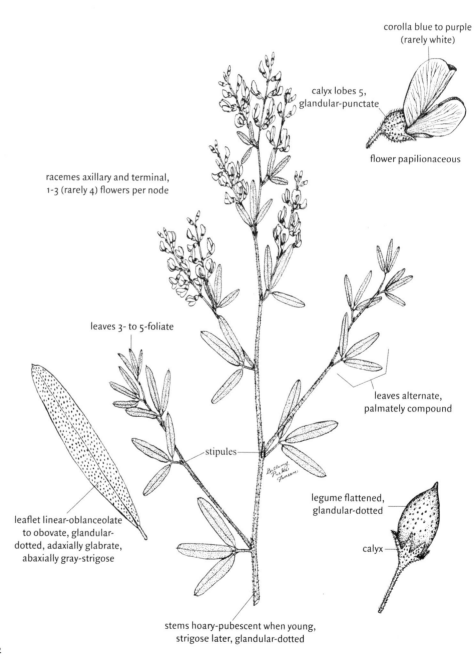

corolla blue to purple (rarely white)

calyx lobes 5, glandular-punctate

flower papilionaceous

racemes axillary and terminal, 1-3 (rarely 4) flowers per node

leaves 3- to 5-foliate

leaves alternate, palmately compound

stipules

leaflet linear-oblanceolate to obovate, glandular-dotted, adaxially glabrate, abaxially gray-strigose

legume flattened, glandular-dotted

calyx

stems hoary-pubescent when young, strigose later, glandular-dotted

Family:	FABACEAE
Species:	*Psoralidium tenuiflorum* (Pursh) Rydb.
Common Name:	Slimflower scurfpea (wild alfalfa, slender scurfpea)
Life Span:	Perennial
Origin:	Native
Season:	Warm

GROWTH FORM

forb (0.2–1.3 m tall); erect, 1 to several stems from a caudex surmounting a deep taproot, much branched; flowers May to September, reproduces from seeds and rarely from rhizomes; above-ground portion disarticulates at the crown following maturity and may tumble with the wind

FLORAL AND FRUIT CHARACTERISTICS

inflorescences: racemes (2–10 cm long), oblong, loose, axillary and terminal, many-flowered, 1–3 (rarely 4) flowers per node; peduncle longer than the subtending leaf

flowers: perfect, papilionaceous; calyx tube campanulate (1.5–3 mm long), lobes 5; lobes glandular-punctate, strigose or villous-hirsute; corolla (4–8 mm long) blue to purple (rarely white)

fruits: legumes (5–9 mm long), ovate or oblong, flattened, with a short and straight beak, glabrous, densely glandular-dotted, often asymmetrical, 1-seeded

VEGETATIVE CHARACTERISTICS

leaves: alternate, palmately compound, usually 3-foliate (sometimes lower leaves 5-foliate); leaflets linear-oblanceolate to obovate (1–5 cm long, 4–12 mm wide), margins entire, glandular-dotted, adaxially glabrate, abaxially gray-strigose; petioles mostly shorter (4–20 mm long) than the leaflets; stipules lanceolate (2–3 mm long)

stems: hoary-pubescent when young, strigose later, glandular-dotted

HISTORIC, FOOD, AND MEDICINAL USES: some Native Americans drank a tea made from stems and leaves for fever, a tea was made from the roots for headache, plants were burned to repel mosquitoes

LIVESTOCK LOSSES: may cause bloat; reported to be poisonous to cattle and horses, but no experimental or circumstantial evidence is available in support of this report

FORAGE VALUE: poor for livestock and wildlife, readily eaten after curing in hay; seeds are an important food for birds and small mammals

HABITAT: dry plains, prairies, and open woods; adapted to a broad range of soils

Gambel oak
Quercus gambelii Nutt.

SYN = *Q. eastwoodiae* Rydb., *Q. gunnisonii* (Torr. & A. Gray) Rydb., *Q. leptophylla* Rydb.,
    Q. nitescens Rydb., *Q. submollis* Rydb.

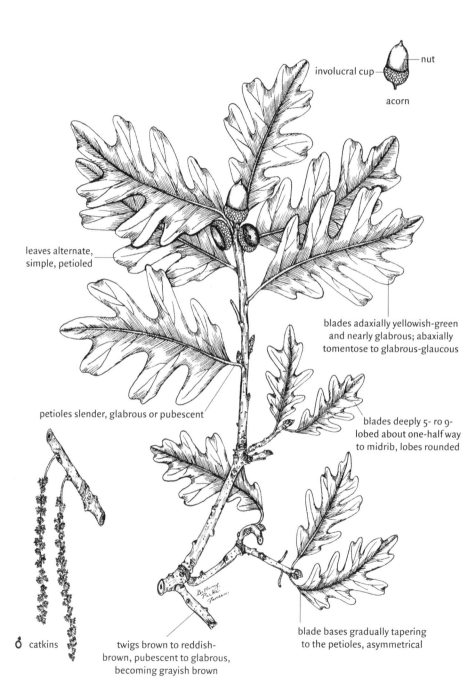

nut

involucral cup

acorn

leaves alternate,
simple, petioled

blades adaxially yellowish-green
and nearly glabrous; abaxially
tomentose to glabrous-glaucous

petioles slender, glabrous or pubescent

blades deeply 5- ro 9-
lobed about one-half way
to midrib, lobes rounded

♂ catkins

twigs brown to reddish-
brown, pubescent to glabrous,
becoming grayish brown

blade bases gradually tapering
to the petioles, asymmetrical

364

Family:	FAGACEAE
Species:	*Quercus gambelii* Nutt.
Common Name:	Gambel oak (encino)
Life Span:	Perennial
Origin:	Native
Season:	Cool

GROWTH FORM

monoecious shrub or small tree (to 20 m tall); crown rounded; grows in dense stands, often forming thickets; flowers March to April, fruits mature in autumn of the first year after flowering, reproduces from seeds

FLORAL AND FRUIT CHARACTERISTICS

inflorescences: staminate flowers in catkins (2.5–4 cm long), pendulous, many-flowered; pistillate flowers 1–3 in reduced catkins at branch apex

flowers: unisexual, apetalous; calyx of staminate flowers 5- or 6-lobed; stamens 5–10; pistillate flowers inconspicuous (3–8 mm long), calyx lobes 6

fruits: acorns solitary or clustered; involucral cup (to 1.5 cm wide), cup encloses one-third to one-half the nut; scales appressed, ovate-acuminate, tomentulose, apex narrowed and rounded, base rounded; nut ovoid to ellipsoid (to 1.5 cm long), 1-seeded

VEGETATIVE CHARACTERISTICS

leaves: alternate, simple; blades highly variable, ovate to obovate or oblong to elliptic (8–16 cm long, 4–7 cm wide), usually deeply 5- to 9-lobed about one-half way to midrib; lobes rounded, middle and lateral lobes the largest; bases gradually tapering to the petiole, asymmetrical; coriaceous; adaxially yellowish-green and nearly glabrous; abaxially tomentose to glabrous-glaucous; petiole slender (7–25 mm long), glabrous or pubescent

stems: twigs slender, brown to reddish-brown, pubescent to glabrous, becoming grayish-brown; bark gray, deeply fissured, scaly

HISTORIC, FOOD, AND MEDICINAL USES: acorns are edible after tannic acid is removed; some Native Americans used ground or powdered acorns to thicken soup and make mush

LIVESTOCK LOSSES: shoots contain tannic and gallic acids, therefore, poisoning of cattle and occasionally sheep may occur from March to April

FORAGE VALUE: fair for all classes of livestock, deer, and porcupines; acorns eaten by livestock and wildlife

HABITAT: valleys, canyons, foothills, and lower mountain slopes in all soil textures

Post oak
Quercus stellata Wangenh.

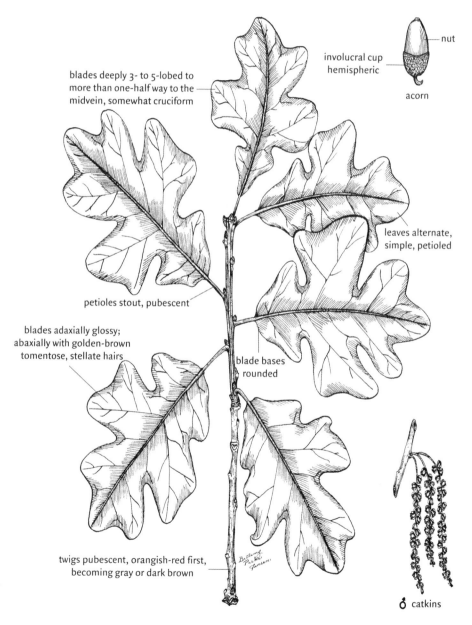

blades deeply 3- to 5-lobed to more than one-half way to the midvein, somewhat cruciform

nut

involucral cup hemispheric

acorn

leaves alternate, simple, petioled

petioles stout, pubescent

blades adaxially glossy; abaxially with golden-brown tomentose, stellate hairs

blade bases rounded

twigs pubescent, orangish-red first, becoming gray or dark brown

♂ catkins

Family:	FAGACEAE
Species:	*Quercus stellata* Wangenh.
Common Name:	Post oak (encino)
Life Span:	Perennial
Origin:	Native
Season:	Warm

GROWTH FORM
monoecious tree or shrub (to 30 m tall); stout limbs, crown rounded or cylindric; flowers March to May, fruits mature in the autumn of the first year after flowering, reproduces from seeds and basal shoots

FLORAL AND FRUIT CHARACTERISTICS
inflorescences: staminate flowers in catkins (8–10 cm long), pendulous, in clusters from lateral buds, many-flowered; pistillate flowers 2 to several, short-stalked or sessile

flowers: unisexual, apetalous; calyx of staminate flowers 5-lobed; lobes (1.2–1.4 mm long) yellow with brown tips, pubescent, acute; stamens 4–6; pistillate flowers inconspicuous

fruits: acorns usually in clusters of 2–4; involucral cup hemispheric, enclosing about one-third (sometimes up to one-half of the nut); scales reddish-brown, hairy-tomentose, closely appressed, apex rounded or acute; nut oval or obovoid-oblong (1–1.5 cm long); 1-seeded

VEGETATIVE CHARACTERISTICS
leaves: alternate, simple; blades broadly obovate to oblong-obovate (10–15 cm long, 7–10 cm wide), deeply 3- to 5-lobed to more than one-half way to the midvein, lobes somewhat cruciform; apex and base rounded; adaxially glossy; abaxially with golden-brown tomentose hairs; hairs stellate; gray pubescence in axils of leaf veins; petiole stout (5–10 mm long), pubescent

stems: twigs pubescent, rigid, orangish-red first, becoming gray or dark brown; bark gray to reddish-brown, fissured with plate-like scales

HISTORIC, FOOD, AND MEDICINAL USES: wood used for furniture, fencing, flooring, fuel, railroad ties, and lumber

LIVESTOCK LOSSES: high content of tannic and gallic acids in the shoots can poison cattle

FORAGE VALUE: poor to fair for cattle, fair for goats; buds, leaves, and twigs eaten by cattle and goats in early spring; acorns eaten by wildlife

HABITAT: open woods, hillsides, and ridges; adapted to a broad range of soil textures; most abundant on dry sites

Broadleaf filaree
Erodium botrys (Cav.) Bertol.

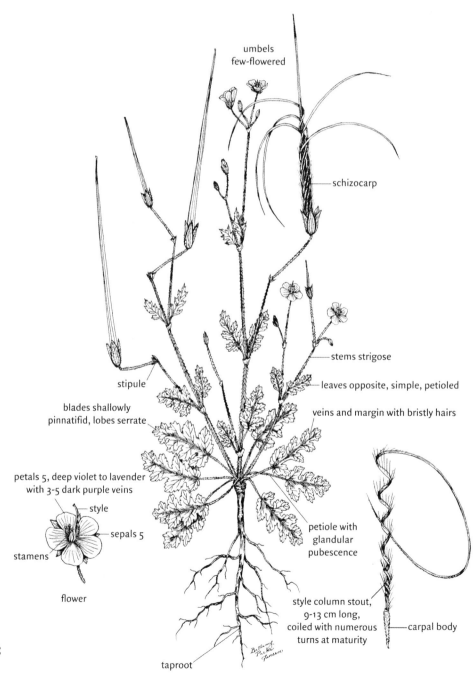

umbels
few-flowered

schizocarp

stems strigose

leaves opposite, simple, petioled

stipule

blades shallowly
pinnatifid, lobes serrate

veins and margin with bristly hairs

petals 5, deep violet to lavender
with 3-5 dark purple veins

style

sepals 5

stamens

flower

petiole with
glandular
pubescence

style column stout,
9-13 cm long,
coiled with numerous
turns at maturity

carpal body

taproot

Family:	GERANIACEAE
Species:	*Erodium botrys* (Cav.) Bertol.
Common Name:	Broadleaf filaree (longbeaked filaree, longbeaked storksbill)
Life Span:	Annual
Origin:	Introduced (from Mediterranean region through Europe)
Season:	Cool

GROWTH FORM

forb (stems 10–90 cm long); few to several stems from a taproot, prostrate to somewhat ascending, first acaulescent then branching from the base; flowers March to May, reproduces from seeds; seeds germinate in winter, forming rosettes, grows rapidly for 3–4 months before slowing significantly

FLORAL AND FRUIT CHARACTERISTICS

inflorescences: umbels axillary or terminal on long peduncles, few-flowered; peduncles glandular-hirsute

flowers: perfect, regular; sepals 5 (7–8 mm long, enlarging in fruit), glandular-pubescent; lobes with 1–2 minute, bristle-like hairs or awns (enlarging in fruit); petals 5, deep violet to lavender, with 3–5 dark purple veins, cuneate (1–1.5 cm long), apex blunt; anther-bearing stamens 5, occurring alternately with 5 sterile stamens

fruits: schizocarp; carpel bodies 5, fusiform (8–10 mm long), pubescent, 2 pits at the base of each carpel body, each subtended by 2 folds forming smaller pits between; style column stout (9–13 cm long), coiled with numerous turns at maturity

VEGETATIVE CHARACTERISTICS

leaves: opposite, simple; blades ovate to oblong-ovate (3–10 cm long), shallowly pinnatifid; lobes serrate, acute; basal leaves sometimes crenate; bristly hairs on veins and margins; petioles (8–24 mm long) with glandular pubescence; stipules present

stems: strigose, much-branched

HISTORIC, FOOD, AND MEDICINAL USES: none

LIVESTOCK LOSSES: has been reported to cause bloat

FORAGE VALUE: good in winter and early spring for cattle, sheep, and wildlife; especially valuable on annual rangeland

HABITAT: pastures, plains, grassy lowlands, foothills, and waste places; adapted to a broad range of soils

Redstem filaree
Erodium cicutarium (L.) L'Hér. *ex* Aiton

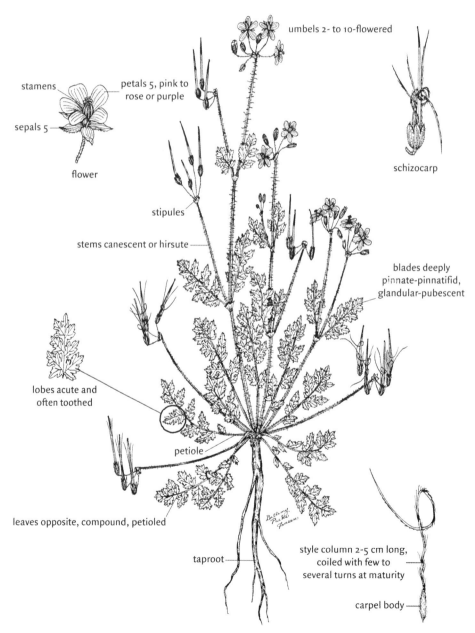

umbels 2- to 10-flowered

stamens

petals 5, pink to rose or purple

sepals 5

flower

schizocarp

stipules

stems canescent or hirsute

blades deeply pinnate-pinnatifid, glandular-pubescent

lobes acute and often toothed

petiole

leaves opposite, compound, petioled

taproot

style column 2-5 cm long, coiled with few to several turns at maturity

carpel body

Family:	GERANIACEAE
Species:	*Erodium cicutarium* (L.) L'Hér. *ex* Aiton
Common Name:	Redstem filaree (alfilaria, alfilerillo, redstem storksbill, heronbill, agujas de pastor)
Life Span:	Annual
Origin:	Introduced (from Mediterranean region through Europe)
Season:	Cool

GROWTH FORM
forb (stems 10–50 cm long); few to several stems from a taproot, first appearing acaulescent as a winter rosette of basal leaves, decumbent to ascending; flowers February to May and occasionally in September, reproduces from seeds; seeds germinate in fall, or is one of the first plants to germinate in spring

FLORAL AND FRUIT CHARACTERISTICS
inflorescences: umbels axillary on long peduncles, 2- to 10-flowered; pedicels (5–20 mm long) glandular pubescent

flowers: perfect, regular (1 cm wide); sepals 5, elliptic (2–6 mm long), with 1 to 2 short, white, bristle-like hairs or awns (0.1–0.5 mm long); petals 5, pink to rose colored or purple, elliptic to obovate (4–8 mm long), ciliate at the base; anther bearing stamens 5, occurring alternately with 5 sterile stamens

fruits: schizocarp; carpel bodies 5, fusiform (4–5 mm long), tardily dehiscent, stiffly pubescent; style column attached (2–5 cm long), coiled with few to several turns at maturity

VEGETATIVE CHARACTERISTICS
leaves: opposite, compound; blades elongate-oblanceolate (3–13 cm long, including the petiole), deeply pinnate-pinnatifid; lobes acute and often toothed; glandular-pubescent; petioles present (8–18 mm long); stipules present

stems: canescent or hirsute, much branched

other: leaves delicate, nearly fern-like

HISTORIC, FOOD, AND MEDICINAL USES: young leaves can be eaten raw or cooked; reputed to contain an antidote for strychnine

LIVESTOCK LOSSES: has been reported to cause bloat

FORAGE VALUE: excellent to good in spring for cattle, sheep, and wildlife

HABITAT: cultivated fields, waste grounds, roadsides, lawns, plains, mesas, and ravines; adapted to a broad range of soil types, most abundant on sandy soils

Wildwhite geranium
Geranium richardsonii Fisch. & Trautv.

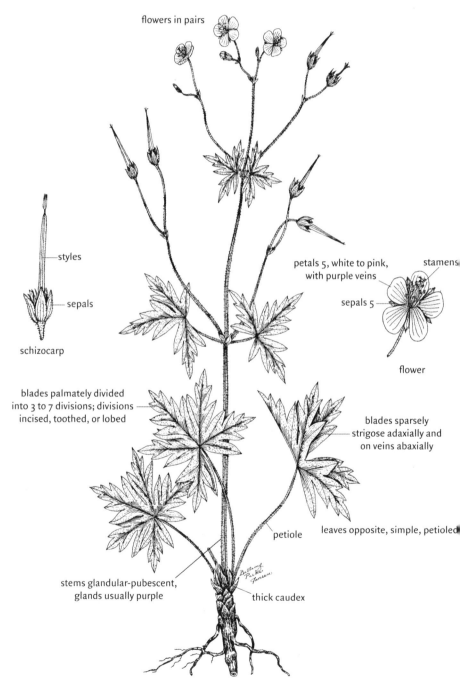

flowers in pairs

styles

sepals

schizocarp

petals 5, white to pink, with purple veins

stamens

sepals 5

flower

blades palmately divided into 3 to 7 divisions; divisions incised, toothed, or lobed

blades sparsely strigose adaxially and on veins abaxially

petiole

leaves opposite, simple, petioled

stems glandular-pubescent, glands usually purple

thick caudex

Family:	GERANIACEAE
Species:	*Geranium richardsonii* Fisch. & Trautv.
Common Name:	Wildwhite geranium (Richardson cranesbill, white cranesbill)
Life Span:	Perennial
Origin:	Native
Season:	Warm

GROWTH FORM

forb (0.2–1 m tall), 1 to few stems, erect or ascending, from a thick branched or unbranched caudex; sometimes rhizomatous; flowers May to August, fruits mature August to September, reproduces from seeds

FLORAL AND FRUIT CHARACTERISTICS

inflorescences: flowers in pairs; pedicellate, pedicels (1–3 cm long) on axillary branches, branches purplish glandular-pilose; gland tips often reddish-purple

flowers: perfect, regular; sepals 5, imbricate and fused at the base, ovate (6–11 mm long), tips setose; petals 5, imbricate, white to pink, with purple veins, obovate to obcordate (1–1.8 cm long), pilose at the base for one-half their length on the inner surface; stamens 10

fruits: schizocarp; carpel bodies 5 (3–4.5 mm long); carpel stylar portions remaining attached at the apex, free portions coiling outward; style branches somewhat pubescent

VEGETATIVE CHARACTERISTICS

leaves: opposite, simple; basal blades palmately divided into 3 to 7 divisions (5–30 cm long, 4–15 cm wide), cleft three-fourths of length or more; main divisions incised, toothed, or lobed; sparsely strigose adaxially and on veins abaxially; petioles long (2–28 cm long); upper blades reduced; petioles shorter (2–12 cm long); stipules present

stems: usually glandular-pubescent, but may be glabrous to rarely strigulose below, the glands usually purple

HISTORIC, FOOD, AND MEDICINAL USES: the Cheyenne Indians pulverized leaves into a powder and snuffed it into their noses to control nosebleeds, roots were also powdered and made into a drink

LIVESTOCK LOSSES: none

FORAGE VALUE: poor to fair for sheep and deer, poor to worthless for cattle, rarely eaten by horses; worthless after reaching maturity

HABITAT: shaded and moist wooded hillsides, ravines, mountains, and foothills; most abundant in rich soils

Wax currant
Ribes cereum Dougl.

SYN = *R. reniforme* Nutt.

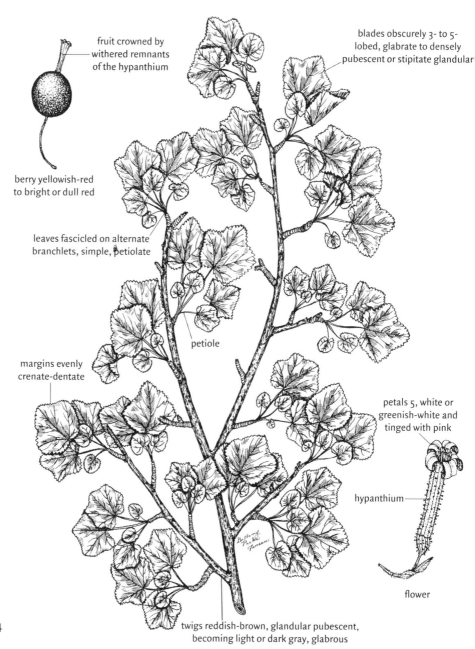

fruit crowned by withered remnants of the hypanthium

blades obscurely 3- to 5-lobed, glabrate to densely pubescent or stipitate glandular

berry yellowish-red to bright or dull red

leaves fascicled on alternate branchlets, simple, petiolate

petiole

margins evenly crenate-dentate

petals 5, white or greenish-white and tinged with pink

hypanthium

flower

twigs reddish-brown, glandular pubescent, becoming light or dark gray, glabrous

Family:	GROSSULARIACEAE
Species:	*Ribes cereum* Dougl.
Common Name:	Wax currant (capulincillo, western redcurrant)
Life Span:	Perennial
Origin:	Native
Season:	Cool

GROWTH FORM

shrub (to 2 m tall); much-branched, erect to spreading; flowers May to July, fruits ripen July to August, reproduces from seeds

FLORAL AND FRUIT CHARACTERISTICS

inflorescences: flowers solitary or in 2- to 8-flowered racemes; inconspicuous, drooping

flowers: perfect, regular; hypanthium tubular (6–9 mm long, 2–3.5 mm wide); calyx narrowly cylindrical, sepals 5; sepals ovate, spreading or reflexed (1.5–3 mm long, 1–1.9 mm wide), white or greenish-white and tinged with pink, pubescent and stipitate glandular; petals 5 (1.5–2 mm long, 1.2–2 mm wide), white to tinged with pink; stamens 5

fruits: berries spherical to ovoid (5–10 mm in diameter), yellowish-red to bright or dull red, smooth to slightly glandular-hairy, few- to several-seeded; crowned by withered remnants of the hypanthium

VEGETATIVE CHARACTERISTICS

leaves: fascicled on alternate branchlets, simple; blades reniform to orbiculate (5–25 mm long, 1–4 cm wide), obscurely 3- to 5-lobed; bases truncate to subcordate; margins evenly crenate-dentate; glabrate to densely pubescent or stipitate glandular, waxy; petiolate (5–20 mm long)

stems: young twigs reddish-brown, glandular pubescent; eventually becoming light or dark gray (sometimes white), glabrous; lenticels transverse; aromatic

HISTORIC, FOOD, AND MEDICINAL USES: some Native Americans ate the tasteless fresh berries or dried the berries to be later eaten with raw mutton and deer fat, plant extracts were used to alleviate stomachache; currently used in jellies

LIVESTOCK LOSSES: none

FORAGE VALUE: poor to fair for livestock, fair to good for wildlife; often abundant, therefore, a rather important browse plant; fruit valuable for many species of birds and small mammals

HABITAT: openings in woods, prairies, open slopes, hills, and ridges; most abundant in dry, rocky or sandy soils

Range ratany
Krameria erecta Willd. *ex* J. A. Schultes

SYN = *K. glandulosa* Rose & Painter, *K. parvifolia* Benth.

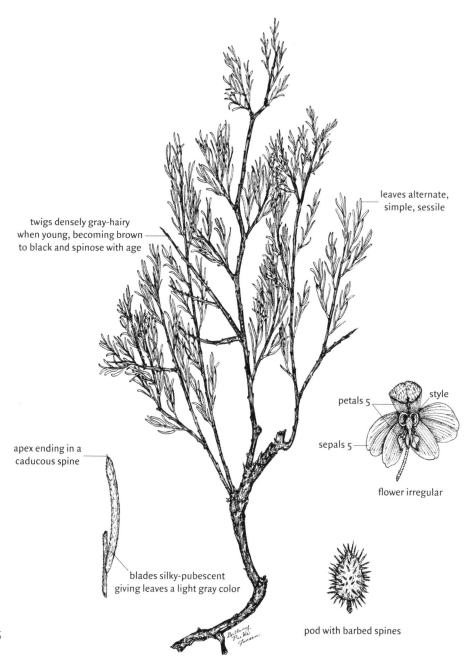

leaves alternate, simple, sessile

twigs densely gray-hairy when young, becoming brown to black and spinose with age

petals 5

style

sepals 5

flower irregular

apex ending in a caducous spine

blades silky-pubescent giving leaves a light gray color

pod with barbed spines

Family:	KRAMERIACEAE
Species:	*Krameria erecta* Willd. *ex* J. A. Schultes
Common Name:	Range ratany (tamichil, little ratany, littleleaf ratany)
Life Span:	Perennial
Origin:	Native
Season:	Cool

GROWTH FORM

shrub (to 70 cm tall); intricately branched, crowns rounded to irregular; flowers April to May and may flower a second time in late summer, reproduces from seeds

FLORAL AND FRUIT CHARACTERISTICS

inflorescences: flowers solitary; terminal and axillary, on glandular-hairy peduncles (4–9 mm long)

flowers: perfect, irregular; sepals 5 (sometimes 4), petal-like, unequal (5–10 mm long), ascending, oblong to oblanceolate, strigose; petals 5, oblong to obovate (2–5 mm long), upper 3 (4–5 mm long) united to form a short claw, remaining 2 reduced (2–3 mm long), nearly orbicular, purple, showy

fruits: pods, globose (6–9 mm long), thick-walled, silky-hairy, bur-like with barbed spines; spines slender (3–4 mm long), purple or red; 1-seeded

other: fragrant flowers

VEGETATIVE CHARACTERISTICS

leaves: alternate, simple; blades linear to oblanceolate (3–15 mm long, 1–2 mm wide), apex ending in a caducous spine; margins entire; silky-pubescent giving leaves a light gray color; sessile; stipules absent

stems: twigs slender, stiff; densely gray-hairy when young, becoming brown to black and spinose with age

other: bluish-green twigs give a bluish tint to the landscape where these plants are abundant

HISTORIC, FOOD, AND MEDICINAL USES: some Native Americans made a decoction from the leaves to use as an eyewash, for diarrhea, and for sores; source for red or brown dye; roots were used in manufacturing ink

LIVESTOCK LOSSES: none

FORAGE VALUE: fair to good for cattle, sheep, goats, and mule deer; furnishes valuable browse but cannot withstand heavy browsing because branches are brittle; bur-like fruits are readily disseminated by livestock, important food for small mammals

HABITAT: foothills, mesas, hillsides, and plains; adapted to a broad range of soil textures; most abundant in dry, gravelly soils

Scarlet globemallow
Sphaeralcea coccinea (Nutt.) Rydb.

SYN = *Malvastrum coccineum* (Nutt.) A. Gray

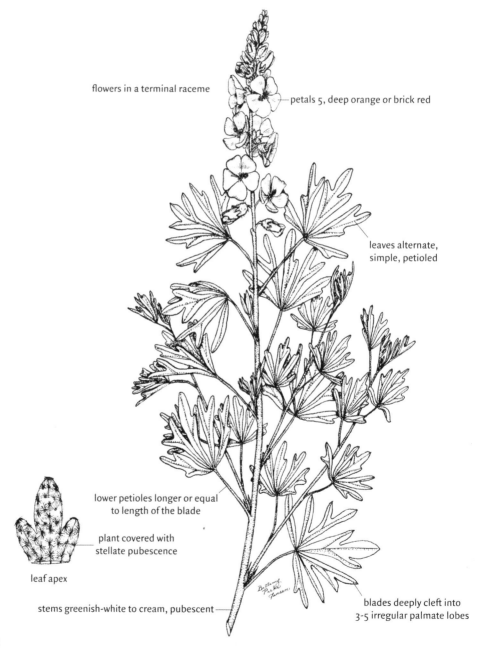

flowers in a terminal raceme

petals 5, deep orange or brick red

leaves alternate,
simple, petioled

lower petioles longer or equal
to length of the blade

plant covered with
stellate pubescence

leaf apex

stems greenish-white to cream, pubescent

blades deeply cleft into
3-5 irregular palmate lobes

378

Family:	MALVACEAE
Species:	*Sphaeralcea coccinea* (Nutt.) Rydb.
Common Name:	Scarlet globemallow (red falsemallow)
Life Span:	Perennial
Origin:	Native
Season:	Warm

GROWTH FORM

forb (stems 10–50 cm long); ascending to decumbent, stems simple or clustered from a woody caudex; flowers April to August, reproduces from seeds; persists in dry periods by shedding leaves

FLORAL AND FRUIT CHARACTERISTICS

inflorescences: racemes; broadly or narrowly ovate (2–10 cm long), terminal

flowers: perfect, regular; calyx conspicuously villous (3–10 mm long), persistent; sepals 5, connate, narrowly triangular to ovate, acuminate; petals 5 (1–2 cm long), deep orange or brick red and drying pinkish, emarginate; stamens numerous

fruits: schizocarp of 10 or more mericarps (3–3.4 mm tall and wide), differentiated into a smooth dehiscent portion at the apex and a roughened indehiscent basal portion; basal portion prominently reticulate, black tuberculate, 1-seeded

VEGETATIVE CHARACTERISTICS

leaves: alternate, simple; blades deltate or suborbicular to ovate in outline (1–6 cm long, wider than long), deeply cleft into 3–5 (rarely more) irregular palmate lobes; segment apex acute to rounded, final segment oblong to obovate or oblanceolate to spatulate; base cuneate; petioles of the lower leaves longer or equal (1–8 cm long) to the length of the blade

stems: decumbent or ascending, greenish-white to cream, pubescent

other: entire plant is covered with stellate pubescence

HISTORIC, FOOD, AND MEDICINAL USES: the Blackfoot Indians chewed these plants and applied the paste to burns, scalds, and external sores as a cooling agent

LIVESTOCK LOSSES: none

FORAGE VALUE: excellent for deer and pronghorn, worthless to fair for domestic livestock; important forage in Southwest but infrequently grazed in the Great Plains; increases in abundance with heavy use and during dry periods

HABITAT: prairies, plains, hills, roadsides, and waste places; adapted to a broad range of soils

Fireweed
Epilobium angustifolium L.

SYN = *Chameron angustifolium* (L.) Holub

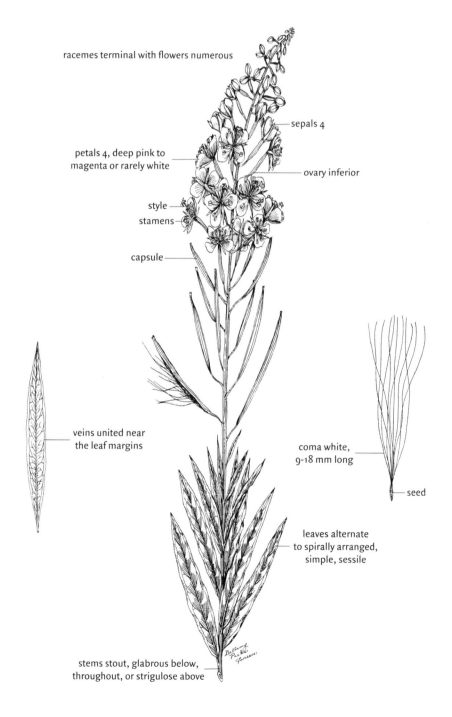

racemes terminal with flowers numerous

sepals 4

petals 4, deep pink to magenta or rarely white

ovary inferior

style

stamens

capsule

veins united near the leaf margins

coma white, 9-18 mm long

seed

leaves alternate to spirally arranged, simple, sessile

stems stout, glabrous below, throughout, or strigulose above

Family:	ONAGRACEAE
Species:	*Epilobium angustifolium* L.
Common Name:	Fireweed (willowherb, blooming Sally, great willowherb)
Life Span:	Perennial
Origin:	Native
Season:	Warm

GROWTH FORM

forb (0.3–2.5 m tall); erect; flowers June to September, reproduces from seeds and rhizomes (up to 8 m long)

FLORAL AND FRUIT CHARACTERISTICS

inflorescences: racemes terminal, elongate, lax, drooping in bud, flowers numerous, bracts resembling leaves

flowers: perfect, irregular; hypanthium absent; sepals 4, narrowly lanceolate (7–16 mm long, 1.6–2.5 mm wide), acute, often tinged in purple, canescent; petals 4, slightly asymmetrical, obovate (8–20 mm long, 6–11 mm wide) tapering to a short claw, deep pink to magenta or rarely white; stamens 8, subequal, shorter than petals; stigma lobes 4, long and slender; ovary inferior

fruits: capsule (3–9 mm long) 4-sided, 4-celled, splitting by 4 valves; ascending to erect; canescent, often purplish; seeds many, fusiform (0.9–1.4 mm long); coma white (9–18 mm long)

VEGETATIVE CHARACTERISTICS

leaves: alternate to spirally arranged, simple; blades narrowly lanceolate to lanceolate (2–20 cm long, 5–40 mm wide); apex acuminate; bases acute to acuminate; margins subentire or obscurely denticulate; midvein strong abaxially, glabrous or with strigulose hairs; veins united near the leaf margins; sessile

stems: stout, generally not branched; glabrous below, throughout, or strigulose above

HISTORIC, FOOD, AND MEDICINAL USES: used as a potherb, young shoots can be cooked like asparagus, young leaves used in salads and steeped for tea, pith of stem can be used to flavor and thicken stews and soups; important honey plant; grown as an ornamental

LIVESTOCK LOSSES: none

FORAGE VALUE: fair to good for sheep; poor to fair for cattle; grazed to a minor extent by horses, deer, and elk; becomes unpalatable with maturity

HABITAT: open woods, along streams, roadsides, and disturbed areas; adapted to dry and moist soils, grows in a broad range of soil types; especially abundant following fire

Pinyon pine
Pinus edulis Engelm.

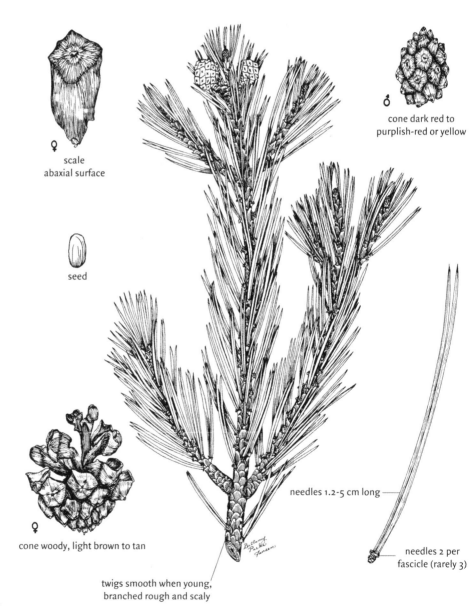

♀
scale
abaxial surface

seed

♂
cone dark red to
purplish-red or yellow

needles 1.2-5 cm long

♀
cone woody, light brown to tan

needles 2 per
fascicle (rarely 3)

twigs smooth when young,
branched rough and scaly

Family:	PINACEAE
Species:	*Pinus edulis* Engelm.
Common Name:	Pinyon pine (pino piñonero, piñon pine, nut pine)
Life Span:	Perennial
Origin:	Native
Season:	Evergreen

GROWTH FORM
monoecious tree (to 15 m tall); crown pyramidal, becoming round-topped with age; flowers April to June, reproduces from seeds

CONE CHARACTERISTICS
cones: unisexual, in clusters at ends of branches; staminate cones ovoid (6–8 mm long), dark red to purplish-red or yellow, in clusters of 20–40; ovulate cones ovoid (2–6 cm long, 2–5 cm wide), purplish to brown at maturity, sessile

mature cones: ovulate cones woody, light brown to tan, scales thickened near apex, incurved lips, unarmed, seeds borne in cavities at the base of middle scales, maturing the first season; seeds ovate (9–16 mm long), thick-shelled, wingless, edible

VEGETATIVE CHARACTERISTICS
leaves: needle-like, in fascicles arranged in spirals; 2 needles per fascicle (rarely 3); needles linear (1.2–5 cm long), often curved upward, aromatic; apex acuminate; margins entire; abaxially rounded, with whitish lines, new growth bluish-green turning yellowish-green; fascicle sheaths mostly decid-uous (4–7 mm long); sessile

stems: twigs smooth when young; branches rough and scaly; bark thin, gray to reddish-brown or nearly black; trunk frequently crooked and twisted; bark irregularly furrowed with small scales, yellowish- to reddish-brown, resinous

HISTORIC, FOOD, AND MEDICINAL USES: seed crop is valuable and is used in making candies, cakes, and cookies; seeds were staple food in some Native American diets and were eaten raw, roasted, or ground into flour; needles were steeped for tea; inner bark served as starvation food for some Native Americans; wood is used for fuel and fence posts

LIVESTOCK LOSSES: none

FORAGE VALUE: worthless to cattle and sheep; seeds important wildlife food for song birds, quail, small mammals, black bears, mule deer, and goats

HABITAT: mountain slopes, foothills, mesas, and plateaus; most abundant on dry and rocky soils

Ponderosa pine
Pinus ponderosa P. & C. Lawson

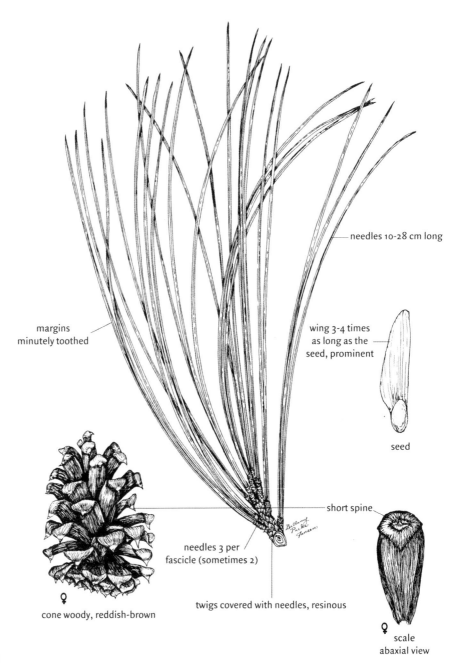

needles 10-28 cm long

margins minutely toothed

wing 3-4 times as long as the seed, prominent

seed

short spine

needles 3 per fascicle (sometimes 2)

♀
cone woody, reddish-brown

twigs covered with needles, resinous

♀
scale abaxial view

Family:	PINACEAE
Species:	*Pinus ponderosa* P. & C. Lawson
Common Name:	Ponderosa pine (pino ponderosa, western yellowpine)
Life Span:	Perennial
Origin:	Native
Season:	Evergreen

GROWTH FORM

tree (to 60 m tall); crown open, pyramidal when immature and becoming round- to flat-topped with age; mature trees generally without branches on the lower trunk; flowers April to June, reproduces from seeds

CONE CHARACTERISTICS

cones: unisexual; staminate cones cylindrical (1.5–2.5 cm long, 6–8 mm wide), yellowish-orange to deep purple, in clusters of 10–20; ovulate cones in clusters or pairs at base of new growth, broadly ovoid (6–15 cm long, 6–9 cm wide), reddish-brown, woody, sessile

mature cones: ovulate cones woody, tip of scales rounded and often with a short spine, 2 seeds per scale, maturing the summer of the second season; seeds dark brown to purplish-mottled (6–7 mm long), with a prominent papery wing 3- to 4-times as long as the seed

VEGETATIVE CHARACTERISTICS

leaves: needle-like, in fascicles arranged in spirals; 3 needles per fascicle (sometimes 2), fascicles clustered or whorled near the tips of the branches; needles linear (10–28 cm long, 1–1.5 mm wide), resinous, aromatic, one-third to one-half round in cross-section; apex abruptly acuminate; margins minutely toothed; yellowish-green to dark green; fascicle sheath dark brown to black; sessile

stems: twigs covered with needles and occasionally with old lanceolate leaf scales, resinous; buds brown; bark rough, thick, with deep fissures, scales reddish-brown; trunk straight

HISTORIC, FOOD, AND MEDICINAL USES: some Native Americans used cones to start fires; one of the principal lumber trees in North America; wood used commercially for boxes, crates, construction, and mill products; sometimes planted as an ornamental and in shelterbelts

LIVESTOCK LOSSES: browsing on needles causes abortion in cattle

FORAGE VALUE: worthless to livestock; seeds are eaten by several species of birds and small mammals; browsed by mule deer, white-tailed deer, and bighorn sheep

HABITAT: rocky hillsides, mountains, and plateaus; grows in moist or dry soils and in a wide range of soil types

Woolly plantain
Plantago patagonica Jacq.

SYN = *P. purshii* Roemer & Schultes

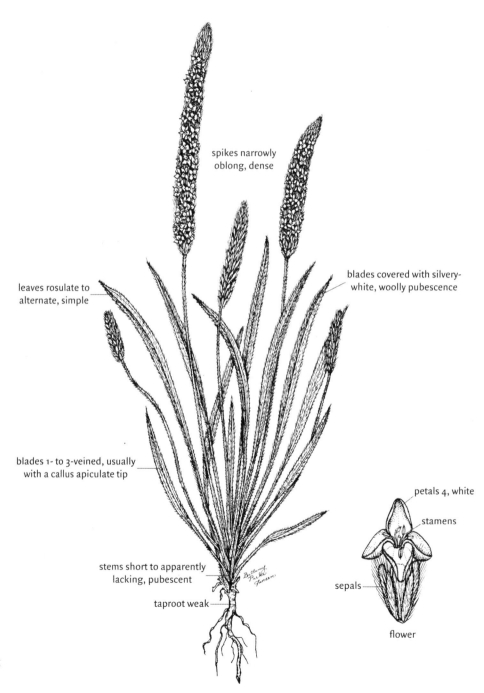

spikes narrowly oblong, dense

blades covered with silvery-white, woolly pubescence

leaves rosulate to alternate, simple

blades 1- to 3-veined, usually with a callus apiculate tip

petals 4, white

stamens

sepals

stems short to apparently lacking, pubescent

taproot weak

flower

Family: PLANTAGINACEAE
Species: *Plantago patagonica* Jacq.
Common Name: Woolly plantain (lanté, woolly indianwheat, tallowweed)
Life Span: Annual
Origin: Native
Season: Cool

GROWTH FORM
forb (3–35 cm tall), appearing acaulescent from a weak taproot; forming winter rosettes; starts growth in late winter, flowers May to August, reproduces from seeds

FLORAL AND FRUIT CHARACTERISTICS
inflorescences: spikes; 1–20 per plant, narrowly oblong (1–15 cm long, less than 1 cm wide), cylindric, dense, borne on scapes (1–20 cm long)

flowers: perfect, regular; sepals narrowly obovate (1.4–2.5 mm long), margins scarious; petals 4 (1–2 mm long), suborbicular to ovate-lanceolate, spreading, white; stamens 4, slightly exserted to included

fruits: capsules (3–4 mm long); breaking apart at or just below the middle; 2-seeded

VEGETATIVE CHARACTERISTICS
leaves: rosulate to alternate, mostly basal, simple; blades of rosettes oblanceolate (5–30 cm long); principal blades linear to oblanceolate (2–20 cm long, 0.5–15 mm wide), acute to acuminate, gradually tapering to the petiolar base, 1- to 3-veined, usually with a callus apiculate tip; margins entire; covered with silvery-white woolly pubescence

stems: short, appear to be lacking, pubescent

HISTORIC, FOOD, AND MEDICINAL USES: some Native Americans chewed and swallowed leaves for internal hemorrhage, leaves were also chewed for toothache

LIVESTOCK LOSSES: none

FORAGE VALUE: good for sheep and poor to fair for cattle and wildlife; a major forage species on lambing ranges, most important in the Southwest where it is often regarded as a spring opportunist, abundance is generally considered an indicator of rangeland deterioration

HABITAT: prairies, pastures, waste places, and roadsides; most abundant in sandy soils

Low larkspur
Delphinium bicolor Nutt.

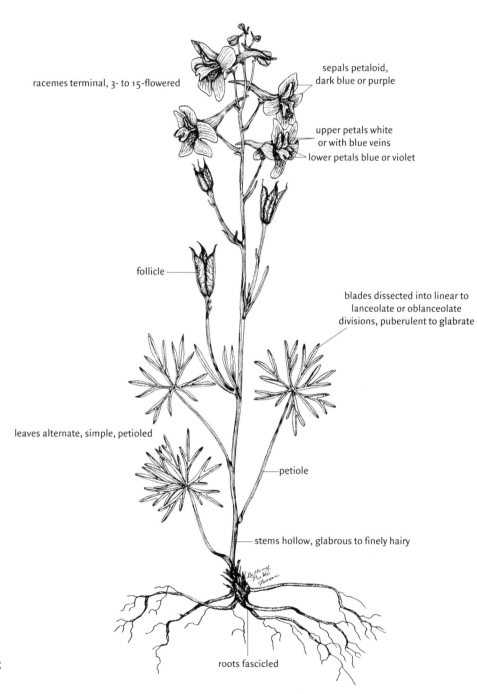

racemes terminal, 3- to 15-flowered

sepals petaloid,
dark blue or purple

upper petals white
or with blue veins

lower petals blue or violet

follicle

blades dissected into linear to
lanceolate or oblanceolate
divisions, puberulent to glabrate

leaves alternate, simple, petioled

petiole

stems hollow, glabrous to finely hairy

roots fascicled

Family: RANUNCULACEAE
Species: *Delphinium bicolor* Nutt.
Common Name: Low larkspur (little larkspur)
Life Span: Perennial
Origin: Native
Season: Cool

GROWTH FORM

forb (10–60 cm tall), erect to weak-stemmed, usually unbranched; from a branching cluster of fascicled roots in early spring; flowers May to July, seeds mature June to July, reproduces from seeds

FLORAL AND FRUIT CHARACTERISTICS

inflorescences: racemes oblong (usually less than 15 cm long), terminal, usually 3- to 15-flowered; pedicels spreading

flowers: perfect, irregular; sepals 5, petaloid, upper sepal with a spur (1–2 cm long); lower sepals (1.5–2 cm long) dark blue or purple (rarely white or pink), lobes wavy; corolla of 2 sets of 2 petals, petals smaller than the sepals; upper pair of petals prolonged at base into the spur, white or with blue veins; lower pair of petals clawed, blue or violet, concealing the stamens, sinuses absent; stamens numerous

fruits: follicles (1.5–2.5 cm long); usually divergent, brown, viscid-pubescent to glabrous, many-seeded

VEGETATIVE CHARACTERISTICS

leaves: alternate, simple; leaves mostly basal to evenly spaced, usually 3–6; blades cuneate to orbicular (2–4 cm wide), dissected into linear to lanceolate or oblanceolate divisions; puberulent to glabrate; lowest petioles long, abruptly reduced above

stems: hollow, glabrous to finely hairy

HISTORIC, FOOD, AND MEDICINAL USES: some Native Americans crushed plants of this genus and applied it to their hair to control lice and other insects; used as an ornamental

LIVESTOCK LOSSES: poisonous to cattle throughout growth cycle, contains alkaloids which act on nervous system, death may result following paralysis of breathing; bloat is common; seeds are the most poisonous part

FORAGE VALUE: fair to good for sheep and some wildlife; palatable to cattle even though other plants are available, unpalatable to horses

HABITAT: open woods, roadsides, hills, and meadows; most abundant on dry soils

Tall larkspur
Delphinium occidentale (S. Wats.) S. Wats.

SYN = *D. cucullatum* A. Nels.

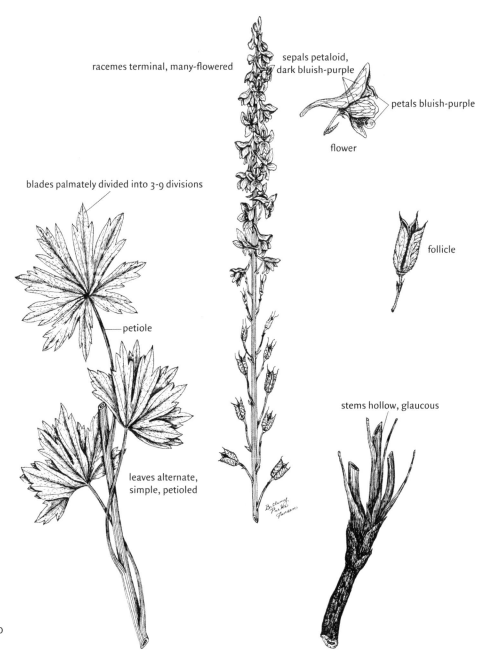

racemes terminal, many-flowered

sepals petaloid, dark bluish-purple

petals bluish-purple

flower

blades palmately divided into 3-9 divisions

follicle

petiole

stems hollow, glaucous

leaves alternate, simple, petioled

Family:	RANUNCULACEAE
Species:	*Delphinium occidentale* (S. Wats.) S. Wats.
Common Name:	Tall larkspur (duncecap larkspur)
Life Span:	Perennial
Origin:	Native
Season:	Warm

GROWTH FORM

forb (0.6–2 m tall); erect from a taproot; growth begins in late spring; flowers July to August, seeds mature from August to September, reproduces from seeds

FLORAL AND FRUIT CHARACTERISTICS

inflorescences: racemes narrowly oblong (usually more than 15 cm long), dense, spicate to loosely paniculate with age, terminal, many-flowered, central axis glandular-hairy

flowers: perfect, irregular; sepals 5, petaloid, dark bluish-purple (rarely white); upper sepal with a spur (9–12 mm long), straight or curved near the tip; lower sepals ovate-oblong (6–12 mm long), apex rounded or acute, light or usually gray-canescent on the back; corolla of 2 sets of 2 petals; upper pair of petals bluish-purple, spur-like; lower pair of petals with broad, wavy-margined lobes, bluish-purple, sinuses 1–2 mm long; stamens numerous

fruits: follicles; short-oblong (9–12 mm long), glabrous to glandular-pubescent, many-seeded

VEGETATIVE CHARACTERISTICS

leaves: alternate, simple; blades cuneate to orbicular (10–18 cm long, 8–12 cm wide), palmately divided into 3–9 divisions, divisions cleft; pubescent to puberulent; petioled

stems: hollow, somewhat straw-colored at the base, glaucous

HISTORIC, FOOD, AND MEDICINAL USES: some Native Americans crushed plants of this genus and applied to their hair to control lice and other insects; used as an ornamental

LIVESTOCK LOSSES: poisonous to cattle which decreases after blossoming, contains alkaloids which act on nervous system, death may result following paralysis of breathing, bloat is common; seeds are the most poisonous part

FORAGE VALUE: fair to good for sheep and some wildlife; palatable to cattle even though other plants are available, seldom eaten by horses

HABITAT: meadows, thickets, stream banks, and open woods; most abundant where the snow pack lasts the longest

Wedgeleaf ceanothus
Ceanothus cuneatus (Hook.) Nutt.

SYN = *C. ramulosus* (Greene) McMinn

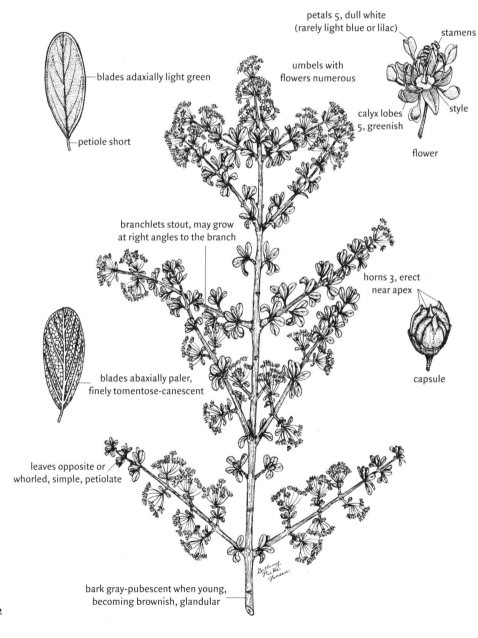

blades adaxially light green

petiole short

petals 5, dull white
(rarely light blue or lilac)

stamens

umbels with
flowers numerous

calyx lobes
5, greenish

style

flower

branchlets stout, may grow
at right angles to the branch

horns 3, erect
near apex

blades abaxially paler,
finely tomentose-canescent

capsule

leaves opposite or
whorled, simple, petiolate

bark gray-pubescent when young,
becoming brownish, glandular

392

Family:	RHAMNACEAE
Species:	*Ceanothus cuneatus* (Hook.) Nutt.
Common Name:	Wedgeleaf ceanothus (narrowleaf buckbrush)
Life Span:	Perennial
Origin:	Native
Season:	Warm

GROWTH FORM

shrub (to 4 m tall); branches rigid and divaricate, may form dense thickets; flowers March to May, reproduces from seeds, seeds germinate best following fire

FLORAL AND FRUIT CHARACTERISTICS

inflorescences: umbels (1–2.5 cm wide) borne on spur-like axillary branchlets; flowers numerous; pedicels slender

flowers: perfect, regular; calyx lobes 5, united below into a cup-like base, bent inward above, somewhat greenish; petals 5, dull white (rarely light blue or lilac), pipe-shaped, long-clawed; stamens 5, opposite the petals; style 3-cleft

fruits: capsules subglobsoe (5–6 mm in diameter), 3-lobed, 3-celled, with 3 erect horns near the apex; separating into 3 carpels each with 1 seed

other: flowers fragrant

VEGETATIVE CHARACTERISTICS

leaves: opposite or whorled, simple; blades oblong or cuneate-obovate to broadly obovate (5–20 mm long); apex obtuse or rounded; bases cuneate; margins entire to finely serrate at apex; coriaceous; adaxially light green, glabrous; abaxially paler, finely tomentose-canescent; petioles short (1–5 mm long)

stems: branchlets stout, may grow at right angles to the branch; bark gray-pubescent when young, becoming brownish; glandular

other: leaves and twigs glandular, producing a balsam-like fragrance

HISTORIC, FOOD, AND MEDICINAL USES: leaves and flowers may be boiled for tea; some Native Americans made an infusion from bark for a tonic; fresh flowers when crushed and rubbed in water to make a perfumed, cleansing lather; branches were used to build fish dams

LIVESTOCK LOSSES: unsubstantiated reports of kidney injury exclusively to male animals browsing this species

FORAGE VALUE: fair browse for sheep, goats, and deer; browsed by cattle only when other forage is unavailable; seeds eaten by small mammals

HABITAT: ridges, slopes, semiarid valleys, and open rocky sites; most abundant in gravelly or other well-drained soils

Fendler ceanothus
Ceanothus fendleri A. Gray

SYN = *C. subericeus* Rydb.

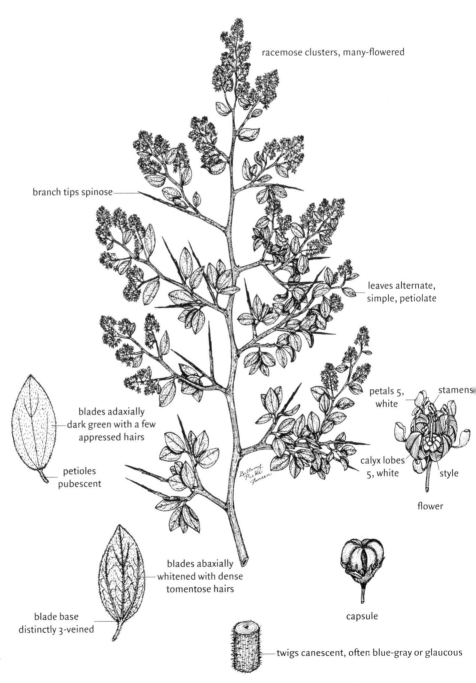

racemose clusters, many-flowered

branch tips spinose

leaves alternate, simple, petiolate

blades adaxially dark green with a few appressed hairs

petioles pubescent

petals 5, white

stamens

calyx lobes 5, white

style

flower

blades abaxially whitened with dense tomentose hairs

blade base distinctly 3-veined

capsule

twigs canescent, often blue-gray or glaucous

Family:	RHAMNACEAE
Species:	*Ceanothus fendleri* A. Gray
Common Name:	Fendler ceanothus (deerbriar, Fendler soapbloom)
Life Span:	Perennial
Origin:	Native
Season:	Warm

GROWTH FORM

shrub (to 1 m tall); stems clustered, loosely branched, forming low thickets; tardily deciduous; flowers April to October, fruits mature June to November, reproduces from seeds

FLORAL AND FRUIT CHARACTERISTICS

inflorescences: racemose clusters terminating main stem and branches, many-flowered

flowers: perfect, regular; calyx lobes 5, white, ovate (1.5 mm long and wide), acute, glabrous, sharply incurved; petals 5, white, spreading, pipe-shaped (1.5–1.8 mm long), clawed; stamens 5, opposite the petals; style 3-cleft (1 mm long), white

fruits: capsules (3–4 mm high, 5–6 mm in diameter); surface rough, brown, 3-lobed, 3-celled, separating into 3 carpels each with 1 seed; capsule lobes slightly keeled, horns absent

VEGETATIVE CHARACTERISTICS

leaves: alternate, simple; blades ovate to elliptic (1–2.5 cm long, 2–10 mm wide); apex acute to rounded, mucronulate; base cuneate, distinctly 3-veined; margins usually entire, occasionally serrate; adaxially dark green with a few appressed hairs; abaxially whitened with dense tomentose hairs; petioles pubescent (3–5 mm long)

stems: twigs canescent, often blue-gray or glaucous; bark of old stems reddish-brown, often with blister-like glands; branch tips spinose

HISTORIC, FOOD, AND MEDICINAL USES: some Native Americans extracted a dye from this species and used the fruits for food; tea can be made from the dried leaves

LIVESTOCK LOSSES: none

FORAGE VALUE: good and important browse for goats and deer, fair to good for cattle and sheep, frequently browsed by horses; highly variable in value between different areas in North America; heavily utilized by porcupines and jackrabbits in summer months

HABITAT: hillsides, open valleys, open woods, and rocky ledges in foothills and mountains; most abundant in well-drained soils

Deerbrush
Ceanothus integerrimus Hook & Arn.

SYN = *C. andersonii* Parry, *C. californicus* Kellogg

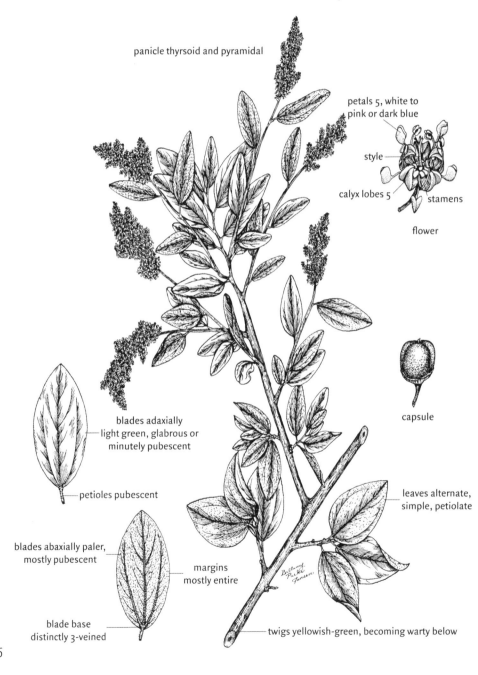

panicle thyrsoid and pyramidal

petals 5, white to pink or dark blue

style

calyx lobes 5

stamens

flower

blades adaxially light green, glabrous or minutely pubescent

petioles pubescent

capsule

leaves alternate, simple, petiolate

blades abaxially paler, mostly pubescent

margins mostly entire

blade base distinctly 3-veined

twigs yellowish-green, becoming warty below

396

Family:	RHAMNACEAE
Species:	*Ceanothus integerrimus* Hook. & Arn.
Common Name:	Deerbrush (bluebrush, mountain lilac, sweet birch, soapbush)
Life Span:	Perennial
Origin:	Native
Season:	Warm

GROWTH FORM

shrub (to 4 m tall, often broader than tall); widely and loosely branched, branches slender, often drooping; crown rounded to irregular; flowers May to July, reproduces from seeds and rootstocks; seeds germinate best following fire

FLORAL AND FRUIT CHARACTERISTICS

inflorescences: panicles thyrsoid and pyramidal (4–15 cm long), terminal, simple or compound

flowers: perfect, regular; calyx lobes 5, united below, white, curved sharply between the petals; petals 5, white to pink or dark blue, pipe-shaped, long-clawed; stamens 5, opposite the petals; style 3-cleft near the apex

fruits: capsules globose to widely ovate, viscid, 3-lobed, 3-celled, separating into 3 carpels each with 1 seed; oblong glands common on the back of each capsule lobe, sometimes indistinct

other: flowers fragrant

VEGETATIVE CHARACTERISTICS

leaves: alternate, simple; blades ovate to oblong-ovate (2–8 cm long, 1–4 cm wide); apex acute to rounded; base rounded, distinctly 3-veined; margins entire or occasionally denticulate near the apex; adaxially light green, glabrous or minutely pubescent; abaxially paler, mostly pubescent; petioles pubescent (3–15 mm long)

stems: twigs slender, often slightly drooping, yellowish-green, becoming warty below

HISTORIC, FOOD, AND MEDICINAL USES: extract from the bark was used by some Native Americans as a tonic and for making a soapy lather, flexible stems were used in basket weaving; valuable honey plant

LIVESTOCK LOSSES: none

FORAGE VALUE: good to excellent for cattle, sheep, goats, and deer; fair to good for horses; important for small mammals; quail and other birds eat the seeds

HABITAT: mountain slopes and ridges; adapted to a broad range of soils, most abundant in well-drained and fertile soils; grows in the open or in partial shade, does not tolerate full shade

Snowbrush
Ceanothus velutinus Dougl. ex Hook.

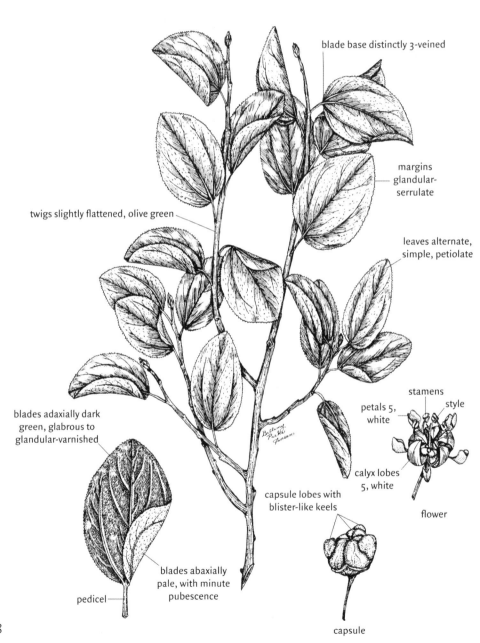

blade base distinctly 3-veined

margins glandular-serrulate

twigs slightly flattened, olive green

leaves alternate, simple, petiolate

blades adaxially dark green, glabrous to glandular-varnished

stamens

petals 5, white

style

calyx lobes 5, white

capsule lobes with blister-like keels

flower

blades abaxially pale, with minute pubescence

pedicel

capsule

398

Family:	RHAMNACEAE
Species:	*Ceanothus velutinus* Dougl. *ex* Hook.
Common Name:	Snowbrush (tobaccobrush, buckbrush, mountain balm)
Life Span:	Perennial
Origin:	Native
Season:	Evergreen

GROWTH FORM

shrub (to 2 m tall); several to many stems from the base, diffusely spreading, often procumbent, forming dense colonies, crown often rounded; flowers May to July, fruits mature in August and September, reproduces from seeds

FLORAL AND FRUIT CHARACTERISTICS

inflorescences: thyrse (3–10 cm long, 3–4 cm wide), 1–5 main branches, each with a lanceolate bract near the base, axis pubescent; on axillary and terminal peduncles (1–4 cm long); usually drying chocolate-brown

flowers: perfect, regular; calyx lobes 5, united below, white, triangular (1.5–2 mm long, 1.5 mm wide), curved sharply between the petals over the flower; petals 5, white, pipe-shaped (2–3 mm long), long-clawed, usually curved downward; stamens 5, opposite the petals; style 3-cleft near the apex

fruits: capsules subglobose (3–4 mm high, 5–6 mm in diameter), sticky-glandular, 3-lobed, 3-celled, separating into 3 carpels each with 1 seed, each capsule lobe with a blister-like keel

VEGETATIVE CHARACTERISTICS

leaves: alternate, simple; blades broadly ovate to elliptic (2.5–8 cm long, 1.5–5 cm wide); apex obtuse; base rounded or subcordate, distinctly 3-veined; margins glandular-serrulate; coriaceous; adaxially dark green, glabrous to glandular-varnished; abaxially pale, with minute pubescence; petiolate (5–21 mm long)

stems: twigs rigid, slightly flattened, olive green, unarmed, puberulent; bark reddish-brown

other: plants produce a strong cinnamon- or balsam-like odor

HISTORIC, FOOD, AND MEDICINAL USES: some Native Americans used leaves as a tobacco substitute

LIVESTOCK LOSSES: none

FORAGE VALUE: generally not considered a browse species, although it is occasionally browsed by deer and elk in winter; goats, sheep, and deer may consume flowers; deer often bed in this shrub

HABITAT: open woods and mountain slopes; most abundant in rocky soils on logged or burned areas

Chamise
Adenostoma fasciculatum Hook. & Arn.

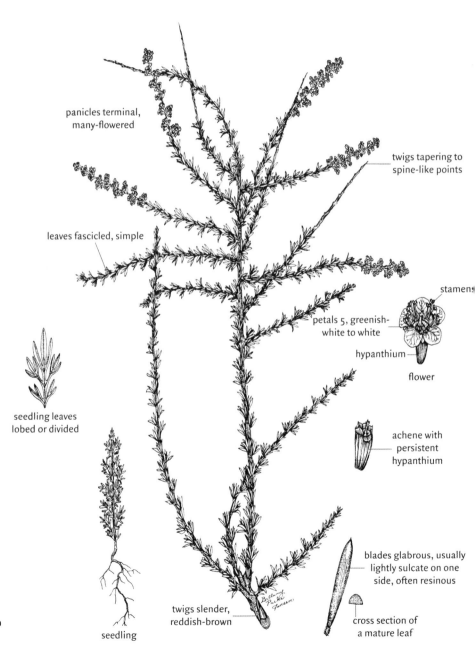

panicles terminal,
many-flowered

twigs tapering to
spine-like points

leaves fascicled, simple

stamens

petals 5, greenish-
white to white

hypanthium

flower

seedling leaves
lobed or divided

achene with
persistent
hypanthium

blades glabrous, usually
lightly sulcate on one
side, often resinous

twigs slender,
reddish-brown

cross section of
a mature leaf

seedling

Family:	ROSACEAE
Species:	*Adenostoma fasciculatum* Hook. & Arn.
Common Name:	Chamise (chamizo, greasewood, hierba del pasmo)
Life Span:	Perennial
Origin:	Native
Season:	Evergreen

GROWTH FORM
shrub (to 3.5 m tall); diffusely branched, forms dense thickets; growth starts in January and ends after flowering in June, reproduces from basal sprouts and seeds, seedlings rapidly establish following fire

FLORAL AND FRUIT CHARACTERISTICS
inflorescences: panicles terminal (4–12 cm long), crowded, many-flowered

flowers: perfect, regular; hypanthium obconical (1–2 mm long), striate with 10 lines, with glands inside margin; sepals 5; petals 5, greenish-white to white, rounded, spreading; stamens 10–15, in groups of 2–3, alternating with the petals; sessile or nearly so

fruits: achenes small, hard, 1 per flower; hypanthium persistent with achene

VEGETATIVE CHARACTERISTICS
leaves: fascicled, simple; blades subulate to spatulate (4–15 mm long); margins entire; glabrous, usually lightly sulcate on one side, often resinous, rigid; stipules minute; seedling leaves lobed or divided

stems: twigs slender, stiff, some tapering to spine-like points, reddish-brown; old bark grayish-brown or reddish-brown, becoming shreddy with age

other: thick canopy and numerous roots prevent rapid runoff of water and soil loss

HISTORIC, FOOD, AND MEDICINAL USES: used by some Native Americans for various ceremonial purposes

LIVESTOCK LOSSES: none

FORAGE VALUE: mature plants are worthless to cattle, sheep, and horses; largely unpalatable; fair browse for deer in winter; seedlings are palatable to cattle and sheep

HABITAT: mountain slopes, ridges, and foothills; chaparral; adapted to a broad range of soils, common on infertile soils; most abundant at elevations of 200–1800 m

Serviceberry
Amelanchier alnifolia (Nutt.) Nutt. *ex* Roemer

SYN = *A. utahensis* Koehne

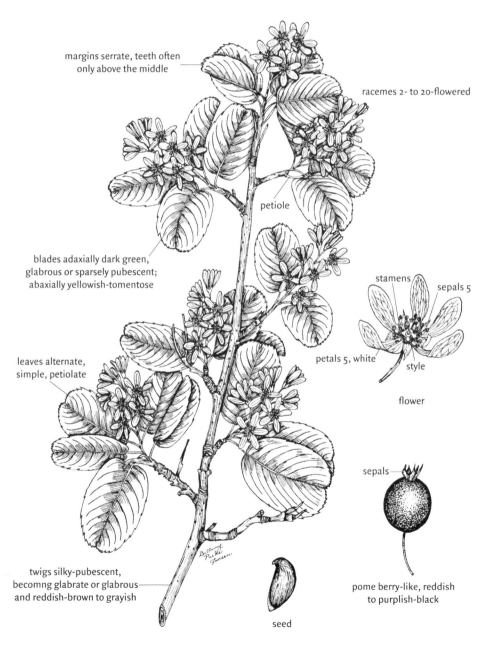

margins serrate, teeth often only above the middle

racemes 2- to 20-flowered

petiole

blades adaxially dark green, glabrous or sparsely pubescent; abaxially yellowish-tomentose

stamens

sepals 5

petals 5, white

style

flower

leaves alternate, simple, petiolate

sepals

twigs silky-pubescent, becomng glabrate or glabrous and reddish-brown to grayish

pome berry-like, reddish to purplish-black

seed

Family:	ROSACEAE
Species:	*Amelanchier alnifolia* (Nutt.) Nutt. *ex* Roemer
Common Name:	Serviceberry (Saskatoonberry, Saskatoon serviceberry, juneberry)
Life Span:	Perennial
Origin:	Native
Season:	Cool

GROWTH FORM

shrub or small tree (to 6 m tall); single or clustered trunks with branches near the base; flowers April to June, fruit matures in July and August, reproduces from seeds and stolons, forming colonies

FLORAL AND FRUIT CHARACTERISTICS

inflorescences: racemes widely ovate to ovate (2–4 cm long), erect, terminal on new growth, 2- to 20-flowered; axis silky-pubescent, becoming glabrous or glabrate

flowers: perfect, regular; hypanthium campanulate (3–4 mm in diameter), browinish-white; sepals 5, triangular (1.5–4 mm long), reflexed at anthesis; petals 5 (5–16 mm long, 2–3 mm wide), white, obovate to spatulate, spreading; stamens 10–20

fruits: pomes berry-like, globose to obovoid (8–11 mm long, 7–9 mm wide), fleshy, reddish to purplish-black, 3- to 6-seeded

other: flowers are ill-scented

VEGETATIVE CHARACTERISTICS

leaves: alternate, simple; blades oval or obovate to oblong (2–6 cm long, 2.5–4 cm wide); apex truncate, rounded, or obtuse; base rounded or subcordate; margins serrate, teeth often only above the middle, main lateral veins entering the teeth; adaxially dark green, glabrous or sparsely pubescent; abaxially yellowish-tomentose; petioles (7–20 mm long); stipules narrowly lanceolate (1–3 mm long), caducous

stems: young twigs silky-pubescent, becoming glabrate or glabrous and reddish-brown to grayish, smooth, rigid

HISTORIC, FOOD, AND MEDICINAL USES: fresh fruit is dry, mealy, and not very palatable; fruit can be used to make jams, pies, and wines; some Native Americans used stems for arrow shafts and tepee stakes

LIVESTOCK LOSSES: none

FORAGE VALUE: young growth is fair to good for livestock, excellent browse for deer and moose; fruit is an important food for small mammals, black bears, and birds; bark is eaten by beavers and marmots

HABITAT: brushy hillsides, open woods, canyons, and creek banks; usually growing in well-drained soils, but occasionally grows around bogs

Curlleaf mountainmahogany
Cercocarpus ledifolius Nutt. *ex* Torr. & A. Gray

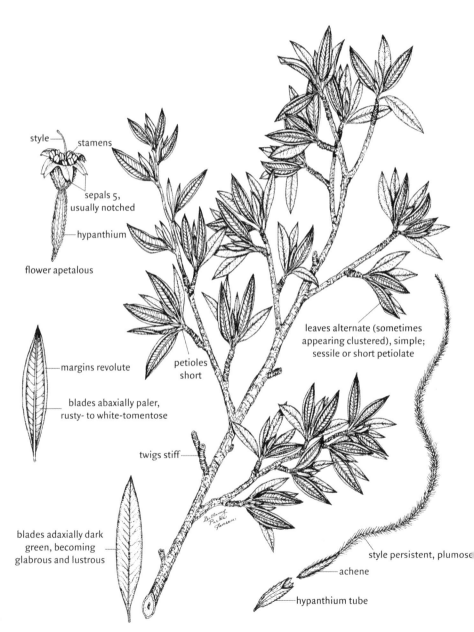

style — stamens

sepals 5,
usually notched

hypanthium

flower apetalous

margins revolute

blades abaxially paler,
rusty- to white-tomentose

petioles
short

leaves alternate (sometimes
appearing clustered), simple;
sessile or short petiolate

twigs stiff

blades adaxially dark
green, becoming
glabrous and lustrous

style persistent, plumose

achene

hypanthium tube

Family:	ROSACEAE
Species:	*Cercocarpus ledifolius* Nutt. *ex* Torr. & A. Gray
Common Name:	Curlleaf mountainmahogany (desert mahogany)
Life Span:	Perennial
Origin:	Native
Season:	Evergreen

GROWTH FORM

shrub or small tree (to 8 m tall); 1 to several trunks; characteristically grows in scattered patches; flowers May to July, reproduces from seeds

FLORAL AND FRUIT CHARACTERISTICS

inflorescences: flowers solitary or in clusters of 2 or 3; sessile or subsessile in axils

flowers: perfect, apetalous; hypanthium tube (3–12 mm long); sepals 5, ovate to triangular (1–2.5 mm long), spreading, usually notched, villous to tomentose outside, glabrous within; stamens 15–30; pistil single, style elongating in fruit

fruits: achenes terete (5–9 mm long), hard, narrow, sharp-pointed; style terminal, elongate (4–8 cm long), exserted, persistent, plumose

VEGETATIVE CHARACTERISTICS

leaves: alternate (sometimes appearing clustered), simple; blades narrowly lanceolate to oblanceolate (1–4 cm long, 5–12 mm wide); apex acute; base cuneate; margins entire, revolute; midvein prominent; coriaceous, resinous, aromatic; adaxially dark green, becoming glabrous and lustrous; abaxially paler, rusty- to whitish-tomentose; sessile or short petiolate (to 6 mm long)

stems: twigs stiff; bark reddish-brown, finely villous or tomentose with white hairs, becoming gray and deeply sulcate

HISTORIC, FOOD, AND MEDICINAL USES: wood is hard and dense (will not float), providing excellent fuel, producing intense heat and burns for long periods; the Gosiute Indians of Utah made bows from this wood

LIVESTOCK LOSSES: none

FORAGE VALUE: fair for sheep and cattle in fall and winter, good for big game in winter, not readily eaten at other times; provides cover and some browse for deer and pronghorn, except in summer

HABITAT: hills, canyons, rocky slopes, and rocky ridges at altitudes of 1200–3000 m; usually on south- and west-facing slopes; adapted to a wide range of soil textures, most abundant in dry coarse-textured soils

True mountainmahogany
Cercocarpus montanus Raf.

SYN = *C. breviflorus* A. Gray

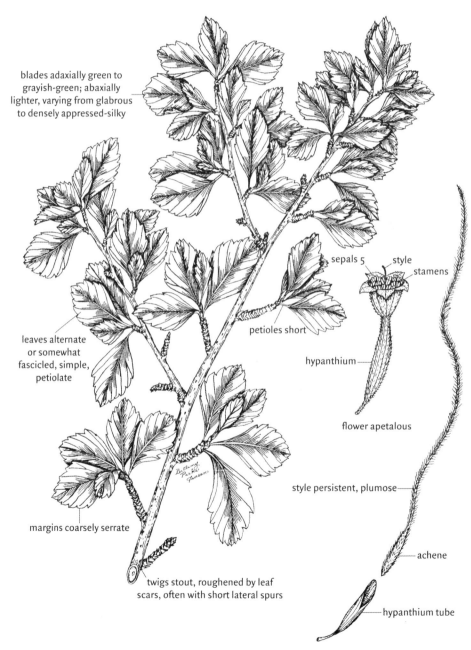

blades adaxially green to grayish-green; abaxially lighter, varying from glabrous to densely appressed-silky

leaves alternate or somewhat fascicled, simple, petiolate

margins coarsely serrate

twigs stout, roughened by leaf scars, often with short lateral spurs

sepals 5

style

stamens

petioles short

hypanthium

flower apetalous

style persistent, plumose

achene

hypanthium tube

Family:	ROSACEAE
Species:	*Cercocarpus montanus* Raf.
Common Name:	True mountainmahogany (palo duro, birchleaf mountainmahogany)
Life Span:	Perennial
Origin:	Native
Season:	Cool

GROWTH FORM

shrub or small tree (to 6 m tall); branches upright or spreading; flowers June to July (some southern varieties flower as late as November), fruits usually mature in August, reproduces from seeds

FLORAL AND FRUIT CHARACTERISTICS

inflorescences: flowers solitary or clustered in groups of 2–3; axillary, some-times crowded in groups of 5–15 on short spur-like branchlets

flowers: perfect, apetalous; inconspicuous; hypanthium tube cylindrical (5–11 mm long); sepals 5 (1–2 mm long, 3–5 mm wide), greenish-yellow, becoming reddish-brown, elongate, hairs spreading-villous or appressed-silky; stamens 22–44

fruits: achenes cylindric-fusiform (8–12 mm long), appressed-silky; style terminal, elongate (5–10 cm long), twisted, exserted, persistent, plumose

VEGETATIVE CHARACTERISTICS

leaves: alternate or somewhat fascicled, simple; blades usually ovate to oval or obovate (2–5 cm long, 1.5–3.5 cm wide); apex acute to rounded; base usually cuneate; margins coarsely serrate; adaxially green to grayish-green; abaxially lighter, varying from glabrous to densely appressed-silky, 3–10 prominent veins; petioles short (3–6 mm long)

stems: twigs stout, rigid, roughened by leaf scars, often with short lateral spurs; bark thin, young branches reddish-brown, older branches gray to brown

HISTORIC, FOOD, AND MEDICINAL USES: some Native Americans used wood from these plants to make tools and war clubs, the Hopis used the bark to make a reddish-brown dye for leather

LIVESTOCK LOSSES: leaves may contain a cyanogenic glycoside, which may cause hydrocyanic acid poisoning

FORAGE VALUE: good to very good for cattle, sheep, and goats; extremely valuable winter browse for deer, elk, and bighorn

HABITAT: rocky bluffs, mountain sides, canyons, rimrock, breaks, and open woodland; most abundant in dry soils

Blackbrush
Coleogyne ramosissima Torr.

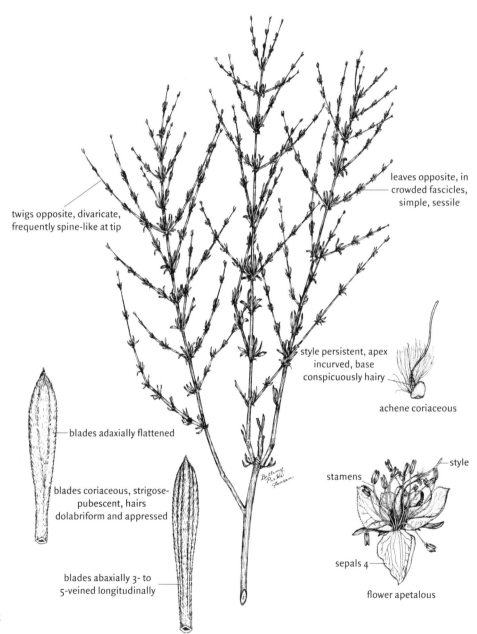

leaves opposite, in crowded fascicles, simple, sessile

twigs opposite, divaricate, frequently spine-like at tip

style persistent, apex incurved, base conspicuously hairy

achene coriaceous

blades adaxially flattened

blades coriaceous, strigose-pubescent, hairs dolabriform and appressed

blades abaxially 3- to 5-veined longitudinally

style

stamens

sepals 4

flower apetalous

Family:	ROSACEAE
Species:	*Coleogyne ramosissima* Torr.
Common Name:	Blackbrush (burrobrush)
Life Span:	Perennial
Origin:	Native
Season:	Evergreen

GROWTH FORM
shrub (to 3 m tall); much-branched; branches opposite, short, rigid; forms nearly pure stands on large areas; flowers March to May, reproduces from seeds

FLORAL AND FRUIT CHARACTERISTICS
inflorescences: flowers solitary; subtended by 1 or 2 pairs of 3-lobed bractlets; terminating short branchlets

flowers: perfect, apetalous; hypanthium tube very short, coriaceous; sepals 4 (8–12 mm long), arranged in 2 pairs, petaloid; adaxially greenish-yellow (sometimes brownish); abaxially green to purplish, strigose; 2 outer sepals lanceolate to oblong (4–7 mm long), inner ones ovate and acute (5–8 mm long), margins scarious; stamens 20–40; ovary enclosed in a tubular sheath

fruits: achenes somewhat compressed (3–5 mm long), glabrous, coriaceous; with a long, twisted, and thread-like stalk (style) from 1 side; style persistent, apex incurved, base conspicuously hairy; hair long, dense

VEGETATIVE CHARACTERISTICS
leaves: opposite, in crowded fascicles, simple; blades linear to oblanceolate (4–15 mm long, 1–5 mm wide); apex obtuse to mucronate; margins entire; coriaceous, strigose-pubescent, hairs dolabriform and appressed; adaxially flattened; abaxially 3- to 5-veined longitudinally; sessile

stems: twigs opposite and divaricate, frequently spine-like at the tip; bark gray to ashy, becoming black with age, striate

other: gray barks turns black when wet

HISTORIC, FOOD, AND MEDICINAL USES: none

LIVESTOCK LOSSES: none

FORAGE VALUE: fair for cattle, sheep, goats, and deer during the winter; survives on heavily used rangeland because of the spiny character of the branches

HABITAT: desert mesas, open plains, and foothills in dry and well-drained soils; most abundant in sandy, gravelly, and rocky soils

Apache plume
Fallugia paradoxa (D. Don) Endl.

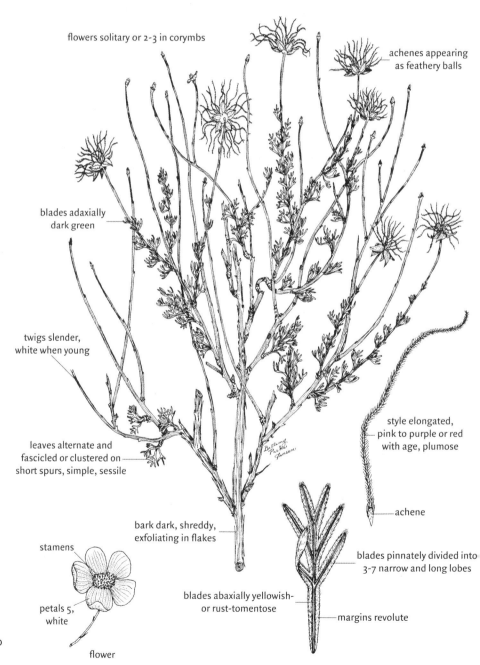

flowers solitary or 2-3 in corymbs

achenes appearing as feathery balls

blades adaxially dark green

twigs slender, white when young

leaves alternate and fascicled or clustered on short spurs, simple, sessile

style elongated, pink to purple or red with age, plumose

achene

bark dark, shreddy, exfoliating in flakes

blades pinnately divided into 3-7 narrow and long lobes

stamens

petals 5, white

blades abaxially yellowish- or rust-tomentose

margins revolute

flower

Family:	ROSACEAE
Species:	*Fallugia paradoxa* (D. Don) Endl.
Common Name:	Apache plume (poñil, feather rose)
Life Span:	Perennial
Origin:	Native
Season:	Cool

GROWTH FORM

shrub (to 2.5 m tall); branchlets slender, straggly, crown irregular, clump-forming; flowers May to October, reproduces from seeds

FLORAL AND FRUIT CHARACTERISTICS

inflorescences: flowers solitary or 2–3 in corymbs; peduncle elongated, nearly leafless

flowers: mostly unisexual, sometimes perfect, regular; hypanthium cupulate; sepals 5, ovate, long-acuminate or trifid alternating with 5 linear-lanceolate or bifid bractlets; corolla showy (2.5–3.5 cm in diameter); petals 5, white, rounded (9–16 mm long), spreading; stamens numerous, in 3 series

fruits: achenes numerous, appearing as feathery balls, obovoid-fusiform (2.5–3.5 mm long); styles terminal, elongated (2.5–4 cm long), pink to purple or red with age, plumose

VEGETATIVE CHARACTERISTICS

leaves: alternate and fascicled or clustered on short spurs, simple; blades cuneate to obovate (1–2 cm long), pinnately divided into 3–7 (sometimes 9) narrow and long lobes; apex obtuse; margins entire and revolute, thick; adaxially dark green; abaxially yellowish- or rust-tomentose; midvein prominent; sessile; stipules triangular, small, persistent

stems: twigs slender; bark white when young, turning dark with age, shreddy, exfoliating in flakes

other: mostly evergreen

HISTORIC, FOOD, AND MEDICINAL USES: some Native Americans used bundles of twigs as brooms and older stems for arrow shafts; a decoction from leaves was used as a supposed growth stimulant for hair; currently used as an ornamental and for erosion control

LIVESTOCK LOSSES: none

FORAGE VALUE: fair for cattle and goats, good winter forage for sheep; furnishes important browse for big game

HABITAT: arroyos, foothills, plains, and mesas in deserts and chaparral; most abundant in rocky or gravelly slopes and alluvial plains

Shrubby cinquefoil
Pentaphylloides fruticosa (L.) O. Schwartz

SYN = *Dasiphera floribunda* (Pursh) Kartesz, *Potentilla fruticosa* L.

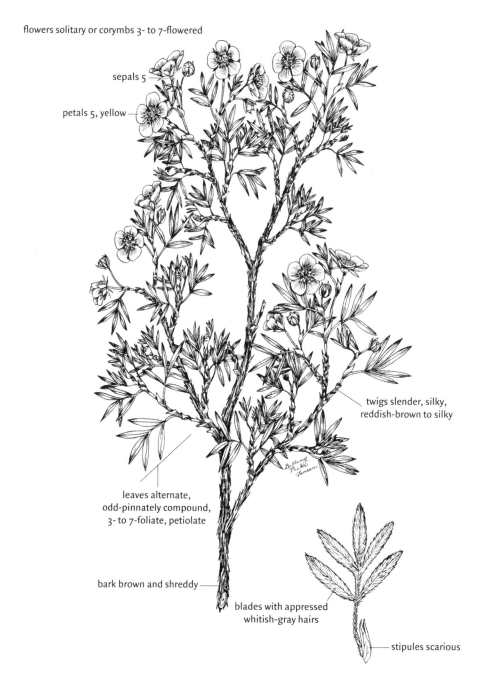

flowers solitary or corymbs 3- to 7-flowered

sepals 5

petals 5, yellow

twigs slender, silky, reddish-brown to silky

leaves alternate, odd-pinnately compound, 3- to 7-foliate, petiolate

bark brown and shreddy

blades with appressed whitish-gray hairs

stipules scarious

Family: ROSACEAE
Species: *Pentaphylloides fruticosa* (L.) O. Schwartz
Common Name: Shrubby cinquefoil (bush cinquefoil, yellow rose)
Life Span: Perennial
Origin: Native
Season: Cool (partially evergreen)

GROWTH FORM
shrub (to 1.5 m tall); usually erect or ascending, much-branched, stems leafy; flowers April to June, seeds mature June to September, reproduces from seeds

FLORAL AND FRUIT CHARACTERISTICS
inflorescences: corymbs 3- to 7-flowered, terminating branches or solitary flowers in leaf axils

flowers: perfect, regular; hypanthium saucer-shaped, long-hairy; sepals 5, ovate (4–7 mm long), acuminate, alternating with 5 lanceolate and pubescent bractlets; petals 5 (5–15 mm long), yellow, nearly orbicular; stamens 15–25; carpels numerous, angular, pubescent

fruits: achenes (1.5–2.5 mm long) pubescent

VEGETATIVE CHARACTERISTICS
leaves: alternate, odd-pinnately compound, 3- to 7-foliate; leaflets crowded, upper 3 leaflets sometimes confluent at base, narrowly elliptic (5–20 mm long), tapering at each end; margins entire and often revolute; appressed whitish-gray hairs, especially abaxially; petioles variable (5–12 mm long); stipules broadly lanceolate to ovate, scarious

stems: twigs slender; bark first silky and reddish-brown to gray, later brown and shreddy

HISTORIC, FOOD, AND MEDICINAL USES: steeped leaves were sometimes used as a tea by some Native Americans and pioneers; popular ornamental with many horticultural varieties; sometimes planted for erosion control

LIVESTOCK LOSSES: none

FORAGE VALUE: poor for cattle, good browse for sheep and goats in the Southwest and intermountain areas; browsed extensively by mule deer, less so by white-tailed deer

HABITAT: alpine meadows, bogs, rocky grounds at higher elevations (2000–3000 m), and subarctic regions; adapted to a broad range of soils, most abundant in moist soils

Chokecherry
Prunus virginiana L.

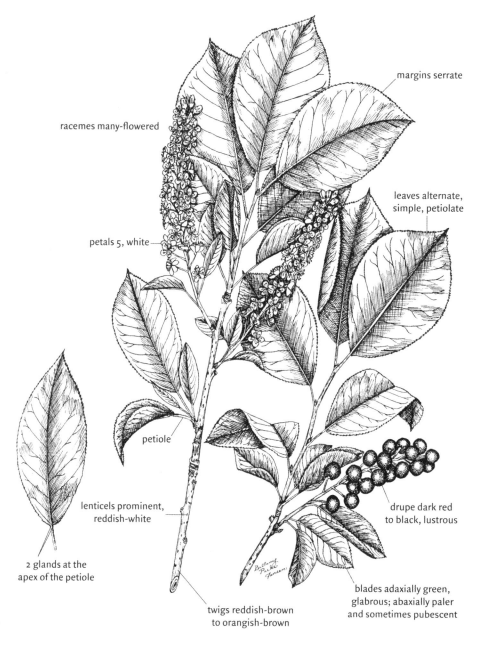

margins serrate

racemes many-flowered

leaves alternate,
simple, petiolate

petals 5, white

petiole

lenticels prominent,
reddish-white

drupe dark red
to black, lustrous

2 glands at the
apex of the petiole

twigs reddish-brown
to orangish-brown

blades adaxially green,
glabrous; abaxially paler
and sometimes pubescent

Family:	ROSACEAE
Species:	*Prunus virginiana* L.
Common Name:	Chokecherry (capulín, western chokecherry, black chokecherry)
Life Span:	Perennial
Origin:	Native
Season:	Cool

GROWTH FORM

shrub or small tree (to 10 m tall); erect with horizontal branches, forming dense colonies, colony crown rounded; flowers April to July, fruits mature July to September; reproduces from seeds, rhizomes, and basal sprouts

FLORAL AND FRUIT CHARACTERISTICS

inflorescences: racemes oblong (5–15 cm long), cylindric, terminal, dense, many-flowered

flowers: perfect, regular; hypanthium 2–3 mm deep; sepals 5, short, obtuse at apex, glandular-lancinate on the margins; corolla 6–9 mm in diameter; petals 5, white, suborbicular (3–5 mm long); stamens 20–30

fruits: drupes globose (6–10 mm in diameter), dark red to black, lustrous, skin thick, flesh juicy, astringent and acidulous

other: flowers fragrant

VEGETATIVE CHARACTERISTICS

leaves: alternate, simple; blades elliptic (2–10 cm long, 1–6 cm wide); apex acuminate; margins serrate; adaxially green, glabrous; abaxially paler, sometimes pubescent; 2 glands at the apex of the petiole (1–2 cm long)

stems: twigs slender, greenish first and then becoming reddish-brown to orangish-brown; bark gray to black; lenticels prominent, reddish-white

HISTORIC, FOOD, AND MEDICINAL USES: fruits are used for jelly; bark is sometimes used as a flavoring agent in cough syrup; some Native Americans used bark extract to cure diarrhea; fruits were used to treat canker sores and added to pemmican; wood was used for arrows, bows, and pipe stems; ornamental

LIVESTOCK LOSSES: may contain toxic quantities of hydrocyanic acid in leaves, stems, and seeds; poisonous to all classes of livestock, poisoning generally occurs when plant is stressed from drought or freezing

FORAGE VALUE: poor to fair for cattle and sheep, twigs are good winter browse for wildlife; fruit is an important food source for wildlife

HABITAT: prairies, fence rows, roadsides, hillsides, and canyons; most abundant in moist soils; adapted to a broad range of soil types

Mexican cliffrose
Purshia stansburiana (Torr.) Henr.

SYN = *Cowania mexicana* D. Don

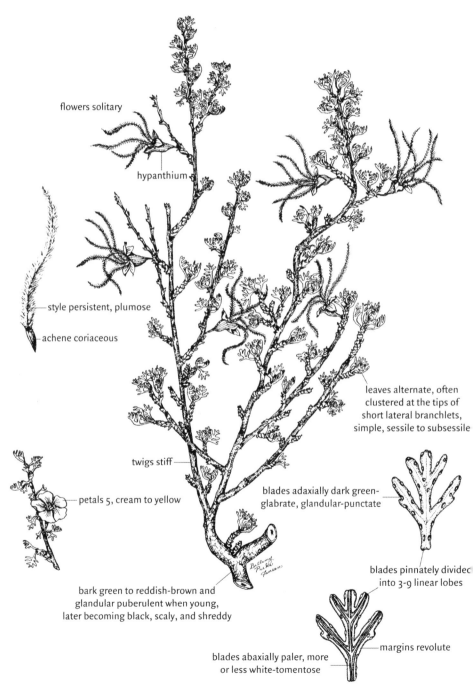

flowers solitary

hypanthium

style persistent, plumose

achene coriaceous

leaves alternate, often
clustered at the tips of
short lateral branchlets,
simple, sessile to subsessile

twigs stiff

petals 5, cream to yellow

blades adaxially dark green-
glabrate, glandular-punctate

blades pinnately divided
into 3-9 linear lobes

bark green to reddish-brown and
glandular puberulent when young,
later becoming black, scaly, and shreddy

margins revolute

blades abaxially paler, more
or less white-tomentose

Family:	ROSACEAE
Species:	*Purshia stansburiana* (Torr.) Henr.
Common Name:	Mexican cliffrose (rosa mexicana, quininebush, Stansbury cliffrose)
Life Span:	Perennial
Origin:	Native
Season:	Evergreen

GROWTH FORM

shrub or small tree (rarely to 8 m tall); much-branched, spreading, crown rounded; flowers April to June, fruits mature September to October, reproduces from seeds

FLORAL AND FRUIT CHARACTERISTICS

inflorescences: flowers solitary; borne at ends of small lateral branches

flowers: perfect, regular; hypanthium tube hemispheric to turbinate (4–6 mm long); sepals 5, broadly ovate, glandular-tomentose; corolla showy (1.2–1.8 cm in diameter); petals 5, cream to yellow, broadly ovate to obovate (7–15 mm long); stamens numerous, in 2 series; pistils 4–10, styles elongating in fruit

fruits: achenes 4–10; body narrowly oblong (4–8 mm long), coriaceous, striate-ribbed, glabrous at maturity; style terminal, elongate (1.2–5 cm long), persistent, silvery, plumose

other: flowers fragrant

VEGETATIVE CHARACTERISTICS

leaves: alternate, often clustered at the tips of short lateral branchlets, simple; blades obovate to narrowly spatulate (6–15 mm long), pinnately divided into 3–9 linear lobes; margins revolute; adaxially dark green-glabrate, glandular-punctate; abaxially paler, more or less white-tomentose, scattered glands; sessile to subsessile

stems: twigs erect, stiff; bark green to reddish-brown and glandular puberulent when young, later becoming dark gray to black, scaly, and shreddy

other: resinous with a strong odor

HISTORIC, FOOD, AND MEDICINAL USES: leaves used by some Native Americans as a medicinal wash for wounds; wood used for arrow shafts; fibers were used for baskets, sandals, ropes, and clothing; cultivated as an ornamental

LIVESTOCK LOSSES: none

FORAGE VALUE: good for cattle and sheep, especially important browse in winter; staple feed for mule deer in some areas

HABITAT: cliffs, hillsides, mesas, and washes; most abundant in dry and rocky soils at altitudes of 1000–2500 m

Antelope bitterbrush
Purshia tridentata (Pursh) DC.

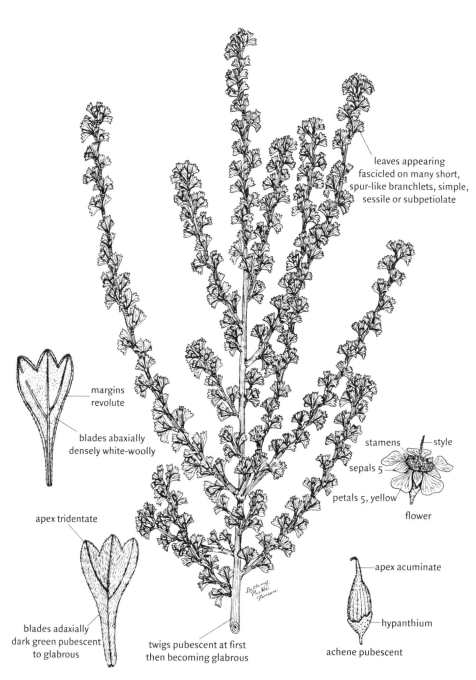

leaves appearing
fascicled on many short,
spur-like branchlets, simple,
sessile or subpetiolate

margins
revolute

blades abaxially
densely white-woolly

stamens — style

sepals 5

petals 5, yellow

flower

apex tridentate

apex acuminate

hypanthium

blades adaxially
dark green pubescent
to glabrous

twigs pubescent at first
then becoming glabrous

achene pubescent

418

Family:	ROSACEAE
Species:	*Purshia tridentata* (Pursh) DC.
Common Name:	Antelope bitterbrush (bitterbrush)
Life Span:	Perennial
Origin:	Native
Season:	Evergreen

GROWTH FORM

shrub (to 3 m tall); late-deciduous to evergreen, much-branched, crown rounded; flowers April to August, fruits mature July to September, reproduces from
seeds; trailing branches often root when in contact with the soil

FLORAL AND FRUIT CHARACTERISTICS

inflorescences: flowers solitary; subsessile; terminating short branchlets

flowers: perfect, regular; hypanthium funnelform (2–5 mm long), tomentose and sometimes glandular; sepals 5 (1.8–3 mm long), ovate to oblong;
petals 5 (4–8 mm long), yellow, spatulate or obovate; stamens 20–25,
exserted; style beak-like (4–6 mm long), stout, persistent

fruits: achenes (7–12 mm long) exceeding the calyx, acuminate at apex,
longitudinally ribbed, pubescent

VEGETATIVE CHARACTERISTICS

leaves: alternate, appearing fascicled, simple; blades cuneate (5–26 mm
long); apex 3-lobed to tridentate; margins entire but revolute; adaxially
dark green, pubescent to glabrous; abaxially densely white-woolly; sessile
or subpetiolate; stipules small, triangular, persistent

stems: twigs gray to brown, with many short, spur-like branchlets, pubescent at first then becoming glabrous; buds small and scaly

HISTORIC, FOOD, AND MEDICINAL USES: occasionally cultivated as an ornamental

LIVESTOCK LOSSES: none

FORAGE VALUE: good for cattle, sheep, and goats especially in late fall and
winter when the ground is snow-covered; usually not eaten by horses; excellent
for mule deer; less valuable to pronghorn, white-tailed deer, and elk; seeds are
important for birds and small mammals; withstands heavy grazing

HABITAT: plains, foothills, mountain slopes, mesas, and open woodland; most
abundant in well-drained sandy, gravelly, or rocky soils

Wild rose
Rosa woodsii Lindl.

SYN = *R. demareei* Palmer

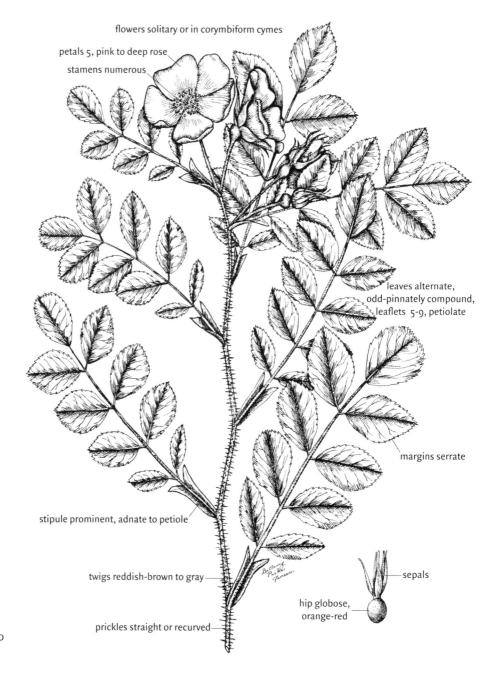

flowers solitary or in corymbiform cymes

petals 5, pink to deep rose

stamens numerous

leaves alternate,
odd-pinnately compound,
leaflets 5-9, petiolate

margins serrate

stipule prominent, adnate to petiole

twigs reddish-brown to gray

sepals

hip globose,
orange-red

prickles straight or recurved

Family:	ROSACEAE
Species:	*Rosa woodsii* Lindl.
Common Name:	Wild rose (rosa silvestre, Woods rose)
Life Span:	Perennial
Origin:	Native
Season:	Cool

GROWTH FORM

shrub (to 1.5 m tall); usually forming thickets; growth starts in early spring, flowers May to July, reproduces from seeds

FLORAL AND FRUIT CHARACTERISTICS

inflorescences: corymbiform cymes or flowers solitary; terminating branches of the current season, cymes few-flowered

flowers: perfect, regular; hypanthium glabrous (3–5 mm in diameter); sepals 5 (1–2 cm long, 1.5–3.5 mm wide at the base), apex attenuate, often dilated, pubescent (rarely glandular), persistent, erect or spreading in fruit; petals 5, obovate (1.5–2.5 cm long), pink to deep rose; stamens numerous

fruits: achenes (3–4 mm long), hairy on one side, 15–35 contained inside an orange-red, globose hip (6–12 mm in diameter) formed by the hypanthium

VEGETATIVE CHARACTERISTICS

leaves: alternate, odd-pinnately compound; leaflets 5–9, elliptic to oval or elliptic-ovate (2–5 cm long, 1–2.5 cm wide); margins sharply serrate; adaxially glabrous and shiny; abaxially puberulent or glandular to glabrous; petiolate; stipules prominent, adnate to the petiole at the base (4–7 mm wide), glandular-pubescent on back

stems: twigs reddish-brown to gray, with straight or recurved prickles; infrastipular prickles more prominent than those at the internodes

HISTORIC, FOOD, AND MEDICINAL USES: Europeans utilized hips as a source of vitamins A and C, rose hip powder used as a flavoring in soups and for making syrup; some Native Americans ate the young shoots as a potherb and steeped leaves for tea; petals are consumed today raw, in salads, candied, or made into syrup; inner bark has been smoked like tobacco; and dried petals were stored for perfume

LIVESTOCK LOSSES: prickles may injure soft tissue

FORAGE VALUE: fair to good for sheep and cattle; good browse for elk and deer; small mammals and birds feed on hips

HABITAT: prairies, open woods, plateaus, dry slopes, ravines, and thickets; in a broad range of soils

Quaking aspen
Populus tremuloides Michx.

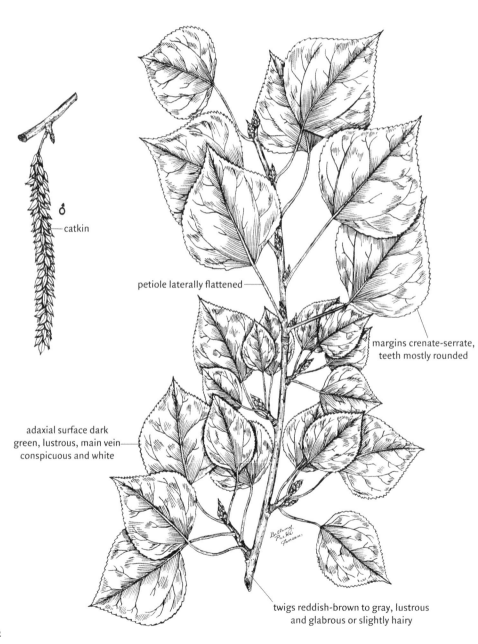

♂

—catkin

petiole laterally flattened—

margins crenate-serrate,
teeth mostly rounded

adaxial surface dark
green, lustrous, main vein—
conspicuous and white

twigs reddish-brown to gray, lustrous
and glabrous or slightly hairy

Family:	SALICACEAE
Species:	*Populus tremuloides* Michx.
Common Name:	Quaking aspen (chopo temblón, trembling aspen, aspen, álamo blanco, alamillo)
Life Span:	Perennial
Origin:	Native
Season:	Cool

GROWTH FORM

dioecious tree (to 30 m tall); trunk erect, long and slender; crown rounded; flowers April to June, fruits ripen May to July, grows rapidly from basal sprouts and root sprouts, seldom reproduces from seeds

FLORAL AND FRUIT CHARACTERISTICS

inflorescences: catkins (2–8 cm long) drooping, dense, rachis thinly pubescent

flowers: unisexual, apetalous; staminate floral disk oblique (1.2–1.9 mm wide), entire, stamens 6–12; scales divided into 3–5 lobes; lobes triangular-lanceolate, acute to acuminate, hairy-fringed; pistillate floral disks crenate

fruits: capsules borne in catkin (which may lengthen to 12 cm); oblong-conic to lance-ovoid (3–6 mm long), light green to brown, thin walled, 2-valved, 3–6 seeds per valve

VEGETATIVE CHARACTERISTICS

leaves: alternate, simple; blades highly variable, ovate to broadly ovate or reniform (2.5–7.5 cm long, 2.5–7 cm wide); apex acute or acuminate; base truncate or rounded; margins crenate-serrate, teeth mostly rounded; adaxially dark green, glabrous, lustrous, main vein conspicuous and white; abaxially pale green, glabrous; petiole slender (4–6.5 cm long), flattened laterally, as long as the blade

stems: twigs slender, reddish-brown to gray, lustrous and glabrous or slightly hairy; trunk bark whitish, soft, becoming dark

other: leaves change to bright yellow or yellowish-orange in fall

HISTORIC, FOOD, AND MEDICINAL USES: bark was used by pioneers and some Native Americans as a fever remedy and for scurvy, contains salicin (similar to the active ingredient in aspirin); a substance similar to turpentine was extracted and used internally as an expectorant and externally as a counterirritant; wood is used for pulp and lumber

LIVESTOCK LOSSES: none

FORAGE VALUE: fair to good for sheep, fair for cattle; twigs, bark, and buds are browsed by horses and wildlife; seeds eaten by birds and small mammals

HABITAT: moist upland woods, mountain slopes, parkland, and stream banks; occurs on nearly all soil types

Beaked willow
Salix bebbiana Sarg.

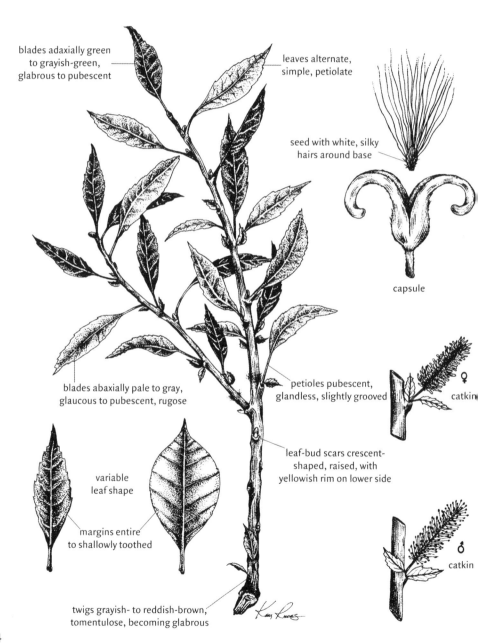

blades adaxially green to grayish-green, glabrous to pubescent

leaves alternate, simple, petiolate

seed with white, silky hairs around base

capsule

blades abaxially pale to gray, glaucous to pubescent, rugose

petioles pubescent, glandless, slightly grooved

♀ catkin

leaf-bud scars crescent-shaped, raised, with yellowish rim on lower side

variable leaf shape

margins entire to shallowly toothed

♂ catkin

twigs grayish- to reddish-brown, tomentulose, becoming glabrous

Family:	SALICACEAE
Species:	*Salix bebbiana* Sarg.
Common Name:	Beaked willow (Bebb willow, longbeaked willow)
Life Span:	Perennial
Origin:	Native
Season:	Cool

GROWTH FORM
dioecious shrub (to 4 m tall); flowers April to May, fruits May to June, reproduces from seeds and basal sprouts

FLORAL AND FRUIT CHARACTERISTICS
inflorescence: catkins appearing with the leaves; staminate catkins (1.5–2.5 cm long) densely flowered on short new stems (usually with 2 leaves); pistillate catkins (1.5–6 cm long) densely flowered on short new stems (usually with 2–4 leaves), persistent after capsule dehiscence

flowers: unisexual, apetalous; staminate scales ovate (1.5–2.5 mm long), blunt, pubescent, yellowish to brown, stamens 2; pistillate scales ovate (2 mm long), pubescent, greenish yellow, apex dark when immature, becoming yellowish brown with age, ovary silky, stigmas 2

fruits: capsules born in catkins, ovoid-conic (5–8 mm long), silky pubescent, splitting into 2 valves; seeds obovoid (1 mm long) with white, silky hairs (5–8 mm long) around base

VEGETATIVE CHARACTERISTICS
leaves: alternate, simple; elliptical to narrowly ovate or obovate (3–6 cm long, 1.5–2.5 cm wide); apex short acuminate to acute; base cuneate; margins entire to shallowly toothed; adaxially green to grayish-green, glabrous to pubescent, somewhat coriaceous; abaxially pale to gray, glaucous to pubescent, rugose; petiole pubescent (5–15 mm long) glandless, slightly grooved

stems: twigs grayish- to reddish-brown (1–1.2 mm wide), flexuous, tomentulose, becoming glabrous; lenticels elliptical, light colored, not obvious; leaf bud scars crescent-shaped, raised, with yellowish rim on lower side; trunk gnarled, bark gray

other: reproduced from stem cuttings for revegetation

HISTORIC, FOOD, AND MEDICINAL USES: some Native Americans made tea from the bark to treat fever and headache

LIVESTOCK LOSSES: none

FORAGE VALUE: fair for cattle and good for sheep; valuable browse for elk, deer, and moose

HABITAT: moist valleys, stream banks, wet meadows, seeps, moist to dry hillsides

Sandbar willow
Salix exigua Nutt.

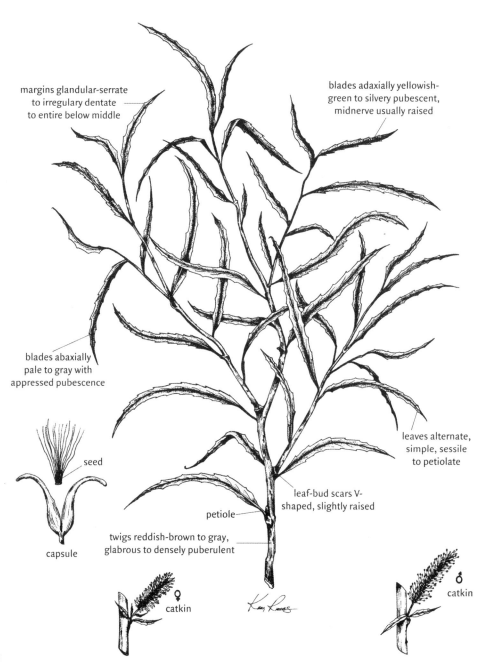

margins glandular-serrate to irregulary dentate to entire below middle

blades adaxially yellowish-green to silvery pubescent, midnerve usually raised

blades abaxially pale to gray with appressed pubescence

seed

capsule

leaves alternate, simple, sessile to petiolate

leaf-bud scars V-shaped, slightly raised

petiole

twigs reddish-brown to gray, glabrous to densely puberulent

♀ catkin

♂ catkin

Family:	SALICACEAE
Species:	*Salix exigua* Nutt.
Common Name:	Sandbar willow (sauz, narrowleaf willow, coyote willow, sauce)
Life Span:	Perennial
Origin:	Native
Season:	Cool

GROWTH FORM

dioecious shrub (to 7 m tall); rhizomatous, stems and branches spreading to form a rounded crown; flowers May to June, fruits June to July, reproduces from seeds and rhizomes

FLORAL AND FRUIT CHARACTERISTICS

inflorescence: catkins; staminate catkins (1.5–2.5 cm long, 5–6 mm wide) emerging after the leaves on new leafy shoots, erect or ascending; pistillate catkins (1.5–8 cm long, 4–5 mm wide) on new shoots, ascending to slightly drooping

flowers: unisexual, apetalous; staminate scales ovate to obovate, acute, densely pubescent on both sides, greenish-yellowish, deciduous, stamens 2; pistillate scales rounded, greenish-yellow, rugose, pubescent on both sides, ovary sparsely pubescent

fruits: capsules born in catkins, capsules ovoid (4–7 mm long, 1.5 mm wide), blunt, glabrous or slightly pubescent, splitting into 2 valves

VEGETATIVE CHARACTERISTICS

leaves: alternate, simple; blade linear to lanceolate (4–12 cm long, 2–10 mm wide); apex acute; base acuminate; margins glandular-serrate to irregularly dentate to entire below middle; adaxially yellowish-green to silvery pubescent, permanently pubescent to becoming glabrous, midnerve usually raised; abaxially pale to gray with appressed pubescence; sessile to petiolate (1–5 mm long)

stems: twigs reddish-brown to gray, flexuous, glabrous to densely puberulent; lenticels elliptic, raised, scattered; leaf-bud scars V-shaped, slightly raised; trunk bark gray-green to brown, smooth, shallowly furrowed with age

other: variable species with at least 2 subspecies

HISTORIC, FOOD, AND MEDICINAL USES: some Native Americans used peeled stems to make baskets; a tea was made from the bark to treat fever and headache

LIVESTOCK LOSSES: none

FORAGE VALUE: fair for cattle and good for sheep; valuable browse for deer, elk, and moose

HABITAT: valleys, stream banks, marshy areas, ditch banks, occasionally on dry ground

Blue penstemon
Penstemon glaber Pursh

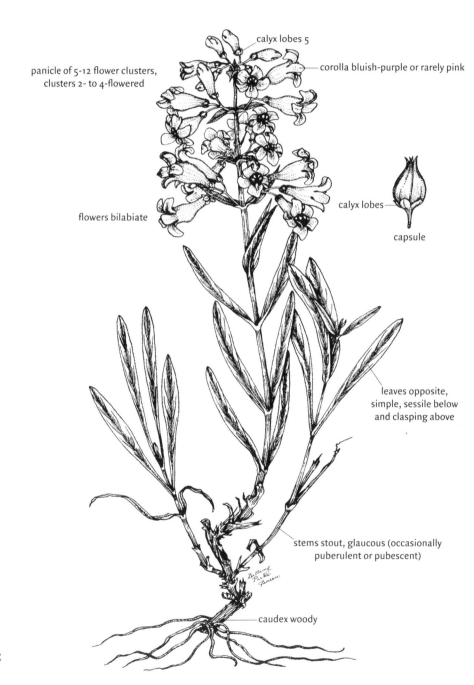

calyx lobes 5

panicle of 5-12 flower clusters, clusters 2- to 4-flowered

corolla bluish-purple or rarely pink

flowers bilabiate

calyx lobes

capsule

leaves opposite, simple, sessile below and clasping above

stems stout, glaucous (occasionally puberulent or pubescent)

caudex woody

Family:	SCROPHULARIACEAE
Species:	*Penstemon glaber* Pursh
Common Name:	Blue penstemon (sawsepal penstemon, smooth beardtongue)
Life Span:	Perennial
Origin:	Native
Season:	Cool

GROWTH FORM

forb (15–80 cm tall); erect to ascending from a woody caudex; flowers June to September, seeds mature August to October, reproduces from seeds

FLORAL AND FRUIT CHARACTERISTICS

inflorescences: panicle (8–30 cm long); flowers in 5–12 clusters, congested, each floral cluster with 2 to 4 flowers

flowers: perfect, bilabiate; calyx lobes 5, rounded or short-acuminate (2–10 mm long, 1.5–4 mm wide), margins scarious and erose, glabrous to puberulent; corolla bluish-purple or rarely pink (2.5–4 cm long), pale internally and lined on the interior with reddish-purple nectar guides, lobes rounded; sterile filament bearded

fruits: capsules 2-valved, ovoid (1–1.5 cm long), thin, pliable, brown at maturity

VEGETATIVE CHARACTERISTICS

leaves: opposite, simple; basal leaves wanting or highly reduced, apex acute to obtuse, occasionally mucronate, sessile or subpetiolate; cauline leaves linear-lanceolate to lanceolate (3–15 cm long, 7–45 mm wide), thick, apex acute to obtuse, margins entire; glabrous (rarely pubescent), glaucous; sessile below and clasping above

stems: 1 to many, stout, glaucous (occasionally puberulent or pubescent)

HISTORIC, FOOD, AND MEDICINAL USES: some Native Americans made a wet dressing from the leaves to treat snakebites, others made tea from the leaves to stop vomiting; also grown as an ornamental

LIVESTOCK LOSSES: may accumulate selenium, but has not been substantiated to cause poisoning

FORAGE VALUE: fair for mule deer and sheep, becomes less palatable with maturity; worthless for cattle; seeds are eaten by birds and small mammals

HABITAT: plains, hills, and mountains; most abundant in sandy to gravelly soils

Spiny hackberry
Celtis pallida Torr.

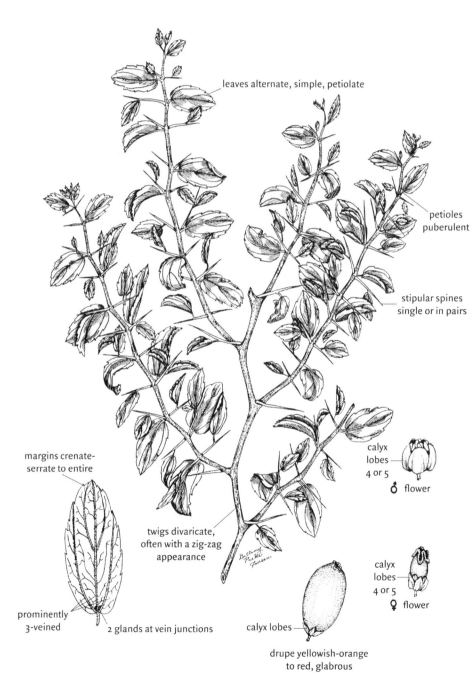

leaves alternate, simple, petiolate

petioles puberulent

stipular spines single or in pairs

calyx lobes 4 or 5

♂ flower

calyx lobes 4 or 5

♀ flower

margins crenate-serrate to entire

twigs divaricate, often with a zig-zag appearance

prominently 3-veined

2 glands at vein junctions

calyx lobes

drupe yellowish-orange to red, glabrous

Family: ULMACEAE
Species: *Celtis pallida* Torr.
Common Name: Spiny hackberry (granjeno, desert hackberry, siempreverde)
Life Span: Perennial
Origin: Native
Season: Warm (semi-evergreen)

GROWTH FORM
polygamous or monoecious shrub (to 6 m tall); densely branched, forming thickets; flowers March to May, reproduces from seeds

FLORAL AND FRUIT CHARACTERISTICS
inflorescences: cymes 2-branched, 3- to 5-flowered, axillary

flowers: unisexual and perfect, apetalous; calyx lobes 4 or 5 (small), greenish-white; stamens as many as the calyx lobes; style absent; stigmas 2, each 2-cleft

fruits: drupes subglobose or ovoid (6–7 mm long, 5–7 mm in diameter), yellowish-orange to red, glabrous; thin-fleshed, mealy, acidulous, edible

VEGETATIVE CHARACTERISTICS
leaves: alternate, simple; blades oblong-ovate to elliptic (2–5.5 cm long, 1–2.5 cm wide), thick; apex rounded to acute; base prominently 3-veined, with 2 glands at vein junctions abaxially; margins crenate-serrate to entire; deep green, puberulent, slightly scabrous; petiole puberulent (2–5 mm long)

stems: twigs divaricate, flexuous, spreading, gray-puberulent, often with a zig-zag appearance; stipular spines (4–25 mm long), straight, stout, single or in pairs, sometimes forming a "V"

HISTORIC, FOOD, AND MEDICINAL USES: wood used for fence posts and firewood; sweet, insipid fruits were ground and eaten by some Native Americans with parched corn or fat; good honey plant

LIVESTOCK LOSSES: spines may cause minor injuries

FORAGE VALUE: worthless for livestock, rarely browsed; poor to fair browse for wildlife; fruits eaten by deer, small mammals, rabbits, quail, and other birds; browsed only when other plants are not available; valuable for erosion control and cover for wildlife

HABITAT: deserts, canyons, mesas, washes, foothills, thickets, brushland, and grassland; most abundant in gravelly or well-drained sandy soils

Creosotebush
Larrea tridentata (Sessé & Moc. *ex* DC.) Coville

SYN = L. *glutinosa* Engelm.

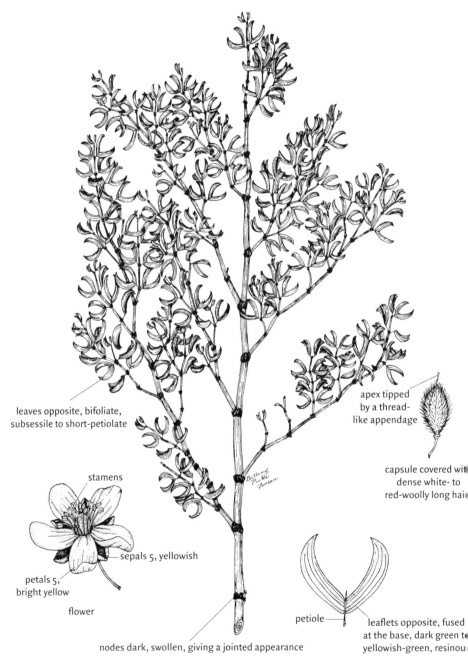

leaves opposite, bifoliate,
subsessile to short-petiolate

apex tipped
by a thread-
like appendage

capsule covered wit▮
dense white- to
red-woolly long hai▮

stamens

sepals 5, yellowish

petals 5,
bright yellow

flower

petiole

leaflets opposite, fused
at the base, dark green t▮
yellowish-green, resinou▮

nodes dark, swollen, giving a jointed appearance

Family: ZYGOPHYLLACEAE
Species: *Larrea tridentata* (Sessé & Moc. *ex* DC.) Coville
Common Name: Creosotebush (gobernadora, hediondilla)
Life Span: Perennial
Origin: Native
Season: Evergreen

GROWTH FORM

shrub (to 3 m tall); no well-defined trunk, numerous limber stems from near ground level; often scattered in nearly pure stands with little variation in size; flowers February to August, reproduces from rhizomes and seeds

FLORAL AND FRUIT CHARACTERISTICS

inflorescences: flowers solitary, axillary

flowers: perfect, regular; sepals 5 (5–8 mm long), unequal, yellowish, silky, early deciduous; petals 5 (7–10 mm long, 5 mm wide), bright yellow, obovate, concave; stamens 10; filaments winged

fruits: capsules globose (7–8 mm long), covered with dense white- to red-woolly long hairs tipped by a thread-like appendage, 5-celled, each cell 1-seeded

VEGETATIVE CHARACTERISTICS

leaves: opposite, bifoliate; leaflets opposite, fused at the base, ovate to oblong or obovate (5–10 mm long, 3–4 mm wide); margins entire; dark green to yellowish-green, glossy, resinous; subsessile to short-petiolate; stipules at base of petiole brown, glandular-hairy

stems: twigs slender, brown; nodes dark, somewhat swollen, conspicuous, giving a jointed appearance

other: foliage emits creosote odor, especially when wet or burned; some plants are estimated to be over 10,000 years old

HISTORIC, FOOD, AND MEDICINAL USES: some Native Americans used decoctions as antiseptics, medicines, and fuel; the Pima used lac from insect larvae found on the plants as glue and to waterproof baskets

LIVESTOCK LOSSES: may cause dermatitis in humans and animals; sheep, especially pregnant ewes, have been reported to die after eating the leaves; seed capsules may contaminate fleece

FORAGE VALUE: worthless to livestock; seldom consumed by livestock or wildlife, although jackrabbits will occasionally eat the leaves

HABITAT: alluvial plains, deserts, mesas, and hillsides; most abundant in sandy and gravelly soils; usually does not occupy saline soils

Glossary

a-	prefix meaning without
abaxial	on the side away from the axis; lower surface
abortion	failure of an organ or structure to develop; early termination of pregnancy resulting in the death of the embryo or fetus
abrupt	changing sharply or quickly, rather than gradually
absent	not present; never developing
acaulescent	stemless; without an above-ground stem or apparently so
acerose	needle-shaped; acicular
achene	a one-seeded, indehiscent fruit with a relatively thin wall in which the seed coat is not fused to the ovary wall
acicular	needle-shaped; acerose
acidic	soil with a low pH (less than 5.5)
acidulous	acid in taste; sour
acorn	a fruit seated in or surrounded by a hard, woody cupule of indurate bracts, such as in the genus *Quercus*
actinomorphic	symmetrical, regular; divisible into equal halves in two or more planes (contrast with zygomorphic)
acuminate	gradually tapering to a sharp point and forming concave sides along the apex; compare with acute
acute	tapering to a pointed apex with more or less straight sides; sharp-pointed; angle less than 90°
adaxial	on the side nearest the axis; upper surface
addicted	having a compulsive need for a habit-forming substance
adnate	attached or grown together; fusion of unlike parts, such as palea and caryopsis in the genus *Bromus*
aggregated	densely clustered
alkali	a soil with a high pH (8.5 or higher) and high exchangeable sodium content (15% or more), normally interferes with the growth of most species
alkaloid	any of numerous nitrogen-containing organic bases, many naturally occurring secondary metabolites in plants; which may be toxic to animals
alluvium	mineral or soil material deposited by running water
alpine	mountainous regions or areas
alternate	located singly at each node; not opposite or whorled
ament	a dense spike or raceme with many small, usually naked, flowers; a catkin such as in *Populus*

androgynous	with both pistillate and staminate flowers, the staminate flowers are borne above the pistillate flowers in the same inflorescence
-angled	suffix meaning a corner
angular	forming an angle; with one or more angles
annual	within one year; applied to plants which do not live more than one year
anterior	in front of; in a flower, the side away from the axis and adjacent to the bract
anther	the part of a stamen in which pollen develops
anthesis	the period of flowering; period during which pollination occurs
antidote	a remedy to counter the effects of a toxic substance
antrorse	directed upwards or forwards; opposed to retrorse
apetalous	without petals
apex	the tip or distal end
apical	relating to the apex
apiculate	ending in an abrupt, sharp point
apomixis	process of setting seed without fusion of gametes
appressed	lying against an organ; flatly pressed against
aquatic	growing in, on, or near water
arching	curved like an arch
arcuate	curved like a bow or arch
aristate	awned or tapering; a very long, narrow apex
armed	having sharp thorns or spines
aromatic	fragrant or having an odor; bearing essential oils
arroyo	a gully or channel; a watercourse in an arid region
articulate	jointed, provided with nodes; separating clearly at maturity
articulation	a joint or point of attachment
ascending	growing or angled upward; obliquely upward
astringent	able to draw together soft tissues; styptic
asymmetrical	not symmetrical; not divisible into equal halves
attenuate	gradually narrowing to a slender apex or base
auricle	ear-shaped lobes, such as those that occur at the base of leaf blades of some grasses
awl-shaped	narrow and sharp pointed; gradually tapering from a narrow base to a pointed apex; subulate
awn	a slender bristle at the end, on the back, or on the edge of an organ; in grasses, the extension of a nerve beyond the leaf-like tissue
awn column	divided portion of the awn below the branches, such as in the genus *Aristida*
awnless	without awns
axil	angle between an organ and its axis
axillary	growing in an axil

axis	the central or main longitudinal support upon which parts are attached
balsam	aromatic and resinous substances from plants containing benzoic or cinnamic acid
banner	upper petal (standard) of the papilionaceous flower in the FABACEAE
barb	a short, rigid projection
barbed	furnished with retrorse projections
bark	exterior covering of a woody stem or root; tissues lying outside the cambium
barren	unproductive sites, usually with shallow soils
basal	located at or near the base of a structure, such as leaves arising from the base of the stem
basifixed	attached by the base (compare with dolabriform)
bead-like	sphere-shaped
beak	a narrow or prolonged tip; a hard point or projection (frequently the remnants of the style base)
bearded	furnished with long, stiff hairs
berry	a fleshy fruit developing from a single pistil with several or many seeds; a fruit that is fleshy or pulpy throughout
bi-	prefix meaning two
bicolored	having two colors
bidentate	having two teeth
biennial	a plant that lives for two years
bifid	one-cleft or two-toothed; applied to the summit of glumes, lemmas, paleas, petals, or leaflets
bifoliate	having two leaves or leaflets
bilabiate	having two lips, as in irregular flowers
bilateral	having two sides
bilocular	having two compartments or locules in the ovary, anther, or fruit
bipinnate	twice pinnate
bitter	a disagreeable taste sensation; one of the four basic taste sensations
bivalved	having two valves
blade	the part of the leaf above the sheath, petiole, or petiolule
bleached	having lost most of the original color
blister	an enclosed raised spot on the surface
bloat	a digestive disturbance of livestock (especially cattle) marked by abdominal swelling due to a build up of gas, potentially fatal
blotched	a spot or mark in an irregular shape
blowout	a depression in the surface of sand or sandy soil caused by wind erosion

blunt	having a point or edge that is not sharp
bodies	parts of the whole
bog	a poorly drained area; wet, spongy ground
borne	attached to or carried by
brackish	salty water, with a saline content less than that of seawater
bract	reduced leaves (frequently associated with the flowers)
bracteate	having bracts
bractleole	a bract borne on a secondary axis
branch	a lateral stem
branchlet	the final, or ultimate, division of the branch
breaks	a rough landform usually caused by water erosion; canyons
bristle	a stiff, slender appendage
bristly	covered with bristles
brittle	easily broken
broom-like	shaped like a broom used for sweeping
browse	twigs, leaves, and other parts of woody plants consumed by herbivores; the act of consuming portions of woody plants
bud	an underdeveloped shoot or flower
bulb	an underground bud with fleshy, thick scales
bulbous	swollen at the base, like a bulb or corm
bulge	a swelling; an abrupt expansion
bur	a rough and prickly covering of a fruit
bushy	resembling a bush; thick and spreading
caducous	early deciduous; falling off early
calcareous	a soil containing sufficient calcium carbonate (often with magnesium carbonate) to effervesce when treated with hydrochloric acid
callous	having the texture of a callus
callus	the indurate downward extension of tissue from the mature lemma in *Nassella*, *Hesperostipa*, *Aristida*, and some other genera; hardened tissue
calyx	the sepals of a flower considered collectively, usually green bracts
campanulate	shaped like a bell
canescent	pale or gray colored because of a dense, fine pubescence
capillary	fine and slender; hair-like, such as a branch or awn
capitate	aggregated into a dense cluster; head-like
capsule	a dry, dehiscent fruit of more than one carpel with more than two seeds
carpel	the modified leaf forming the ovary, may be compound or simple
caryopsis	the fruit or grain of grasses; a small, dry, indehiscent fruit having a single seed with a thin, closely adherent pericarp

catalyst	a substance that increases the speed of a reaction
cathartic	a substance causing the evacuation of the bowels
catkin	a dense spike or raceme with many small, usually naked, flowers; an ament such as in the genus *Populus*
caudate	bearing a slender tail-like projection or appendage
caudex	a short, usually woody, vertical stem located just below the soil surface
caulescent	having a stem
cauline	pertaining to the stem or belonging to the stem
-celled	suffix meaning cavity or individual unit
central	situated at, in, or near the center
cespitose, caespitose	tufted, several or many stems in a close tuft, such as bunch-grasses
chaff	bracts subtending a flower (usually small, membranaceous, and dry)
channeled	deeply grooved
chaparral	a vegetation type with dense thickets of woody plants adapted to dry summers and wet winters
chartaceous	having the texture of writing paper and usually not green
ciliate	fringed with hairs on the margin
ciliolate	fringes with minute hairs on the margins
clasping	one organ or tissue partially or totally wrapped around a second
clavate, claviform	club-shaped, widening toward the apex
claw	the narrowed base of some sepals and petals
cleft	divided into teeth or divisions that extend halfway or more to the midvein or nerve
cleistogamous	applied to flowers or florets fertilized without opening, such as in some species of *Leptochloa* and *Nassella*
club-shaped	widening toward the apex
clump	a dense cluster
cluster	a number of similar tissues or organs growing together; a bunch
coarse	composed of relatively large parts; not fine textured or structured
cobwebby	a tuft of tangled, fine hairs
coil	one or more loops
collar	the area on the abaxial side of a leaf at the junction of the blade and sheath
colony	a group of plants of the same species growing in close association with each other; all members of the group may have originated from a single plant
column	the lower portion of the awn of grasses
coma	a tuft of hairs

comb-like	pectinate; with narrow, closely set, and divergent segments like the teeth in a comb
compact	having a small, dense structure
compound	composed of several parts united into a single structure
compressed	flattened laterally
comprised	to be made up of
concave	hollowed inward like the inside of a bowl
cone	a cluster of scales on an axis, scales may be persistent or deciduous
confluent	merging; coming together, blended together as one
congested	overcrowded; full
conic, conical	cone-shaped with the point of attachment at the broad end
connate	fusion of like parts, such as petals to form a corolla tube
conspicuous	obvious; easy to notice
constricted	drawn together; appearing to be tightly held
contaminate	the introduction of unwanted materials causing a reduction of value or use
contorted	bent; twisted
contracted	inflorescences that are narrow or dense, frequently spike-like
convex	rounded on the surface like the bottom or exterior of a bowl
convolute	rolled longitudinally
cool-season	a category of plants that grow during the cool portions of the year
copious	an abundance
cordate	heart-shaped, with rounded lobes and a sinus at the base
coriaceous	leathery in texture
corm	the bulb-like base of a stem, usually fleshy and underground
corolla	all of the petals considered collectively
corymb	a simple racemose inflorescence that is flat-topped; an indeterminate inflorescence
corymbiform	having the form, but not necessarily the structure, of a corymb
cottony	having the texture of cotton
cotyledon	a leaf of the embryo of a seed; the seed leaf
counterirritant	a substance or act that overcomes an irritation
craving	to want greatly
creeping	continually spreading; a shoot or horizontal stem that roots at the nodes
crenate	having rounded teeth; scalloped margins
crenulate	diminutive of crenate
creosote	an aromatic mixture of phenolic compounds obtained from the distillation of wood tar, commonly used to preserve wood
crescent	a shape having a convex edge and a concave edge
crested	with an elevated ridge or appendage on the top or back
cross-section	cut at a right angle to the main axis; transverse

crowded	pressed close together; a number of structures in a small space
crown	persistent base of a herbaceous perennial; the shape of the foliage of a shrub or tree; the tuft of hairs at the summit of the lemma in some grasses such as in the genus *Nassella*
cruciform	cross-shaped
culm	the hollow or pithy jointed stem or stalk of a grass, sedge, or rush
cumulative	increasing by successive additions
cuneate	wedge-shaped; triangular with the narrow end at the point of attachment
cup, cupule	the cup-like structure at the base of a fruit such as acorn in the genus *Quercus*
cupulate	shaped like a cup
cure	drying, as in standing herbage or hay; relief or recovery from a disease
curled	formed in the shape of curves or spirals
cuspidate	bearing a sharp, firm, and elongated point at the apex
cyanogentic	capable of producing cyanide
cylindric, cylindrical	shaped like a cylinder
cyme	a convex or flat-topped flower cluster with the central flower the first to open; a determinate inflorescence
cymose	resembling a cyme or bearing cymes
deciduous	not persistent, but falling away in less than one year
decoction	an extract obtained by boiling the plant material
decumbent	curved upward from a horizontal or inclined base, with only the end ascending
decurrent	extending downward from the point of attachment
dehiscent	opening at maturity along a definite suture
delicate	fine structure or texture
deltate, deltoid	triangular; shaped like the Greek letter delta
dense	crowded
dentate	with pointed, coarse teeth spreading at right angles to the margin
denticulate	diminutive of dentate
depressed	flattened from above; pressed down
dermatitis	inflammation of the skin
dichotomous	branching repeatedly in pairs
diffuse	open and much-branched, loosely branching
digitate	several members arising from one point at the summit of a support, like the fingers arising from the hand as a point of origin
dilated	enlarged; expanded; widened

dimorphic	two types of leaves, flowers, or other structures on the same plant
dioecious	unisexual flowers on separate plants; pistillate and staminate flowers on separate plants
disarticulating, disarticulation	separating at maturity at a node or joint
discoid	resembling a disk; in the ASTERACEAE, with all the flowers of a head tubular and perfect
dished	shaped like a dish
disk	an outgrowth of the receptacle that surrounds the base of the ovary or ovaries
disk flowers	a regular flower of the ASTERACEAE
dissected	deeply divided into numerous parts
disseminate	to disperse or spread
distal	remote from the place of attachment
distant	to be separated by space
distichous	conspicuously two-ranked leaves, leaflets, or flowers
distinct	clearly evident; separate; apart
disturbance	alteration or destruction of the vegetative cover
disturbed sites	areas on which the vegetative cover has been altered or destroyed
diuretic	a substance causing an increase in the flow of urine
divaricate	widely and stiffly divergent
divergent	widely spreading
divided	separated or cut into distinct parts by inclusions extending to near the base or midrib
division	one of the parts of the whole
dolabriform	hatchet-shaped; attached in the middle (compare with basifixed)
dominant	a species of plant that controls the character of the vegetation
dormancy	an inactive state; period during which plants are not active, such as in winter
dorsal	relating to the back of an organ; opposite the ventral side
dotted	marked with small spots
downy	a soft, fine pubescence
droop	to hang downward
drupe	a fleshy fruit, indehiscent, usually a single seed with a stony endocarp (e.g. a cherry)
dull	lacking brilliance or luster; not shiny
dumbbell	a structure that is narrow in the middle and large on both ends
dwarf	less than normal size
dye	a pigment or other substance used to color other items such as cloth

edible	fit to be eaten
ellipsoid	a solid body circular in cross-section and elliptic in long-section
elliptic, elliptical	shaped like an ellipse; narrowly pointed at the ends and widest in the middle
elongate	narrow, the length many times the width or thickness
emarginate	having a shallow notch at the tip
embedded	enclosed in a supporting structure or organ; imbedded
emetic	a substance causing vomiting
entire	whole; with a continuous margin
enveloped	enclosed within
ephedrine	a crystalline alkaloid extracted from members of the genus *Ephedra*
erect	upright; not reclining or leaning
ergot	fungus disease of grasses
erose	irregularly notched at the apex; appearing gnawed or eroded
even-	prefix meaning number of structures divisible by two
evergreen	woody plants that retain their leaves throughout the year
evident	obvious; distinct; easily seen
exceed	greater than; larger than
exfoliate	shedding in flakes or thin layers
expanded	increased or extended
expectorant	a substance that promotes the discharge of mucous from the respiratory tract
exposed	open to view
exserted	protruding or projecting beyond; not included
extensive	having a wide or considerable range or spread
extract	to separate or remove; material that has been separated
faint	lacking distinctness
falcate	curved or shaped like a sickle
famine	an extreme scarcity of food
fan-shaped	shaped like a segment of a circle with the point of attachment at the narrow end
fascicle	a small bundle or cluster, such as needles of members of the genus *Pinus* in clusters of two to five
feathery	having the texture or appearance of feathers
felty	closely matted with intertwining hairs; having the texture or appearance of felt
fern	a vascular plant with highly divided, delicate leaves
fertile	capable of producing fruit; does not refer to stamen presence or absence in grasses
fibrous	consisting of or containing mostly fibers; commonly used to describe branching root systems (compare with taproot)

filament	the stalk of a stamen supporting the anther; a thread-like structure
filiform	thread-like; long and very slender
firm	hard; resisting distortion when pressure is applied; indurate
fissure	a deep groove
flake	a thin, flattened piece or layer
flanked	to be situated at either side of a structure
flat, flattened	having the major surfaces essentially parallel and distinctly greater than the minor surfaces
fleshy	pulpy; succulent
flexible	capable of being easily bent or flexed; pliant
flexuous	bent alternately in opposite directions; a wavy form
floccose	covered with long, soft, fine hairs that are loosely spreading, these hairs rub off easily
floret	lemma and palea with included flower of the POACEAE; also disk flowers of the ASTERACEAE
floriferous	flower-bearing
folded	a part or organ that is doubled over or laid over
foliage	plant material that is mainly leaves
-foliate	suffix pertaining to or consisting of leaflets (i.e., three-foliate means that the leaves are made up of three leaflets)
follicle	a dry, dehiscent fruit splitting along one suture; a small closed or nearly closed cavity
foothill	a region at the base of a mountain range
forb	herbaceous plants other than grasses and grass-like plants
fragrant	having a sweet or delicate odor
free	not attached to other organs
fringed	having a border consisting of hairs or other structures
frond	a large, divided leaf; a fern leaf
fruit	ripened ovary (pistil); the seed bearing organ
funnelform	shaped like a funnel
furrowed	bearing longitudinal grooves or channels; sulcate
fused	attached
fusiform	shaped like a spindle
geniculate	bent abruptly, like a knee (awns or plant bases may be bent in this manner)
glabrate, glabrescent	nearly glabrous or becoming so with age
glabrous	without hairs
gland	a protuberance or depression that appears to secrete a fluid
glandular	supplied with glands
glaucous	covered with a waxy coating that gives a blue-green color; possessing a waxy surface that easily rubs off

globose	nearly spherical in shape
glomerule	a dense cluster; a dense, head-like cyme
glossy	having a surface luster; shiny
glumes	the pair of bracts at the base of a spikelet in grasses
glutinous	with a firm, sticky substance covering the surface
glycoside, glucoside	organic compounds that yield a sugar and another substance upon hydrolysis; may be found in plants and may be toxic to animals
gnarled	twisted and deformed; knotty
granulate	covered with small grain-like particles
grass	monocotyledonous herbaceous plants
grass-like	herbaceous plants similar in appearance to grasses such as sedges and rushes
graze	to consume growing and/or standing grass or forb herbage; to place animals on pastures to enable them to consume the herbage
groove	a long, narrow channel or depression; sulcus
growing point	apical tissue; vegetative bud
gully	a trench eroded in the land surface by running water
harsh	a texture disagreeable to the touch; rough; unpleasant
head	a dense cluster of sessile or nearly sessile flowers on a short axis; an inflorescence type
hemispheric	shaped like one-half of a sphere
herb, herbaceous	not woody; dying each year or dying back to the crown
herbage	above-ground material produced by herbaceous plants; vegetation that is available for consumption by grazing animals
herringbone	pattern made up of two rows of parallel lines with adjacent rows slanting in reverse directions
hip	fruit consisting of the fleshy floral tube surrounding the mature ovaries, such as in the genus *Rosa*
hirsute	with straight, rather stiff hairs
hispid	with stiff or rigid hairs; bristly hairs
hoary	covered with fine gray or white pubescence
hollow	unfilled space; empty
hooked	curved or bent like a hook
horizontal	parallel to the plane of the earth
horn	an exserted appendage
hyaline	thin and translucent or transparent
hydrocyanic acid	an aqueous solution of hydrogen cyanide that is poisonous
hypanthium	a ring or cup around the ovary formed by a fusion of the bases of sepals, stamens, and petals; a modified receptacle
ill-scented	an unpleasant odor

imbedded	enclosed in a supporting structure or organ; embedded
imbricate	overlapping (like shingles on a roof)
impact	to fix firmly as if by packing or wedging
impenetrable	an area or object that cannot be entered; commonly applied to thick vegetation that grazing or browsing animals cannot enter
improper grazing	animal utilization of herbaceous material that causes a significant decline in the plant community
incised	cut sharply, irregularly, and more or less deeply
included	not exserted nor protruding
inconspicuous	not easily seen; not evident
incurved	curved toward the center
indehiscent	not opening, staying closed at maturity; not splitting
indistinct	not easily seen; not sharply outlined or separable
indurate	hard
indusium	a thin epidermal outgrowth on a fern leaf that covers the sorus
inferior	lower; below; less elevated in position
inflated	swollen or expanded; puffed up; bladdery
inflexed	turned in at the margins
inflorescence	the arrangement of flowers on an axis subtended by a leaf or portion thereof
infra-	a prefix meaning below
infrastipular	below the stipules
infructescence	the arrangement of fruits on an axis
infusion	liquid resulting from soaking or steeping material in water without boiling
inrolled	curved or rolled toward the central axis of the structure
insipid	lacking taste or savor
intermingled	intermixed; mixed together
internerves	spaces between the nerves
internode	the part of a stem between two successive nodes
interrupt	to break the uniformity; to come between two similar objects or structures
intricate	having many complex parts or elements
introduced	not native to North America
intrude	to place, thrust, or force between
involucre	a whorl or circles of bracts below the flower or spikelet cluster
involute	rolled inward from the edges, the upper surface within
irregular	asymmetrical; not equal in similar parts
jam, jelly	food made by boiling fruit and sugar to a thick consistency
joint	the section of a stem or culm from which a leaf or branch arises; a node on a grass culm
jointed	possessing nodes or articulations
junction	place at which two structures or organs join

446

juvenile	young; not mature or fully developed
keel	the sharp fold or ridge at the back of a compressed sheath, blade, glume, lemma, or palea of POACEAE; the united lower petals of FABACEAE
keeled	ridged, like the keel of a boat
knot	the base of a woody branch enclosed in the stem from which it arises
lac	a resinous substance secreted by an insect; once used as glue
lacerate	appearing torn at the edge or irregularly cleft
lacinate	deeply cut into narrow segments
lanate	woolly with long intertwined, curly hairs
lancelinear	shaped like a narrow lance
lanceolate	rather narrow, tapering to both ends, widest below the middle
lateral	belonging to or borne on the side
lax	loose; open and spreading
leafless	without leaves
leaflet	a division of a compound leaf
leafy	with many leaves
leathery	resembling leather in texture and appearance
ledge	a narrow, flat surface on the face of a mountain, hill, or canyon
legume	FABACEAE fruit composed of a single carpel, but with two sutures and dehiscing at maturity along the sutures
lemma	abaxial bract of the floret that subtends the grass flower and palea
lenticel	a slightly raised, lens-shaped area on the surface of a woody stem
lichen	an association of a fungus living symbiotically with an algae on rocks or vascular plants
life span	the length of time a plant will live
ligulate	in the ASTERACEAE, referring to flowering heads solely composed of the flat, strap-shaped flowers on the margin (ray flowers) of the disk; flowers of the head that are strap-like
ligule	in the POACEAE, the appendage, membrane, or ring of hairs on the adaxial side of a leaf at the junction of the sheath and blade; in the ASTERACEAE, the strap-shaped corolla of a ray flower
limb	the expanded portion of a sympetalous corolla; the strap-shaped corolla of a ray flower in the ASTERACEAE
limber	flexible; supple
linear	long and narrow with parallel sides
lobe	the projecting part of an organ with divisions less than one-half the distance to the base or midvein, usually rounded or obtuse
locular	having locules or compartments

locule	a cavity of an ovary, fruit, or anther
loment	a jointed fruit, constricted and breaking apart between the seeds
loose	not arranged tightly together
lustrous	reflecting light evenly without glitter or sparkle
margin	an edge; border
marsh	an area of wet soil; a swamp
mat	a tangled mass of plants growing close to the soil surface and generally rooting at the nodes
mature	fully developed
meadow	moist, level, lowland on which grasses dominate
mealy	surfaces flecked with a lighter color which may rub off
membrane	a thin, soft, and pliable tissue
membranous, membranaceous	thin, opaque, not green; like a membrane
mericarp	a one-seeded portion of a schizocarp; a portion of a dry dehiscent fruit that splits away as if separate
-merous	a suffix referring to the number of parts
mesa	a relatively flat-topped elevated land surface
mesic	characterized by moderately moist conditions; neither dry nor wet
micro-	a prefix meaning small
midnerve, midrib, midvein	the central or principal vein of a leaf or bract
minute	small
monoecious	plants with male and female flowers at different locations on the same plant; all flowers unisexual
mottled	marked with spots or blotches
mouth	the opening of any tube, canal, or cavity (i.e., the top of a leaf sheath in the POACEAE)
mucilage	a gelatinous substance similar to a plant gum
mucro	a short, sharp-pointed tip; a very short awn in the POACEAE
mucronate	tipped with a short, slender, sharp point or awn
mucronulate	tipped with a very small mucro
mush	ground meal boiled in water
naked	uncovered; lacking pubescence; lacking enveloping structures
native	occurring in North America before settlement by Europeans
nauseous	sickening; extremely unpleasant
neck	junction of two structures, such as the junction of the lemma and awn of plants in the genus *Nassella*
nectar guides	lines inside of flowers thought to serve as guides to pollinating insects

needle	a slender needle-shaped leaf such as those in the genus *Pinus*
nerve	the vascular bundles or veins of leaves, culms, glumes, paleas, lemmas, or other organs
neuter	lacking stamens and pistil
nitrates	a compound of nitrogen accumulated by some plants that can cause poisoning if consumed by animals
nodding	inclined somewhat from the vertical; drooping
node	points along the stem where leaves are borne; a joint in a stem or inflorescence
nodulose	with minute knobs or nodules
notch	gap; a V-shaped indentation
nut	an indehiscent, dry, one-seeded fruit
nutlet	a small, usually one-seeded, hard fruit that is indehiscent; a small nut
ob-	a prefix meaning inversely
obconic, obconical	inversely cone-shaped with the attachment at the broad end rather than the narrow end
obcordate	inversely cordate or heart-shaped with the attachment at the point
obdeltoid	inversely triangular-shaped with the attachment at the point of the triangle rather than along the side
oblanceolate	inversely lanceolate with the broadest portion near the apex
oblate	flattened at the poles
oblique	having the axis not perpendicular to the base; neither perpendicular nor parallel
oblong	longer than broad, with sides nearly equal
obovate	opposite of ovate with the widest part toward the far end; egg shaped with the widest part above the middle
obovoid	opposite of ovoid with the attachment at the narrower end
obscure	inconspicuous; not easily seen
obturbinate	inversely turbinate; conical; shaped like a top
obtuse	shape of an apex, with an angle greater than 90°
obvious	easily seen
odd-	number not evenly divisible by two
opposite	borne across from one another at the same node
orbiculate, orbicular	nearly circular in outline
origin	place where the species originally occurred
ornamental	a plant cultivated for its beauty rather than for agronomic use
oval	broadly elliptic
ovary	the expanded basal part of the pistil that contains the ovules
ovate, ovoid	shaped like an egg with the broadest portion toward the base
overflow sites	land that water flows across occasionally

overlap	to extend over and cover part of an adjacent structure
ovulate	bearing ovules
ovule	immature seed located in the ovary; the egg containing part of the ovary
oxalate	a group of chemical compounds that may cause poisoning in animals; salts of oxalic acid
paired	two, together
palatable	acceptable in taste and texture for consumption
pale	not bright; dim; deficient in color
palea	the adaxial bract of a floret, two-nerved; the upper bract subtending a disk flower in the ASTERACEAE
paler	lighter in color
palmate	with three or more lobes, nerves, or leaflets arising from a common point
panicle	inflorescence with a main axis and rebranched branches
paniculiform	having the shape of a panicle
papery	having the texture of writing paper
papilionaceous	a flower type in the FABACEAE having a banner petal, two wing petals, and two partially fused to fused keel petals
papilla	a small, rounded bump or projection
papillose	having minute papillae
pappus	a group of hairs, scales, or bristles that crown the summit of the achene in the ASTERACEAE flower; considered to be a modified calyx
parch	to toast with dry heat
parkland	a vegetation type characterized by scattered clumps of trees in prairie
pectinate	comb-like; divided into numerous narrow segments
pedicel	the stalk of a spikelet or single flower in an inflorescence
pedicellate	having a pedicel
pedicelled	borne on a pedicel
peduncle	the stalk of a flower cluster or spikelet cluster
pellucid	translucent or transparent
peltate	shield-shaped, especially with reference to a leaf attached to the petiole on the lower surface
pemmican	a concentrated food made by some Native Americans consisting of powdered meat mixed with melted fat and occasionally dried fruit
pendant, pendulous	suspended or hanging downward; drooping
perennial	lasting more than two years; applied to plants or plant parts which live more than two years
perfect	applied to flowers having both stamens and pistil

perianth	a floral envelope consisting of the calyx and corolla
pericarp	the fruit wall; wall of a ripened ovary
perigynium	an inflated sac that encloses the achene in the genus *Carex*
perispore	the membrane surrounding a spore
persistent	remaining attached
petal	a part or member of the corolla, usually brightly colored
petaloid	petal-like
petiolar	growing from the petiole; pertaining to the petiole
petiolate	with a petiole
petiole	the stalk of a leaf blade
photosensitization	hypersensitivity of the skin to sunlight due to the ingestion of photodynamic compounds from certain plants
phyllary	a bract of the involucre at the outside of the head of flowers in the ASTERACEAE
pigmented	colored with a substance
pilose	with long soft, straight hairs
pilosulous	minutely pilose
pinna	primary division of a pinnate leaf
pinnate	having two rows of lateral divisions along a main axis (like barbs of a feather)
pinnatifid	deeply cut in a pinnate manner, but not cut entirely to the main axis
pinole	finely ground flour made from parched corn; any of various flours resembling pinole and ground from seeds of other plants
pipe-shaped	shaped like a pipe used to smoke tobacco
pistil	a combination of the stigma, style, and ovary; the female reproductive organ of a flower
pistillate	applied to flowers bearing pistils only; unisexual flowers
pit	a small depression in a surface
pith	soft, spongy material located in the center of a stem
placenta	the structure by which the ovule is attached to the wall of the ovary
plains	flat to rolling land usually covered with vegetation dominated by grasses
planoconvex	flat on one side and convex on the other
plateau	an extensive area of land with a relatively flat surface raised sharply above the adjacent land at least on one side; tableland
plume	an arrangement of hairs that resembles a feather
plumose	feathery, with long pubescence or pinnately arranged bristles
plump	rounded, full
pod	a dry dehiscent fruit splitting along two sutures
pollination	the process of the transfer of pollen from the anther to the stigma
polygamodioecious	monoecious plants having some perfect flowers

polygamous	with bisexual and unisexual flowers on the same plant
pome	an inferior, indehiscent, many-seeded fruit in which the receptacle forms the outer, fleshy portion (e.g., an apple)
potherb	plants that are boiled before being eaten
prairie	an extensive tract of level to rolling land with vegetation composed of dominate grasses together with forbs, shrubs, and grass-like plants
prickle	a small, sharp outgrowth of the bark or epidermis
primary	first
primary branch	branch arising directly from the main inflorescence axis
procumbent	prostrate; lying flat on the ground; trailing but not taking root
prominent	readily noticeable; projecting out beyond the surface
prostrate	lying flat on the ground; procumbent
prow	the projecting front part
puberulent	diminutive of pubescent
pubescent	covered with short, soft hairs
pulp	the soft, succulent portion of a fruit
punctate	having dots, usually with small glandular pits
pungent	sharp and penetrating odor; firm- or sharp-pointed
pustular, pustulate	having small eruptions or blisters
pyramidal	shaped like a pyramid
pyriform	pear-shaped
quinine	a bitter crystalline alkaloid used as a tonic or to treat malaria
raceme	an inflorescence in which the spikelets or flowers are pediceled on a rachis
racemose	raceme-like branch of the inflorescence
rachilla	a small axis; applied especially to the axis of a spikelet
rachis	the axis of a spike, spicate raceme, or raceme inflorescence or pinnately compound leaf
radiate	term used to describe the ASTERACEAE flower arrangement with the marginal flowers ligulate and the disk or central flowers tubular; spreading from a common center
radiating	spreading from a common center
-ranked	rows or series
ravine	a narrow, steep-sided valley caused by water erosion (larger than a gully)
ray	ligulate flowers in the ASTERACEAE; the branch of an umbel
receptacle	the upper end of the stem of a plant to which the flowering parts are attached
recurved	curved backwards
reduced	smaller than normal; not functional

reflexed	bent or turned downward abruptly
regular	having structures of the flower, especially the corolla, of similar shape and equally spaced about the center of the flower
remote	widely spaced
reniform	kidney-shaped
repel	to force away
resinous	producing any of numerous viscous substances such as resin or amber
restoration	returning the contour of the land and the vegetation to its original condition
reticulate	in the form of a network; netted as many leaf veins in dicots
retrorse	pointing backward toward the base
retuse	with a slight notch at a rounded apex
revegetation	replacing current vegetation or starting vegetation on denuded land
revolute	rolled under along the margin toward the abaxial surface
rhizomatous	having rhizomes
rhizome	an underground stem with nodes, scale-like leaves, and internodes
rhombic	having the shape of a four-sided figure with opposite sides parallel and equal but with two of the angles oblique
right-of-way	usually vegetated land along roads, highways, railroad tracks, pipelines, or transmission lines
rigid	firm; not flexible
robust	healthy; full-sized
rootstock	underground stem; rhizome
rosette	a basal, usually crowded, whorl of leaves
rosulate	in the form of or resembling a rosette
rough	not smooth; surface marked by inequalities
rudiment	imperfectly developed organ or part, usually non-functional
rudimentary	underdeveloped
rugose	wrinkled surface
rumen	the large first compartment of the stomach of a ruminant animal
sac	a pouch or bag-like cavity
sagittate	arrowhead-shaped with the lobes turned downward (see *Balsamorhiza* leaf bases)
salicin	a crystalline glucoside found in the bark and leaves of some trees and used as a medicine
saline	a nonsodic soil containing sufficient soluble salts to impair its productivity
sap	the fluid contained in a plant

saponin	any of various glucosides found in plants and marked by the property of producing soapy lather
scaberulous	slightly roughened
scabridulous	minutely roughened
scabrous	rough to the touch; short, angled hairs requiring magnification for observation
scale	reduced leaves at the base of a shoot or a rhizome; a thin chaff-like portion of the bark of woody plants; a thin, flat structure
scaly	having scales
scape	a leafless peduncle arising from the ground or basal whorl of leaves and bearing one or more flowers (see *Taraxacum*)
scapiform	scape-like but not entirely leafless
scapose	bearing a flower or flowers on a scape or resembling a scape
scar	a mark on the stem where a leaf, bud, flower, or fruit was formerly attached
scarious	thin, dry, membranous, not green
schizocarp	a dry fruit consisting usually of two carpels which, when mature, split apart forming one-seeded halves (mericarps)
sclerotium	in fungi, such as ergot, a hardened compact mass of mycelium which gives rise to the fruiting bodies; replaces the caryopsis in ergot-infested POACEAE
scours	diarrhea in livestock
scurfy	covered with minute scales or specialized mealy hairs
season-long	throughout one season; frequently used to describe grazing during the growing season
secretion	materials such as resins, mucilages, gums, oils, and nectar on the exterior of plant parts
seed	a ripened ovule
seep	a place where water oozes slowly to the land surface
segment	a part of a structure that may be separated from the other parts
selenium	a nonmetallic element that is frequently extracted from the soil by plants and may be poisonous to animals in relatively large quantities
sepal	a member of the calyx bracts, usually green
septate	divided by one or more partitions
series	a group with an order of arrangement; in the ASTERACEAE the number of rows of bracts in the involucre
serrate	saw-toothed margins, with teeth pointing toward the apex
serrulate	minutely serrate
sessile	without a pedicel or stalk
setaceous	bristle-like hairs
setose	covered with bristles; hispid
sheath	the lower part of a leaf that encloses the stem

sheathing	forming a sheath, such as the base of a grass leaf blade when it surrounds the culm
shedding	casting off parts
shiny	lustrous; possessing a sheen
shoot	a young stem or branch
showy	attractive, such as a large colorful flower; striking appearance
shred	long, narrow strip; fibrous exfoliating bark
shredding	detaching in long, narrow strips
shrub	a low-growing woody plant; bush with one to many trunks
silicle	a short capsule of two carpels, about at long as broad, in the BRASSICACEAE
silique	a long, slender capsular fruit of two carpels, a type of capsule in the BRASSICACEAE
silky	fine, lustrous, long hair; resembling silk in appearance or texture
silvery	lustrous and gray or white; having the luster of silver
simple	not branched; not compound; single
sinuate	strongly wavy margins
sinus	indentation between two lobes or segments
slough	a place in which shallow water stands for most or all of the year
snuff	to draw forcibly into the nostrils
sod-forming	creating a dense mat
solitary	alone; one by itself
sorus, sori	a cluster of sporangia on the surface of a fern leaf
sparingly	meager; not dense
sparse	scattered; opposite of dense
spathe	a modified sheathing bract of the inflorescence; a modified leaf sheath
spatulate	shaped like a spatula, being broader above than below
spherical, spheroid	having the shape of a sphere or ball; a globular body
spicate	spike-like
spiciform	shaped like a spike
spike	an unbranched inflorescence in which the spikelets or flowers are sessile on a rachis (central axis)
spike-like	having the appearance of a spike
spikelet	the unit of inflorescence in grasses usually consisting of two glumes, one or more florets, and a rachilla
spine	a stiff, pointed outgrowth that is usually woody
spinescent	bearing a spine; terminating in a spine
spinose	having spines
split	divided lengthwise
splotch	a blotch or spot; blending of a spot and blotch
sporangium	a spore-bearing sac or case

spore	reproductive cell from meiotic division in a sporangium
spot grazing	heavy, repeated grazing of localized areas
sprig	a vegetative shoot that is planted to establish a new plant
sprout	the shoot of a plant; especially the first from a root or a germinating seed
spur	any slender, hollow projection of a flower (see *Delphinium* flowers)
squarrose	having spreading and recurved parts, such as bracts surrounding an inflorescence (see *Grindelia*)
stalk	the supporting structure of an organ
stamen	the pollen-producing structure of a flower; typically an anther borne at the apex of a filament
staminate	flower containing only stamens, unisexual flowers
stasis	slowing down or cessation of flow or a process
steep	to soak in water at a temperature under the boiling point
stellate	star-shaped, usually referring to hairs with many branches from the base; type of hair
stem	the portion of the plant bearing nodes, leaves, and buds
sterile	without functional pistils, may or may not bear stamens
sticky	covered with an adhesive-like substance
stiff	not easily bent; rigid
stigma	the portion of the pistil that receives the pollen
stipe	in general, a stalk or stem that supports an organ
stipitate	borne on a stalk or stipe
stipules	appendages, usually leaf-like, occurring in pairs, one on either side of the petiole base; may be modified to spines
stolon	a horizontal, above-ground, modified propagating stem with nodes, internodes, and leaves
stoloniferous	bearing stolons
stone	the hard, inner portion of a drupe that contains the seed
stout	sturdy, strong, rigid
straggly	scattered irregularly
stramineous	straw-colored; straw-like in texture
striate	marked with slender, longitudinal grooves or lines; appearing striped
strict	narrow, with close, upright branches
strigose	rough with short, stiff hairs or bristles
strigulose	minutely strigose
striking	attractive; showy
stringy	resembling or consisting of fibrous material
strychnine	a poisonous alkaloid
stylar	of or pertaining to a style
style	the slender, elongated portion of the pistil that bears the stigma at its apex

sub-	a prefix to denote somewhat, slightly, or in less degree
subshrub	a suffrutescent perennial plant
subtend	to underlie; located below
subulate	shaped like an awl
succulent	fleshy and juicy
suffrutescent	slightly shrubby; having a slightly woody or shrubby base
suffruticose	plants herbaceous above and woody at the base
sulcate	having grooves or furrows
sulcus	a groove or furrow
summit	the top or apex
superior	higher; above; more elevated in position
surmount	directly on top of
surpass	exceed
suture	a line or seam marking the union of two parts; the line of dehiscence of a fruit or capsule
swale	a low-lying depression in the land
swollen	enlarged
sympetalous	petals united at or near the base
tapering	regularly narrowing toward one end
taproot	the primary root of a plant that grows directly downward and gives rise to lateral branches
tar	a brown or black viscous, odorous organic substance
tardily	late
tawny	pale brown or dirty yellow
teeth	pointed lobes or divisions
tepal	segment of a perianth that is not differentiated into a calyx and corolla; a sepal or petal
terete	cylindric and slender; circular in cross-section
terminal	borne at or belonging to the extremity or summit
tetrahedral	a geometric form with four faces
thicket	dense growth of shrubs or small trees
throat	an opening; the orifice of a corolla or calyx
thyrse	a flower cluster of racemosely arranged cymes organized into an elongate panicle
thyrsoid, thrysiform	resembling a thyrse
tiller	a shoot from an adventitious bud at the base of a plant
tinged	slightly colored
tip	apex
tomentose	a surface covered with matted and tangled hairs
tomentulose	finely or slightly tomentose
tomentum	a covering of dense, woolly hairs
tonic	a beverage that refreshes, stimulates, restores, or invigorates

tooth	a pointed projection or division
trailing	prostrate and creeping, but not rooting
translucent	semitransparent; transmitting light rays only partially
transverse	at right angles to the long axis; crosswise; cross-section
tri-	a prefix meaning three
triangular	having three edges and three angles
trichome	an epidermal hair or bristle
tridentate	three-toothed, such as *Artemisia tridentata* leaves
trifid	three-cleft
trifoliate	having three leaflets, such as *Medicago polymorpha*
trigonous	having three angles
truncate	ending abruptly; appearing to be cut off at the end
trunk	the main stem of a tree or shrub
tubercle	a small projection from the surface of an organ or structure
tuberculate	furnished with small projections
tubular	having the shape of a tube, such as the corolla of some flowers
tuft	cluster; bunch
tumble	to roll over and over as when blown by the wind
turbinate	top-shaped; inversely conical
twig	a small branch of a tree or shrub
ultimate	smallest subdivisions
umbel	a simple flat-topped or rounded inflorescence with pedicels radiating from a common point
umbellate	resembling an umbel
unarmed	without thorns or spines
undulate	strong wavy in a perpendicular plane
unilateral	arranged on or directed toward one side
unisexual	said of flowers containing only stamens or only pistils
united	two or more wholly or partially combined parts
unpalatable	not desirable for food; not readily eaten
urceolate	shaped like an urn
urn	hollow and contracted near the mouth; urceolate
utricle	a small one-seeded fruit with a thin wall, dehiscing by the breakdown of the thin wall
valve	one portion of a compound ovary; part of a pod or capsule
variety	a category or taxonomy below the species and subspecies level
vein	a single branch of the vascular system of a plant
velutinous	velvety vestiture
velvety	soft and smooth like velvet; velutinous
verticel	a whorl or level of branching
verticillate	whorls; arranged in verticels
vestige	a rudimentary structure

vestiture	not glabrous; any covering on a surface making it other than glabrous
villous	with long, soft macrohairs; similar to pilose, but with a higher density of hairs
viscid	sticky or clammy
warm-season	a category of plants that grow during the warmer portions of the year
wart	a growth or large blister on the epidermis, resembling a wart on an animal
washes	areas of active water erosion
wavy	margin with small, regular lobes; undulating surface or margin
weak	frail; not stout nor rigid; partially or incompletely
webbed	bearing fine, tangled hairs
whorl	a cluster of several branches or leaves around the axis arising from a common node
wing	a thin projection or border
wiry	being thin and resilient
withered	appearing shriveled and shrunken
woolly	covered with long, entangled soft hairs
wrinkle	small ridges and/or furrows on a surface
xerophyte	a plant adapted to a dry habitat
zig-zag	a series of short, sharp bends
zygomorphic	irregular; divisible into equal halves in only one plane (contrast with actinomorphic)

Authorities

Aiton	William Aiton (1731–1793), English botanist, royal gardener at Kew, England
Allred	Kelly Allred (1949–), New Mexico State University agrostologist
Allred & Gould	Kelly Allred (1949–), New Mexico State University agrostologist, and Frank Walton Gould (1913–1981), Texas A&M University agrostologist
Asch. & Schweinf.	Paul Friedrich August Ascherson (1834–1913), German botanist, and Georg August Schweinfurth (1836–1925), German botanist
Barkw.	Mary Elizabeth Barkworth (1941–), Utah State University agrostologist
Barkw. & D. R. Dewey	Mary Elizabeth Barkworth (1941–), Utah State University agrostologist, and Douglas R. Dewey (1929–1993), U.S. Department of Agriculture and Utah State University cytogeneticist
Baum & Findlay	Bernard Rene Baum (1937–), French-born Canadian botanist, and Judy N. Findlay (1941–), Canadian research technician
Beal	William James Beal (1833–1924), Michigan State University agrostologist
Beauv.	Ambroise Maris Francois Joseph Palisot de Beauvois (1752–1820), French naturalist
Beetle	Alan Ackerman Beetle (1913–2003), University of Wyoming taxonomist
Benth.	George Bentham (1800–1884), English taxonomist at the British Museum
Bertol.	Antonio Bertoloni (1775–1869), Italian professor of botany
Bieb.	Baron Friedrich August Marschall von Bieberstein (1768–1826), German explorer (Russia and the Caucasus)
Biehler	Johann Friedrich Theodor Biehler (1785–18??), German botanist
Bierner	Mark William Bierner (1946–), Southwest Texas State University botanist
Bigel.	Jacob Bigelow (1787–1879), professor of botany, Boston
Bisch.	Gottlieb Wilhelm Bischoff (1797–1854), German professor
Blake	Sidney Fay Blake (1892–1959), U.S. Department of Agriculture scientist
Boland.	Henry Bolander (1831–1897), German-born, state botanist for California

Boott	Francis Boott (1805–1887), American botanist and authority on *Carex*
Britt.	Nathaniel Lord Britton (1859–1934), director-in-chief of the New York Botanical Garden
Britt. & Rusby	Nathaniel Lord Britton (1859–1934) director-in-chief of the New York Botanical Garden, and Henry Hurd Rusby (1855–1940), dean of the New York College of Pharmacy and active collector, especially in South America
Buckl.	Samuel Botsford Buckley (1809–1884), naturalist and state geologist of Texas
Bush	Benjamin Franklin Bush (1858–1937), amateur Missouri botanist
Cav.	Antonio José Cavanilles (1745–1804), Spanish botanist
Chase	Mary Agnes Merrill Chase (1869–1963), agrostologist and custodian of grasses with the U.S. National Herbarium, Washington DC
Cockerell	Theodore Dru Alison Cockerell (1866–1948), professor of biological sciences in New Mexico and Colorado
Collotzi	Albert William Collotzi (1936–1997), Utah botanist
Coville	Fredrick Vernon Coville (1867–1937), curator of the U.S. National Herbarium
DC.	Augustin Pyramus de Candolle (1778–1841), Swiss botanist and professor of botany
Dewey. D. R.	Douglas R. Dewey (1929–1993), U.S. Department of Agriculture and Utah State University cytogeneticist
Dietr. D.	David Nathanael Friedrich Dietrich (1799–1888), German botanist
Don, D.	David Don (1799–1841), English botanist
Dougl.	David Douglas (1798–1834), Scottish botanical collector in northwestern United States
Dun.	Michel Felix Dunal (1789–1856), French botanist
Du Roi	John Philipp Du Roi (1741–1785), German physician and specialist in the taxonomy of American trees
Eastw.	Alice Eastwood (1859–1953), Canadian-born curator of botany at the California Academy of Sciences
Eaton	Amos Eaton (1776–1842), American botanist, produced first botanical manual in United States with descriptions in English
Elmer	Adolph Daniel Edward Elmer (1870–1942), American botanist
Endl.	Stephan Friedrich Ladislaus Endlicher (1804–1849), Austrian botanist
Engelm.	George Engelmann (1809–1884), physician and botanist in St. Louis, Missouri
Fern.	Merritt Lyndon Fernald (1873–1950), plant geographer and systematist, director of the Gray Herbarium, Harvard University

Fern. & J. F. Macbr.	Merritt Lyndon Fernald (1873–1950), plant geographer and systematist, director of the Gray Herbarium, Harvard University, and James Francis Macbride (1897–1976), taxonomist at the Gray Herbarium
Fisch.	Friedrich Ernst Ludwig von Fischer (1782–1854), director of St. Petersburg, Russia, botanical garden
Fisch. & Trautv.	Friedrich Ernst Ludwig von Fischer (1782–1854), director of St. Petersburg, Russia, botanical garden, and Ernst Rudolph von Trautvetter (1809–1889), Russian botanist
Fourn.	Eugene Pierre Nicolas Fournier (1834–1884), physician in Paris and amateur botanist
Gaertn.	Joseph Gaertner (1732–1791), German botanist
Geyer	Carl Andreas Geyer (1809–1853), Austrian botanist who collected in northwestern United States
Gould	Frank Walton Gould (1913–1981), Texas A&M University agrostologist
Gray, A.	Asa Gray (1810–1888), professor of botany at Harvard University
Greene	Edward Lee Greene (1842–1915), American botanist at the University of California
Griffiths	David Griffiths (1867–1935), British-born, U.S. Department of Agriculture agronomist
Hack.	Eduard Hackel (1850–1922), Austrian agrostologist
Harv. & A. Gray	William Henry Harvey (1811–1866), Irish botanist, and Asa Gray (1810–1888), professor of botany at Harvard University
Heller	Amos Arthur Heller (1867–1944), Pennsylvania botanist
Henr.	Jan Theodoor Henrard (1881–1974), Dutch pharmacist, conservator of Rijksherbarium, Netherlands
Herter	Wilhelm Gustav Herter (1884–1958), German botanist who resided in Uruguay after 1924
Hitchc., A. S.	Albert Spear Hitchcock (1865–1935), U.S. Department of Agriculture agrostrologist
Hitchc., C. L.	Charles Leo Hitchcock (1902–1986), botanist at the University of Washington, Seattle
Holub	Josef Ludwig Holub (1930–1999), Czech botanist
Hook.	Sir William Jackson Hooker (1785–1865), director of the Royal Botanical Gardens at Kew, England
Hook. & Arn.	Sir William Jackson Hooker (1785–1865), director of the Royal Botanical Gardens at Kew, England, and George Arnott Walker Arnott (1799–1868), Scottish botanist
Howell, J. T.	John Thomas Howell (1903–1994), California botanist
Host	Nicolaus Thomas Host (1761–1834), Austrian physician and botanist
Huds.	William Hudson (1730–1793), British botanist and apothecary
Isely	Duane Isely (1918–2000), Iowa State University taxonomist

Jacq.	Nikolaus Joseph Baron von Jacquin (1727–1817), Austrian taxonomist
Kartesz	John T. Kartesz (1948–), University of North Carolina taxonomist and author
Kellogg	Albert Kellogg (1813–1887), physician and botanist from San Francisco, California
Koehne	Bernhard Adalbert Emil Koehne (1848–1914), German botanist
Kuhn	Friedrich Adalbert Maximilian Kuhn (1842–1894), German fern taxonomist
Kunth	Carl Sigismund Kunth (1788–1850), German botanist
L.	Carolus Linnaeus (1707–1778), Swedish botanist and author of *Species Plantarum*, which is the starting point for botanical nomenclature
Lag.	Mariano Lagasca y Segura (1776–1839), Spanish professor
Lam.	Jean Baptiste Antoine Pierre Monnet de Lamarck (1744–1829), French botanist
Lawson, P. & C.	Peter Lawson (17??-1820), and Sir Charles Lawson (1794–1873), the son, Scottish nurserymen
Ledeb.	Carl Friedrich von Ledebour (1785–1851), professor and author from Dorpat, Estonia
Less.	Christian Friedrich Lessing (1809–1862), German physician and ASTERACEAE specialist
Leyss.	Freidrich Wilhelm von Leysser (1731–1815), German botanist and author of *The Flora of Halle*
L'Hér.	Charles Lous L'Heritier de Brutelle (1746–1800), French magistrate and botanist
Lindl.	John Lindley (1799–1865), English professor of botany
Link	Johann Heinrich Friedrich Link (1767–1851), German professor of natural science, Berlin
Löve, A.	Askell Löve (1916–1994), Icelandic botanist who worked in North America
Löve, A. & D.	Askell Löve (1916–1994), Icelandic botanist who worked in North America and Doris Benta Maria (Mrs. Askell) Löve (1918–2000), Swedish-born Icelandic botanist
Löve. A. & D., & Kapoor	Askell Löve (1916–1994), Icelandic botanist who worked in North America, Doris Benta Maria (Mrs. Askell) Löve (1918–2000), and Brij Mohan Kapoor (1936–), Indian-born botanist
Macoun	John Macoun (1831–1920), Irish-born, Canadian botanist
Malte	Oscar Malte (1880–1933), chief botanist of the National Herbarium of Canada
McClure	Floyd Alonzo McClure (1897–1970), U.S. Department of Agriculture botanist and Smithsonian research associate
McMinn	Howard Earnest McMinn (1891–1963), professor of botany in California

Meeuse, A. D. J., & Smit	Adrianus Dirk Jacob Meeuse (1914–), and J. Smit (1934–), Danish botanists
Melderis	Aleksandre Melderis (1909–1986), European taxonomist and author
Merr.	Elmer Drew Merrill (1876–1956), director of the New York Botanical Garden and Arnold Arboretum, Harvard University
Mey. C. A.	Carl Anton von Meyer (1795–1855), director of the St. Petersburg, Russia, botanical garden
Michx.	Andre Michaux (1746–1802), French botanist and explorer of North America
Moench	Conrad Moench (1744–1802), German botanist
Moq.	Christian Horace Benedict Alfred Moquin-Tandon (1804–1863), French botanist
Muhl.	Gotthilf Heinrich Ernest Muhlenberg (1753–1815), German-educated Lutheran minister and pioneer botanist in Pennsylvania
Nash	George Valentine Nash (1864–1921), agrostologist and head gardener at the New York Botanical Garden
Nees	Christian Gottfried Daniel Nees von Esenbeck (1776–1858), German botanist
Nels., A.	Aven Nelson (1859–1952), professor of botany and president of the University of Wyoming
Nesom & Baird	Guy L. Nesom (1945–), Botanical Research Institute of Texas botanist, and Gary I. Baird (1955–), taxonomist at Brigham Young University-Idaho
Nevski	Sergei Arsenjevic Nevski (1908–1938), Russian agrostologist
Nutt.	Thomas Nuttall (1786–1859), English-American naturalist who collected in western United States
Opiz, P.	Philipp Maxmilian Opiz (1787–1858), Bohemian botanist
Pall.	Peter Simon Pallas (1741–1811), German botanist
Palmer	Ernest Jesse Palmer (1875–1962), collector for the Arnold Arboretum, Harvard University, and Missouri Botanical Garden
Parry	Charles Christopher Parry (1823–1890), botanist with the Mexican Border Survey
Payne	Willard William Payne (1937–), botanist at the University of Illinois
Pers.	Christiaan Hendrick Persoon (1761–1836), Dutch-born, South African and French botanist
Phil.	Rudolf Amandus Philippi (1808–1904), Chilean botanist
Pohl	Richard Walter Pohl (1916–1993), taxonomist specializing in POACEAE at Iowa State University
Poir.	Jean Louis Marie Poiret (1755–1834), French botanist
Porsild	Morton Pedersen Porsild (1872–1956), Danish botanist

Porter & Coult.	Thomas Conrad Porter (1822–1901), professor of botany at Lafayette College, Pennsylvania, and John Merle Coulter (1851–1928), professor of botany at the University of Chicago
Pott	Johann Friedrich Pott (1738–1805), German botanist and physician
Presl	Karel Boriwag Presl (1794–1852), professor of natural history in Prague
Pursh	Fredrick Traugott Pursh (1774–1820), German author and botanical collector in North America
Raf.	Constantin Samuel Rafinesque (1783–1840), Constantinople-born, pioneer naturalist in Kentucky
Richter	Karl Richter (1855–1891), Austrian botanist
Ricker	Percy Leroy Ricker (1878–1973), agronomist with the U.S. Department of Agriculture who specialized in the POACEAE and FABACEAE
Roemer	Johann Jacob Roemer (1763–1819), Swiss botanist
Roemer & Schultes	Johann Jacob Roemer (1763–1819), Swiss botanist, and Joseph August Schultes (1773–1831), Austrian botanist
Roof, J. B.	James B. Roof (1910–1983), California botanist
Rose & Painter	Joseph Nelson Rose (1862–1928), botanist with the U.S. Department of Agriculture and National Herbarium, and William Hunt Painter (1835–1910), British clergyman and botanist
Roth	Albrecht Wilhelm Roth (1757–1834), German physician and botanist
Rydb.	Per Axel Rydberg (1860–1931), Swedish-born botanist at the University of Nebraska and New York Botanical Garden
Sarg.	Charles Sprague Sargent (1841–1927), botanist at Harvard University
Scheele	Georg Heinrich Adolf Scheele (1808–1864), German botanist who described plants collected in Texas
Schrad.	Heinrich Adolph Schrader (1767–1836), German botanist and professor
Schultes, J. A.	Joseph August Schultes (1773–1831), Austrian botanist and professor
Schum, K.	Karl Moritz Schumann (1851–1904), German botanist
Schwartz, O.	Otto Karl Anton Schwartz (1900–1983), botanist at Friedrich Schiller University and Jena Botanical Garden, Germany
Scribn.	Frank Lamson Scribner (1851–1938), agrostologist with the U.S. Department of Agriculture
Scribn. & Merr.	Frank Lamson Scribner (1851–1938), agrostologist with the U.S. Department of Agriculture, and Elmer Drew Merrill (1876–1956), director of the New York Botanical Garden and Arnold Arboretum, Harvard University

Scribn. & J. G. Smith	Frank Lamson Scribner (1851–1938), agrostologist with the U.S. Department of Agriculture, and Jared Gage Smith (1866–1925), American agrostologist with the U.S. Department of Agriculture
Scribn. & Williams	Frank Lamson Scribner (1851–1938), agrostologist with the U.S. Department of Agriculture, and Thomas Albert Williams (1865–1900), agrostologist with the U.S. Department of Agriculture
Sennen & Pau	Padre Sennen (Etienne Marcellin Granie-Blanc) (1861–1937), French-born clergyman, plant collector, and taxonomist in Ecuador, and Carlos Pau y Espanola (1857–1937), Spanish botanist
Sessé & Moc.	Martin de Sessé y Lacasta (1751–1808), Spanish botanist and director of the botanical garden in Mexico City, and José Mariano Mociño (1757–1820), Mexican physician
Shinners	Lloyd Herbert Shinners (1918–1971), Canadian-born botanist, professor of botany at Southern Methodist University
Simonkai	Lajos von Simonkai (Simkovics) (1851–1910), Hungarian phyto-geographer
Sims	John Sims (1749–1831), English botanical editor
Smith, J. G.	Jared Gage Smith (1866–1925), American agrostologist with the U.S. Department of Agriculture
Spreng.	Curt Polykarp Joachim Sprengel (1766–1833), professor of botany at Halle, Germany
Steud.	Ernst Gottlieb Steudel (1783–1856), German physician and authority on the POACEAE
Stokes	Susan Gabriella Stokes (1868–1954), Rocky Mountain naturalist and San Diego high school teacher
Sw.	Olof Peter Swartz (1760–1818), Swedish botanist
Swezey	Goodwin Deloss Swezey (1851–1934), plant collector and professor of astronomy at the University of Nebraska
Thurb.	George Thurber (1821–1890), American botanist with Mexican Boundary Survey
Torr.	John Torrey (1796–1873), American physician and professor of botany
Torr. & Frém.	John Torrey (1796–1873), American physician and botanist, and John Charles Frémont (1813–1890), soldier, explorer, and presidential candidate
Torr. & A. Gray	John Torrey (1796–1873), American physician and botanist, and Asa Gray (1810–1888), professor of botany at Harvard University
Trel.	William Trelease (1857–1945), professor of botany at the University of Illinois
Trin.	Carl Bernhard von Trinius (1778–1844), Russian physician, poet, and authority on grasses

Trin. & Rupr.	Carl Bernhard von Trinius (1778–1844), Russian physician, poet, and authority on grasses, and Franz Joseph Ruprecht (1814–1870), Austrian-born Russian physician and botanist
Vasey	Geroge Vasey (1822–1893), English-born botanist and curator of the U.S. National Herbarium
Vent.	Etienne Pierre Ventenat (1757–1808), French professor of botany
Vitman	Fulgencio Vitman (1728–1806), Italian clergyman and botanist
Walt.	Thomas Walter (1740–1789), British-American botanist
Wangenh.	Friedrich Adam Julius von Wangenheim (1749–1800), German forester
Ward, G. H.	George Henry Ward (1916–), Illinois botanist
Wats., S.	Sereno Watson (1826–1892), curator of the Gray Herbarium, Harvard University
Weber	Frédéric Albert Constantin Weber (1830–1903), French botanist and member of a French expedition in Mexico
Willd.	Carl Ludwig Willdenow (1765–1812), German botanist
Wood	Alphonso W. Wood (1810–1881), author of first American book to employ dichotomous keys
Woot. & Standl.	Elmer Otis Wooton (1865–1945), professor of biology at New Mexico State University, and Paul Carpenter Standley (1884–1963), curator, U.S. National Herbarium

Selected References

Abrams, Leroy. 1951. Illustrated flora of the Pacific States. Volume 3. Stanford University Press, Stanford, California.

Abrams, Leroy, and Roxana Stinchfield Ferris. 1960. Illustrated flora of the Pacific States. Volume 4. Stanford University Press, Stanford, California.

Agricultural Research Service. 1970. Selected weeds of the United States. Agricultural Research Service, United States Department of Agriculture, Washington, D.C.

Alanís Flores, Glafiro J., Marcela González Alvarez, Marco Antonio Guzmán Lucio, and Gerónimo Cano Cano. 1995. Flora representativa de Chipinque: Arboles y arbustos. Universidad Autonoma de Nuevo Leon, Monterrey.

Albee, B. J., L. M. Shultz, and S. Goodrich. 1988. Atlas of the vascular plants of Utah. Utah Museum of Natural History, Salt Lake City.

Alley, Harold P., and Gary A. Lee. 1969. Weeds of Wyoming. Bulletin 498. Agricultural Experiment Station, University of Wyoming, Laramie.

Allred, Kelley W. 1982. Describing the grass inflorescence. Journal of Range Management 35:672–675.

Andersen, Berniece A. Undated. Desert plants of Utah. Extension Circular 376. Cooperative Extension Service, Utah State University, Logan.

Andersen, Berniece A., and Arthur H. Holmgren. Undated. Mountain plants of northeastern Utah. Circular 319. Cooperative Extension Service, Utah State University, Logan.

Babcock, Ernest Brown. 1947. The genus *Crepis*. Part two. Systematic treatment. University California Press, Los Angeles.

Bare, Janet E. 1979. Wildflowers and weeds of Kansas. Regents Press of Kansas, Lawrence.

Barkley, T. M., Editor. 1977. Atlas of the flora of the Great Plains. Iowa State University Press, Ames.

Barkley, T. M., Editor. 1986. Flora of the Great Plains. University Press of Kansas, Lawrence.

Beetle, Alan Ackerman. 1960. A study of sagebrush: The section *Tridentatae* of *Artemisia*. Bulletin 368. Agricultural Experiment Station, University of Wyoming, Laramie.

Beetle, Alan Ackerman. 1977. Noteworthy grasses from Mexico. Phytologia 37:317–407.

Beetle, Alan Ackerman, Elizabeth Manrique Forceck, Víctor Jaramillo Luque, M. Patricia Guerrero Sánchez, J. Alejandro Miranda Sánchez, Irama Núñez Tancredi, and Aurora Chimal Hernández. 1987. Las Gramineas de Mexico. Volume 2. COTECOCA, Mexico City.

Beetle, Alan Ackerman, Elizabeth Manrique Forceck, Javier Alejandro Miranda Sánchez, Víctor Jaramillo Luque, Aurora Chimal Hernández, Angélica Maria Rodriguez Rodreguez. 1991. Las Gramineas de Mexico. Volume 3. COTECOCA, Mexico City.

Beetle, Alan Ackerman, and Donald J. Gordon. 1991. Gramineas de Sonora. COTE-COCA, Hermosillo, Sonora.

Beetle, Alan Ackerman, and Kendall L. Johnson. 1982. Sagebrush in Wyoming. Bulletin 779. Agricultural Experiment Station, University of Wyoming, Laramie.

Beetle, Alan Ackerman, and Morton May. 1971. Grasses of Wyoming. Research Journal 39. Agricultural Experiment Station, University of Wyoming, Laramie.

Beetle, Alan Ackerman, Rafael Guzmán Mejía, Víctor Jaramillo Luque, Matilde P. Guerrero Sánchez, Elizabeth Manrique Forceck, Aurora Chimal Hernández, Celia Shariff Bujdud, and Irama Núñez Tancredi. 1983. Las Gramineas de Mexico. Volume 1. COTECOCA, Mexico City.

Beetle, Alan Ackerman, Javier Alejandro Miranda Sánchez, Víctor Jaramillo Luque, Angélica Maria Rodriguez Rodreguez, Laura Aragón Melchor, Martha Aurora Vergara Batalla, Aurora Chimal Hernández, and Oscar Dominguez Sepúlverda. 1995. Las Gramineas de Mexico. Volume 4. COTECOCA, Mexico City.

Benson, L. D., and R. A. Darrow. 1981. The trees and shrubs of southwestern deserts. University of New Mexico Press, Albuquerque.

Bentley, H. L. 1898. Grasses and forage plants of central Texas. Bulletin 10. Division of Agrostology. United States Department of Agriculture, Washington, D.C.

Best, Keith F., Jan Looman, and J. Baden Campbell. 1971. Prairie grasses. Publication 1413. Canada Department of Agriculture, Saskatchewan.

Blackwell, Will H. 1990. Poisonous and medicinal plants. Prentice Hall, Englewood Cliffs, New Jersey.

Booth, W. E., and J. C. Wright. 1959. Flora of Montana. Part 2. Dicotyledons. Department of Botany and Microbiology, Montana State University, Bozeman.

Box, T. W., G. M. Van Dyne, and N. E. West. 1966. Syllabus on range sources of North America. Department of Range Science, College of Forestry and Renewable Resources, Colorado State University, Fort Collins.

Brummitt, Richard K., and C. Emma Powell. 1992. Authorities of plant names. Royal Botanic Gardens, Kew, England.

Budd, A. C. 1957. Wild plants of the Canadian prairies. Publication 983. Canada Department of Agriculture, Saskatchewan.

Burbidge, Nancy T., and Surrey W. L. Jacobs. 1984. Australian grasses. Angus and Robertson Publishers, Sidney.

Campbell, J. B., R. W. Lodge, and A. C. Budd. 1956. Poisonous plants of the Canadian prairies. Publication 900. Canada Department of Agriculture, Ottawa.

Campbell, Robert S., and Wesley Keller. 1973. Range resources of the southeastern United States. Special Publication Number 21. American Society of Agronomy, Madison, Wisconsin.

Cassady, J. T. 1951. Bluestem range in the piney woods of Louisiana and east Texas. Journal of Range Management 4:173–177.

Cheeke, Peter R., and Lee R. Shull. 1985. Natural toxicants in feeds and poisonous plants. AVI Publishing Company, Westport, Connecticut.

Christiansen, Paul, and Mark Müller. 1999. An illustrated guide to Iowa prairie plants. University or Iowa Press, Iowa City.

Clark, Lewis J. 1973. Wild flowers of British Columbia. Gray's Publications Limited, Sidney, British Columbia.

Clausen, J., David D. Keck, and William Hiesey. 1940. Experimental studies on the nature of species. I. Effect of varied environments on western North American plants, the *Achillea millefolium* complex. Carnegie Institute Publication 520:296–324.

Coon, Nelson. 1979. Using plants for healing. Rodale Press, Emmaus, Pennsylvania.

Copple, R. F., and A. E. Aldous. 1932. The identification of certain native and naturalized grasses by their vegetative characters. Technical Bulletin 32. Agricultural Experiment Station, Kansas State College, Manhattan.

Copple, R. F., and C. P. Pase. 1967. A vegetative key to some common Arizona range grasses. Research Paper RM-27. Rocky Mountain Forest and Range Experiment Station, Forest Service, United States Department of Agriculture, Washington, D.C.

Correll, Donovan S., and Marshall C. Johnston. 1970. Manual of the vascular plants of Texas. Texas Research Foundation, Renner.

Cronin, Eugene H., and Darwin B. Nielson. 1979. The ecology and control of rangeland larkspurs. Bulletin 499. Agricultural Experiment Station, Utah State University, Logan.

Cronquist, Arthur. 1980. Vascular flora of the southeastern United States. Volume 1. Asteraceae. University of North Carolina Press, Chapel Hill.

Cronquist, Arthur, Arthur H. Holmgren, Noel H. Holmgren, James R. Reveal, and Patricia K. Holmgren. 1977. Intermountain flora. Volume 6. The monocotyledons. The New York Botanical Garden, Bronx.

Cronquist, Arthur, Arthur H. Holmgren, Noel H. Holmgren, James R. Reveal, and Patricia K. Holmgren. 1984. Intermountain flora. Volume 4. Subclass Asteridae. The New York Botanical Garden, Bronx.

Cronquist, Arthur, Arthur H. Holmgren, Noel H. Holmgren, James R. Reveal, and Patricia K. Holmgren. 1989. Intermountain flora. Volume 3, Part B. Fabales (by Rupert C. Barneby). The New York Botanical Garden, Bronx.

Cronquist, Arthur, Arthur H. Holmgren, Noel H. Holmgren, James R. Reveal, and Patricia K. Holmgren. 1994. Intermountain flora. Volume 5. Asterales. The New York Botanical Garden, Bronx.

Cronquist, Arthur, Noel H. Holmgren, and Patricia K. Holmgren. 1997. Intermountain flora. Volume 3, Part A. Subclass Rosidae. The New York Botanical Garden, Bronx.

Davis, Ray J. 1952. Flora of Idaho. William C. Brown Company, Dubuque, Iowa.

Dayton, William A. 1931. Important western browse plants. Miscellaneous Publication 101. United States Department of Agriculture, Washington, D.C.

Densmore, Frances. 1974. How Indians use wild plants. Dover Publications, Incorporated, New York.

Diggs, George M., Jr., Barney L. Lipscomb, and Robert J. O'Kennon. 1999. Illustrated flora of north central Texas. Botanical Research Institute of Texas, Fort Worth.

Dollahite, J. W., G. T. Householder, and B. J. Camp. 1966. Oak poisoning in livestock. Bulletin 1049. Agricultural Experiment Station, Texas A&M University, College Station.

Durrell, L. W. 1951. Halogeton—a new stock-poisoning weed. Circular 170-A. Agricultural Experiment Station, Colorado State College, Fort Collins.

Durrell, L. W., and I. E. Newsom. 1939. Colorado's poisonous and injurious plants. Bulletin 445. Agricultural Experiment Station, Colorado State College, Fort Collins.

Dyksterhuis, E. J. 1948. The vegetation of the Western Cross Timbers. Ecological Monographs 18:325–376.

Elias, Thomas S. 1980. Trees of North America. Van Nostrand Reinhold Co., New York.

Elmore, Francis H. 1976. Shrubs and trees of the Southwest uplands. Southwest Parks and Monuments Association, Tucson, Arizona.

Emboden, William. 1979. Narcotic plants. Macmillian Publishing Company, Incorporated, New York.

Eppas, Alan C. 1976. Wild edible and poisonous plants of Alaska. Publication 28. Cooperative Extension Service, University of Alaska, Fairbanks.

Evers, Robert A., and Roger P. Link. 1972. Poisonous plants of the Midwest and their effects on livestock. Special Publication 24. College of Agriculture, University of Illinois, Urbana-Champaign.

Farrar, John Laird. 1995. Trees of the northern United States and Canada. Iowa State University Press, Ames.

Farrar, Jon. 1990. Wildflowers of Nebraska and the Great Plains. Nebraska Game and Parks Commission, Lincoln.

Featherly, H. I. 1938. Grasses of Oklahoma. Technical Bulletin 3. Agricultural Experiment Station, Oklahoma Agricultural and Mechanical College, Stillwater.

Fernald, Marritt Lyndon. 1950. Gray's manual of botany. American Book Company, New York.

Flora of North America Editorial Committee. 1993. Volume 2. Pteridophytes and Gymnosperms. Oxford University Press, New York.

Flora of North America Editorial Committee. 1993. Volume 3. Magnoliophyta: Magnoliidae and Hamamelidae. Oxford University Press, New York.

Forest Service. 1937. Range plant handbook. Forest Service, United States Department of Agriculture, Washington, D.C.

Frankton, Clarence. 1955. Weeds of Canada. Canada Department of Agriculture, Ottawa.

Gates, Frank C. 1937. Grasses in Kansas. Volume 55, Number 220-A. Kansas State Board of Agriculture, Topeka.

Gay, Charles W., Don Dwyer, Chris Allison, Stephan Hatch, and Jerry Schickedanz.

472

1980. New Mexico range plants. Circular 374. Cooperative Extension Service, New Mexico State University, Las Cruces.

Gleason, Henry A. 1952. Illustrated flora of the northeastern United States and adjacent Canada. Volume 1. The Pteridophyta, Gymnospermae, and Monocotyledoneae. The New York Botanical Garden, Bronx.

Gleason, Henry A. 1952. Illustrated flora of the northeastern United States and adjacent Canada. Volume 2. The Chloripetalous Dycotyledoneae. Lancaster Press, Incorporated, Lancaster, Pennsylvania.

Gleason, Henry A. 1952. Illustrated flora of the northeastern United States and adjacent Canada. Volume 3. The Sympetalous Dicotyledoneae. Lancaster Press, Incorporated, Lancaster, Pennsylvania.

Gould, Frank W. 1951. Grasses of southwestern United States. Biological Sciences Bulletin Number 7. University of Arizona, Tucson.

Gould, Frank W. 1975. Texas plants: a checklist and ecological summary. Miscellaneous Publication 585. Agricultural Experiment Station, Texas A&M University Press, College Station.

Gould, Frank W. 1978. Common Texas grasses. Texas A&M University Press, College Station.

Gould, Frank W., and R. Moran. 1981. The grasses of Baja California, Mexico. Memoir 12. San Diego Society of Natural History.

Gould, Frank W., and Robert B. Shaw. 1983. Grass systematics. Texas A&M University Press, College Station.

Grinnell, George Bird. 1962. The Cheyenne Indians, their history and ways of life. Volume 2. Cooper Square Publisher, Incorporated, New York.

Hafenrichter, A. L., Lowell A. Mullen, and Robert L. Brown. 1949. Grasses and legumes for soil conservation in the Pacific Northwest. Miscellaneous Publication 678. United States Department of Agriculture, Washington, D.C.

Hallsten, Gregory P., Quentin D. Skinner, and Alan A. Beetle. 1987. Grasses of Wyoming. Research Journal 202. Agricultural Experiment Station, University of Wyoming, Laramie.

Harrington, H. D. 1954. Manual of the plants of Colorado. Sage Books, Denver.

Harrington, H. D. 1967. Edible native plants of the Rocky Mountains. University of New Mexico Press, Albuquerque.

Harrington, H. D., and L. W. Durrell. 1944. Key to some Colorado grasses in vegetative condition. Technical Bulletin 33. Agricultural Experiment Station, Colorado State College, Fort Collins.

Harrington, H. D., and L. W. Durrell. 1957. How to identify plants. The Swallow Press Incorporated, Chicago.

Harris, James G., and Melinda Woolf Harris. 2001. Plant identification terminology. Spring Lake Publishing, Spring Lake, Utah.

Hart, Jeff, and Jacqueline Moore. 1976. Montana-native plants and native people. Montana Historical Society, Helena.

Hatch, Stephan L. 2002. Gould's grasses of Texas. Department of Rangeland Ecology and Management, Texas A&M University, College Station.

Hatch, Stephan L., K. N. Gandhi, and L. E. Brown. 1990. Checklist of the vascular plants of Texas. MP 1655. Agricultural Experiment Station, Texas A&M University, College Station.

Hatch, Stephan L., and Jennifer Pluhar. 1993. Texas range plants. Texas A&M University Press, College Station.

Hatch, Stephan L., Joseph L. Schuster, and D. Lynn Drawe. 1999. Grasses of the Texas gulf prairies and marshes. Texas A&M University Press, College Station.

Hayes, Doris W., and George A. Garrison. 1960. Key to important woody plants of eastern Oregon and Washington. Agriculture Handbook 148. Forest Service, United States Department of Agriculture, Washington, D.C.

Hermann, F. J. 1966. Notes on western range forbs. Agriculture Handbook 293. Forest Service, United States Department of Agriculture, Washington, D.C.

Hermann, F. J. 1970. Manual of the carices of the Rocky Mountains and Colorado Basin. Agriculture Handbook 374. Forest Service, United States Department of Agriculture, Washington, D.C.

Hignight, K. W., J. K. Wipff, and Stephan L. Hatch. 1988. Grasses (Poaceae) of the Texas Cross Timbers and prairies. Miscellaneous Publication 1657. Agricultural Experiment Station, Texas A&M University, College Station.

Hilken, Thomas O., and Richard F. Miller. 1980. Medusahead (*Taeniantherum asperum* Nevski): A review and annotated bibliography. Station Bulletin 644. Agricultural Experiment Station, Oregon State University, Corvallis.

Hitchcock, A. S. 1951. Manual of the grasses of the United States. Revised by Agnes Chase. Miscellaneous Publication 200. United States Department of Agriculture, Washington, D.C.

Hitchcock, C. Leo. Undated. A key to the grasses of Montana. John S. Swift Company, Incorporated, St. Louis.

Hitchcock, C. Leo, and Arthur Cronquist. 1973. Flora of the Pacific Northwest. University of Washington Press, Seattle.

Holmgren, Arthur H. 1965. Handbook of vascular plants of the northern Wasatch. Nature Press Company, Palo Alto, California.

Holmgren, Arthur H., and James L. Reveal. 1966. Checklist of the vascular plants of the Intermountain Region. Research Paper INT-32. Forest Service, United States Department of Agriculture, Washington, D.C.

Hosie, R. C. 1973. Native trees of Canada. Canadian Forest Service, Department of Environment, Ottawa.

Hulten, Eric. 1968. Flora of Alaska and neighboring territories. Stanford University Press, California.

Humphrey, Robert R. 1958. Arizona range grasses. Bulletin 298. Agricultural Experiment Station, University of Arizona, Tucson.

Instituto Nacional de Estadistica. 1995. Catálogo de Herbario INEGI. Volume 1. Geografia e Informatica, Mexico City.

Instituto Nacional de Estadistica. 1995. Catálogo de Herbario INEGI. Volume 2. Geografia e Informatica. Mexico City.

Instituto Nacional de Estadistica. 1995. Catálogo de Herbario INEGI. Volume 3. Geografia e Informatica. Mexico City.

Isely, Duane. 1990. Vascular flora of the southeastern United States. Volume 3, Part 2. Leguminosae (Fabaceae). University of North Carolina Press, Chapel Hill.

Jepson, Willis L. 1951. Flowering plants of California. University of California Press, Berkeley.

Johnson, James R., and Gary E. Larson. Grassland plants of South Dakota and the Northern Great Plains. Bulletin 566. Agricultural Experiment Station, South Dakota State University, Brookings.

Johnson, W. M. 1964. Field key to the sedges of Wyoming. Bulletin 419. Agricultural Experiment Station, University of Wyoming, Laramie.

Jones, F. B. 1975. Flora of the Texas Coastal Bend. Mission Press, Corpus Christi, Texas.

Judd, B. Ira. 1962. Principal forage plants of Southwestern ranges. Station Paper 69. Rocky Mountain Forest and Range Experiment Station, Forest Service, United States Department of Agriculture, Washington, D.C.

Kartesz, John T. 1994. A synonymized checklist of the vascular flora of the United States, Canada, and Greenland. Volume 1. Checklist. The Biota of North America Program, University of North Carolina Press, Chapel Hill.

Kartesz, John T. 1994. A synonymized checklist of the vascular flora of the United States, Canada, and Greenland. Volume 2. Thesaurus. The Biota of North America Program, University of North Carolina Press, Chapel Hill.

Kearney, Thomas H., and Robert H. Peebles. 1960. Arizona flora. University of California Press, Berkeley.

Keim, F. D., G. W. Beadle, and A. L. Frolik. 1932. The identification of the more important prairie grasses of Nebraska by their vegetative characteristics. Research Bulletin 65. Agricultural Experiment Station, University of Nebraska, Lincoln.

Keeler, Richard F., Kent R. Van Kampen, and Lynn F. James. 1978. Effects of poisonous plants on livestock. Academic Press, New York.

Kinch, Raymond C., Leon Wrage, and Raymond A. Moore. 1975. South Dakota weeds. Cooperative Extension Service, South Dakota State University, Brookings.

Kingsbury, John M. 1964. Poisonous plants of the United States and Canada. Prentice-Hall, Incorporated, Englewood Cliffs, New York.

Knobel, Edward. 1980. Field guide to the grasses, sedges and rushes of the United States. Dover Publishing, Incorporated, New York.

Langman, I. K. 1964. A selected guide to the literature on the flowering plants of Mexico. University of Pennsylvania Press, Philadelphia.

Larson, Gary. 1993. Aquatic and wetland vascular plants of the northern Great Plains. General Technical Report RM-238. Forest Service, United States Department of Agriculture, Washington, D.C.

Leithead, Horace L., Lewis L. Yarlett, and Thomas N. Shiftlet. 1971. 100 native forage grasses in 11 southern states. Agriculture Handbook 389. Soil Conservation Service, United States Department of Agriculture, Washington, D.C.

LeSueur, H. 1945. The ecology of the vegetation of Chihuahua, Mexico, north of Parallel 28. Publication 4521. University of Texas Press, Austin.

Lewis, Walter H. 1977. Medical botany. John Wiley & Sons, New York.

Little, Elbert L., Jr. 1971. Atlas of United States trees. Volume 1. Conifers and important hardwoods. Miscellaneous Publication 1146. United States Department of Agriculture, Washington, D.C.

Little, Elbert L., Jr. 1976. Atlas of United States trees. Volume 3. Minor western hardwoods. Miscellaneous Publication 1314. United States Department of Agriculture, Washington, D.C.

Lommasson, Robert C. 1973. Nebraska wild flowers. University of Nebraska Press, Lincoln.

Looman, J., and K. F. Best. 1979. Budd's flora of the Canadian prairie provinces. Publication 1662. Research Branch Agriculture, Canada.

Marsh, C. D. 1929. Stock-poisoning plants of the range. Bulletin 1245. United States Department of Agriculture, Washington, D.C.

Martin, S. Clark. 1975. Ecology and management of southwestern semi-desert grass-shrub ranges: The status of our knowledge. Research Paper RM-156. Forest Service, United States Department of Agriculture, Washington, D.C.

Martin, William C., and Charles R. Hutchins. 1980. A flora of New Mexico. Volume 1. J. Cramer, Germany.

Martin, William C., and Charles R. Hutchins. 1981. A flora of New Mexico. Volume 2. J. Cramer, Germany.

May, Morton. 1960. Key to the major grasses of the Big Horn Mountains—based on vegetative characters. Bulletin 371. Agricultural Experiment Station, University of Wyoming, Laramie.

McArthur, E. Durant, and Bruce L. Welch. 1984. Proceedings—symposium on the biology of *Artemisia* and *Chrysothamnus*. General Technical Report INT-200. Forest Service, United States Department of Agriculture, Washington, D.C.

McKean, William T. 1976. Winter guide to central Rocky Mountain shrubs. Colorado Department of Natural Resources, Denver.

McKell, Cyrus M., James P. Blaisdell, and Joe R. Goodin. 1972. Wildland shrubs—their biology and utilization. General Technical Report INT-1. Forest Service, United States Department of Agriculture, Washington, D.C.

Moreno, N. P. 1984. Glosario botanico ilustrado. Instituto National de Investigaciones Xalapa, Veracruz, Mexico.

Morris, H. E., W. E. Booth, G. F. Payne, and R. E. Stitt. Undated. Important grasses on Montana ranges. Bulletin 470. Agricultural Experiment Station, Montana State College, Bozeman.

Morton, Julia F. 1974. Folk remedies of the low country. E. A. Seemann Publishing, Incorporated, Miami, Florida.

Moser, L. E., D. R. Buxton. and M. D. Casler. 1996. Cool-season forage grasses. Agronomy Monograph 34. American Society of Agronomy, Madison, Wisconsin.

Mozingo, Hugh H. 1987. Shrubs of the Great Basin. University of Nevada Press, Reno.

Muenscher, Walter Conrad. 1939. Poisonous plants of the United States. Macmillan Publishing Company, Incorporated, New York.

Munson, T. V. 1883. Forest and forest trees of Texas. American Journal of Forestry 1:433–451.

Munz, Philip A., and David D. Keck. 1959. A California flora. University of California Press, Berkeley.

Nebraska Statewide Arboretum. 1982. Common and scientific names of Nebraska plants. Publication 101. Nebraska Statewide Arboretum, Lincoln.

Nelson, Ruth Ashton. 1968. Wild flowers of Wyoming. Bulletin 490. Cooperative Extension Service, University of Wyoming, Laramie.

Owensby, Clinton E. 1980. Kansas prairie wildflowers. Iowa State University Press, Ames.

Pammel, L. H., Carleton R. Ball, and F. Lamson-Scribner. 1904. The grasses of Iowa. Part 2. Iowa Geological Survey, Des Moines.

Parker, Karl G., Lamar R. Mason, and John F. Vallentine. Undated. Utah grasses. Extension Circular 384. Cooperative Extension Service, Utah State University, Logan.

Parker, Kittie F. 1972. Arizona weeds. University of Arizona Press, Tucson.

Parks, H. B. 1937. Valuable plants native to Texas. Bulletin 551. Agricultural Experiment Station, Texas A&M University, College Station.

Pavlick, Leon. 1995. Bromus L. of North America. Royal British Columbia Museum, Victoria, British Columbia.

Phillips, Jan. 1979. Wild edibles of Missouri. Missouri Department of Conservation, Jefferson City.

Phillips Petroleum Company. 1963. Pasture and range plants. Phillips Petroleum Company, Bartlesville, Oklahoma.

Pool, Raymond J. 1971. Handbook of Nebraska trees. Bulletin 32. Conservation Survey Division, University of Nebraska, Lincoln.

Porter, C. L. 1960. Wyoming trees. Circular 164R. Cooperative Extension Service, University of Wyoming, Laramie.

Powell, A. Michael. 1988. Trees and shrubs of Trans-Pecos, Texas. Big Bend Natural History Association, Incorporated, Big Bend.

Powell, A. Michael. 1994. Grasses of the Trans-Pecos and adjacent areas. University of Texas Press, Austin.

Preston, Richard J., Jr. 1976. North American trees. Iowa State University Press, Ames.

Reed, P. B., Jr. 1988. National list of plant species that occur in wetlands: Texas, United States Fish and Wildlife Service, United State Department of Interior, Washington, D.C.

Richardson, Alfred. 1995. Plants of the Rio Grande Delta. University of Texas Press, Austin.

Rickett, Harold William. 1966. Wild flowers of the United States. McGraw-Hill Book Company, New York.

Runkel, Sylvan T., and Dean M. Roosa. 1989. Wildflowers of the tallgrass prairie. Iowa State University Press, Ames.

Rydberg, Per Axel. 1932. Flora of the prairies and plains of central North America. Hafner Publishing Company, New York.

Rydberg, Per Axel. 1954. Flora of the Rocky Mountains and adjacent plains. Hafner Publishing Company, New York.

Rzedowski, J. 1978. Vegetacion de Mexico. Editorial Limusa, Mexico City.

Sampson, Arthur W. 1924. Native American forage plants. John Wiley & Sons, New York.

Sampson, Arthur W., and Agnes Chase. 1927. Range grasses of California. Bulletin 430. University of California, Berkeley.

Sampson, Arthur W., Agnes Chase, and Donald W. Hedrick. 1951. California grasslands and range forage grasses. Bulletin 724. Agricultural Experiment Station, University of California, Berkeley.

Saunders, Charles Francis. 1976. Edible and useful wild plants. Dover Publications Incorporated, New York.

Scoggan, H. J. 1979. The flora of Canada. Part 2, Pteriodophyta, Gymnospermae and Monocotyledoneae. Museums of Canada, Ottawa.

Scoggan, H. J. 1979. The flora of Canada. Part 3, Dicotyledoneae (Saururaceae to Violaceae). Museums of Canada, Ottawa.

Scoggan, H. J. 1979. The flora of Canada. Part 4, Dicotyledoneae (Loasaceae to Compositae). Museums of Canada, Ottawa.

Shantz, H. L., and R. Zon. 1936. The natural vegetation of the United States. Pages 1–29. In Atlas of American Agriculture Part 1, Section E. United States Department of Agriculture, Washington, D.C.

Shaw, R. B., and J. D. Dodd. 1976. Vegetative key to the Compositae of the Rio Grande Plains of Texas. Miscellaneous Publication 1274. Agricultural Experiment Station, Texas A&M University, College Station.

Silveus, W. A. 1933. Texas grasses. Clegg Company, San Antonio.

Simpson, Benny J. 1996. A field guide to Texas trees. Gulf Publishing Company, Houston.

Smeins, Fred E., and Robert B. Shaw. 1978. Natural vegetation of Texas and adjacent areas: 1675–1975 bibliography. Miscellaneous Publication 1399. Agricultural Experiment Station, Texas A&M University, College Station.

Smith, James Payne, Jr. 1977. Vascular plant families. Mad River Press, Incorporated, Eureka, California.

Soil Conservation Service. 1965. Important native grasses for range conservation in Florida. Soil Conservation Service, United States Department of Agriculture, Washington, D.C.

Sperry, O. E., J. W. Dollahite, G. O. Hoffman, and B. J. Camp. 1977. Texas plants poisonous to livestock. Cooperative Extension Service, Texas A&M University, College Station.

State of Nebraska. 1962. Nebraska weeds. Bulletin 101-R. Weed and Seed Division, Department of Agriculture and Inspection, Lincoln.

Stechman, John V. 1977. Common western range plants. Vocational Education Productions, California Polytechnic State University, San Luis Obispo.

Stephens, H. A. 1973. Woody plants of the north central plains. University Press of Kansas, Lawrence.

Stevens, O. A. 1963. Handbook of North Dakota plants. North Dakota Institute of Regional Studies, Fargo.

Steyermark, Julian A. 1963. Flora of Missouri. Iowa State University Press, Ames.

Stoddart, L. A., A. H. Holmgren, and C. W. Cook. 1949. Important poisonous plants of Utah. Special Report Number 2. Agricultural Experiment Station, Utah State University, Logan.

Stubbendieck, James. 1994. Rangeland plants. Pages 559–574, In Encyclopedia of Agricultural Science, Academic Press, New York.

Stubbendieck, James, and Elverne C. Conard. 1989. Common legumes of the Great Plains. University of Nebraska Press, Lincoln.

Stubbendieck, James, Geir Y. Friisoe, and Margaret R. Bollick. 1995. Weeds of Nebraska and the Great Plains. Nebraska Department of Agriculture, Lincoln.

Stubbendieck, James, and Thomas A. Jones. 1996. Other cool-season grasses. Pages 765–780, In Cool-season forage grasses. Agronomy Monograph 34. American Society of Agronomy, Madison, Wisconsin.

Stubbendieck, James, James T. Nichols, and Charles H. Butterfield. 1989. Nebraska range and pasture forbs and shrubs. Extension Circular 89–118. Cooperative Extension Service, University of Nebraska, Lincoln.

Stubbendieck, James, James T. Nichols, and Kelly K. Roberts. 1985. Nebraska range and pasture grasses. Extension Circular 85–170. Cooperative Extension Service, University of Nebraska, Lincoln.

Sutherland, David M. 1975. A vegetative key to Nebraska grasses. Pages 283–316. In Prairie: A multiple view. University of North Dakota Press, Grand Forks.

Texas Forest Service. 1963. Forest trees of Texas. Texas Forest Service, College Station.

Thilenius, John F. 1975. Alpine range management in the western United States—principles, practices, and problems: The status of our knowledge. Forest Service Research Paper RM-157. United States Department of Agriculture, Washington, D.C.

Tidestrom, I. 1925. Flora of Utah and Nevada. United States Natural Herbarium Contributions 25:1–665.

Tsvelev, N. N. 1984. Grasses of the Soviet Union. Russian Translation Series A.A. Balkema, Rotterdam.

Turner, B. L. 1959. The legumes of Texas. University of Texas Press, Austin.

Vallentine, John F. 1967. Nebraska range and pasture grasses. Extension Circular 76–170. Cooperative Extension Service, University of Nebraska, Lincoln.

Van Bruggen, Theodore. 1976. The vascular plants of South Dakota. Iowa State University Press, Ames.

Vasey, George. 1891. Illustrations of North American Grasses. Volume 1. Grasses of the Southwest. United States Department of Agriculture, Washington, D.C.

Villarreal-Q, J. A. 1983. Malezas de Buenavista Coahuila. Universidad Autonoma Agraria "Antonio Narro". Buenavista, Saltillo, Mexico.

Vines, Robert A. 1960. Trees, shrubs and woody vines of the southwest. University of Texas Press, Austin.

Wagner, Warren L., and Earl F. Aldon. 1978. Manual of the saltbushes (*Atriplex* spp.) in New Mexico. General Technical Report RM-57. Forest Service, United States Department of Agriculture, Washington, D.C.

Wasser, C. H. 1982. Ecology and culture of selected species useful in revegetating disturbed lands in the West. Fish and Wildlife Service, United States Department of Interior, Washington, D.C.

Waterfall, U. T. 1962. Keys to the flora of Oklahoma. Oklahoma State University Press, Stillwater.

Watson, L., and M. J. Dallwitz. 1988. Grass genera of the world. The Australian National University, Canberra.

Watson, L., and M. J. Dallwitz. 1992. Grass genera of the world. CAB International, University Press, Cambridge.

Weiner, Michael A. 1980. Earth medicine—earth food: plant remedies, drugs, and natural foods of the North American Indians. Macmillian Publishing Company, Incorporated, New York.

Western Regional Technical Committee W-90. 1972. Galleta: Taxonomy, ecology, and management of *Hilaria jamesii* on western rangelands. Bulletin 487. Agricultural Experiment Station, Utah State University, Logan.

Williams, Kim. 1977. Eating wild plants. Mountain Press Publishing Company, Missoula, Montana.

Winward, A. H. 1980. Taxonomy and ecology of sagebrush in Oregon. Station Bulletin 642. Agricultural Experiment Station, Oregon State University, Corvallis.

Yatskievych, George. 1999. Flora of Missouri. Missouri Botanical Garden Press, St. Louis.

Index

Page references for the scientific names of species appear in boldface.